Selected Titles in This Series

A Station Favorable to the Pursuits of Science:
Primary Materials in the History of Mathematics at the United States Military Academy

History of Mathematics • Volume 18

A Station Favorable to the Pursuits of Science:
Primary Materials in the History of Mathematics at the United States Military Academy

Joe Albree
David C. Arney
V. Frederick Rickey

American Mathematical Society
London Mathematical Society

Editorial Board

American Mathematical Society	London Mathematical Society
George E. Andrews	David Fowler, Chair
Bruce Chandler	Jeremy J. Gray
Karen Parshall, Chair	S. J. Patterson
George B. Seligman	

1991 *Mathematics Subject Classification*. Primary 01A70;
Secondary 01A74, 01A90, 00A15.

This book is not endorsed by the U. S. Army or the Department of Defense. Opinions expressed by the authors employed by the Department of Defense are those of the authors and not those of the U. S. Army or Department of Defense.

A list of photograph credits and acknowledgments is included at the beginning of this volume.

Library of Congress Cataloging-in-Publication Data

Albree, Joe, 1939–
   A station favorable to the pursuits of science : primary materials in the history of mathematics at the United States Military Academy / Joe Albree, David C. Arney, V. Frederick Rickey.
      p. cm. — (History of mathematics, ISSN 0899-2428 ; v. 18)
   Includes bibliographical references.
   ISBN 0-8218-2059-1 (alk. paper)
   1. Mathematics—Bibliography—Catalogs. 2. United States Military Academy. Library—Catalogs. 3. Mathematics—Study and teaching (Higher)—New York (State)—West Point—History. I. Arney, David C. II. Rickey, V. Frederick, 1941– III. Title. IV. Series.
   Z6654.2 .A38 1999
   [QA37.2]
   016.51—dc21
                                                                    99-051659
                                                                    CIP

Contents

Photo Credits

The American Mathematical Society gratefully acknowledges the kindness of these institutions in granting the following photographic permissions:

West Point Museum Collections, United States Military Academy

Cover Art:
Portrait of Sylvanus Thayer as a Young Officer by Arthur Dawson, n.d.
Portrait of Charles Davies by H. F. Gray, painted: 1875.
Portrait of Edwin Church by Daniel Huntington, painted: 1874.
Engraving, "View of West Point" by Robert Havell, dated 1848.
Lithograph, "Building for the Library & Philosophic Apparatus of the Military Academy", n.d. (c. 1850)

West Point Library Special Collections and Archives

30 photographs appearing in Appendix 2, pp. 238–268. Courtesy of the Directorate of Information Management, United States Military Academy, West Point, NY.

Preface

The phrase "A station favorable to the pursuits of science" in the title of this volume is from a letter written by Jonathan Williams (the Academy's first Superintendent and grandnephew of Benjamin Franklin) to William Franklin (son of Benjamin Franklin), October 22, 1807, wherein Williams describes West Point and his intellectual debts to Benjamin Franklin.

The United States Military Academy (USMA) was founded at West Point on March 16, 1802. The reflection and self examination, which will be a natural part of the Bicentennial of the Academy over the coming years, will provide appropriate opportunities to focus on the history and heritage of mathematics at West Point. There will be obvious interest in how mathematics affected the evolution of the Academy, the development of the values and talents of the Academy's graduates, technological innovation in America's Army, and the development of the standards for the military profession. In algebra, geometry, calculus, descriptive geometry, mechanics, surveying, mathematics education, and perhaps other mathematical disciplines, USMA professors and graduates have generated major changes across the United States, in colleges and universities, and more generally in the growth and development of the country. From textbooks to curricular designs to the professionalization of engineering, technology, and science in America, the many influences and contributions of Academy mathematicians point to a rich and significant history of mathematics at West Point. This volume is intended to launch the bicentennial reflection on USMA mathematics and to provide a resource for scholars examining this history, especially as it relates to the books in the mathematics collection at the Academy.

At the turn of the nineteenth century, both mathematics and the new United States of America were entering a time of self definition. Mathematics was in various stages of separation from mechanics, astronomy, and engineering, and the process which would lead to "pure mathematics" had begun. The United States had finally decided to establish the means for educating the leadership for a standing army, had begun in earnest the toils of westward expansion, and was developing a truly independent society.

Mathematically, in 1802, the United States was still a weak, lost colony of the English. However, less than a decade after the War of 1812, the young nation began to develop its own character in mathematics and mathematics education. It was "the time when American science got its start."[1] Academy Superintendent Sylvanus

---

[1]George H. Daniels, 1968. *American Science in the Age of Jackson.* New York: Columbia University Press. p. 7.

Thayer brought numerous French and British mathematics textbooks and treatises to West Point and adapted the pedagogy and curriculum of the École polytechnique of Paris to the American culture. In mathematics, Professor Charles Davies (USMA 1815) expressed what was to become the new American philosophy, "to unite ... the scientific discussions of the French, with the practical methods of the English: that theory and practice, science and art, may mutually aid and illustrate each other."[2] Davies' series of textbooks, which were collectively called the "Course of Mathematics" became the best-selling college and high school mathematics texts during the middle of the nineteenth century. Mathematics has always been central to the program at USMA, and just as the Academy served the military needs of the emerging nation, mathematics at West Point provided leadership in the formation of American scientific and educational communities in the nation's colleges and technological institutions.

These ideas were in the minds of the authors as we cataloged the mathematics collection of the West Point Library and wrote our introductory chapters about the Academy, its library, its people, and its books. We hope that this volume is both informative and valuable for scholarly pursuits. We hope that it is a catalyst for academic and mathematical scholarly pursuits associated with West Point's bicentennial celebration.

---

[2]Charles Davies, 1835. *Elements of Algebra Translated from the French of M. Bourdon.* Revised Edition. Philadelphia: A. S. Barnes and Co. p. iv.

Acknowledgments

Fortunate circumstances, as is often the case, inspired this project. As late as 1987, no one really knew the scope and quality of the mathematics collection at West Point. David Cameron, Head of the Academy's Mathematics Department at the time, encouraged co-author Arney to determine whether West Point truly had a noteworthy collection of mathematics books. Arney sought help from co-author Albree, who had recently completed cataloging another collection. Albree and Arney first met at the 1988 AMS Centennial Meeting in Providence, RI, to forge a cataloging plan. George Rosenstein, visiting the Academy for the academic year 1988-89, provided inspiration and initial help. Then, Albree and Arney were joined by co-author Rickey, when the latter visited West Point for the 1989-90 academic year, and together they began the process of discovering, cataloging and investigating volumes. Over the next ten years, their research and analysis progressed and excitement grew as the authors amassed additional important books along with information about these books and the entire collection.

The people who were most helpful in this project are the librarians and staff members of the Special Collections Division of the USMA Library. We would especially like to thank Alan Aimone, rare book librarian of the Special Collections Division, whose detailed knowledge of the West Point Library and its collections and whose advice and encouragement has been of immense help over the past decade. Others in the library have helped us in many ways and we wish to thank the following for all that they have done to bring this project to fruition: Nicholas Batapalia, Gladys Calvetti, Suzanne Christoff, Dawn Crumpler, Carol Koenig, Pat Maher, Elaine McConnell, Charlyn Richardson, and Judith Sibley. This group was always helpful and courteous and took care of all our special requests. We thank the USMA Library for the photographs contained in this volume and on the cover. We deeply appreciate the support given by the USMA Class of 1950 who, through the West Point Association of Graduates, established and restored the books of the "Thayer Collection." This collection is a treasured resource, and it was the foundation for the collection described in this volume.

We also received help from many members of the Department of Mathematical Sciences, USMA. Many faculty of the department share our excitement over the collection and its potential use in educating cadets and faculty. Special thanks to the following West Point mathematicians for their help and encouragement: Jeffrey Arndt, Robert Ball, Warren Chellman, William Ebel, Edward Franklin, Alan Geishecker, Dick Jardine, Jack Kloeber, Gary Krahn, David Lavery, Michael Linnington, George Mitroka, Mike Phillips, Jack Robertson, George Rosenstein,

James Tattersall, Martin Vozzo, and Louis Yuengert. We have also received valuable advice from several other mathematicians and historians, including Joseph Arkin, Otto Bekken, James Blake, Steven Daniell, Ivor Grattan-Guinness, Joe Hill, George Pappas, John Robertson, and Carlton Woods. We are especially grateful to Uta Merzbach and Karen Parshall for their personal support of our efforts, for their helpful advice, and for their expert and careful editing of our work.

We are grateful to the American Mathematical Society (AMS), its staff, and its editorial board for the History of Mathematics Series for their support of our efforts to edit and publish this bibliography. Publications Staff member Christine Thivierge was very helpful in coordinating the final stages of our efforts with the AMS. When we began our work on this project, we had no idea of its vast size and scope. As it grew, we were thankful for all the help we received. This project could not have been completed without the encouragement and support of the following special administrators and leaders: David Cameron, Daniel Christman, Roy Flint, Gerald Galloway, Frank Giordano, Howard Graves, Kenneth Hedman, Fletcher Lamkin, and John Martin. We thank Sue Arney for her help in typing and Ruth Hyde for proofreading. This was truly a team effort. Our sincere apologies to anyone we have inadvertently failed to mention.

We know there are errors in this volume. We just don't know where they are. We take full responsibility. Our only consolation is that their discovery will lead to further advances in the history of mathematics and mathematics education. In that regard, we hope this book will encourage scholars to visit and study this remarkable resource.

WEST POINT, NEW YORK AND MONTGOMERY, ALABAMA
August 1999

CHAPTER 1

Introduction

1. Science and the Formation of USMA

On December 18, 1816, Brevet Lieutenant Colonel William McRee (USMA 1805) reported that he and Brevet Major Sylvanus Thayer (USMA 1808) had just shipped approximately one thousand specially selected books from Paris to the Military Academy at West Point. These volumes ranged from encyclopedias to mathematics texts, to works on the art and science of war, to books on natural philosophy, politics, engineering, and other scholarly subjects [Boynton 1863; Molloy 1975].[1] With this shipment, the collection of primary sources in the history of mathematics at West Point was auspiciously inaugurated.

George Washington had requested, in 1783, that the new nation establish "academies, one or more, for the Instruction of the Art Military" [Ambrose 1966, p. 9], and similar calls were heard from the Revolutionary War to the end of the century. As the French Revolution and Napoleonic Wars brought instability to national life in Europe, the feeling of vulnerability spread to the United States, and on May 7, 1794, Congress created the new rank of "cadet" to strengthen the army's Corps of Artillery and Engineering [Anonymous 1904]. This Corps, which had been stationed at West Point since the Revolution, was the army's most technical branch, and these cadets were to be its junior officers in training [Ambrose 1966, p. 11]. There were strongly ambivalent opinions about the very existence much less the size of a standing professional army in the early years of the young republic as it was beginning to seek its way militarily and politically. The army, such as it was, offered both the cadets and the junior officers stationed at West Point a mixture of formal and informal instruction in mathematics, science, fortifications, gunnery, and other subjects during the last years of the eighteenth century. By 1780, West Point had a library, in its own building, "the first federal library in the United States" [Norton 1986, p. 3]. A fire in 1796 destroyed what was called the academic building; the fate of whatever library collection there was is unknown [Anonymous 1904, pp. 213, 214].

On March 16, 1802, in "An Act Fixing the Military Peace Establishment of the United States," Congress authorized the creation of a Corps of Engineers, separate from the artillery. This Corps, "when so organized shall be stationed at West Point, in the state of New York, and shall constitute a military academy" [Peters 1845, p. 132]. The United States Military Academy dates its foundation from the passage of this law.

---

[1] References listed at the end of this Introduction are enclosed in square brackets [ ] and references to works in the bibliography itself are enclosed in curved brackets { }.

President Thomas Jefferson appointed Major Jonathan Williams the first Commander of the Corps of Engineers, and as such, the first Superintendent of the Academy in 1802. Williams, Benjamin Franklin's grandnephew, completed his education in London under Franklin's direction in the early 1770s and then assisted him in France during the years after the American Revolution. On their journey home together in 1785, Williams and Franklin collaborated on research involving ocean water temperatures and the Gulf Stream, which ultimately resulted in Williams' book *Thermometrical Navigation* (1799). Even though Williams was commissioned in the engineers and artillery in 1801, he was always more of a scientist than a military man [Zuersher 1974].

From the Academy's inception, there have been tensions between scientific and scholarly interests, on the one hand, and military activities on the other. The first faculty appointments went to Captain Jared Mansfield, May 3, 1802, and Captain William A. Barron, July 6, 1802. During its first decade, however, the Academy had no standards for admission, no academic structure or program and no specifications for graduation. Cadets might enter the Academy at any time and they could be ordered to other army duties without regard to their progress, academic or military [Molloy 1975].[2]

"[T]o supplement the educational and scientific activities of the Corps of Engineers and the United States Military Academy," and inspired by the American Philosophical Society, Williams organized the United States Military Philosophical Society (USMPS) on November 12, 1802 [Forman 1945, p. 273]. In Williams' view, "The art of Fortification, however simple it may seem to a superficial beholder, is based on Mathematical, Mechanical, and Philosophical principles ..." [Zuersher 1974, p. 103]. To buttress this endeavor, Williams recruited Thomas Jefferson as what he called a "perpetual patron" and he also convinced Dewitt Clinton, James Madison, and many other prominent men of the day to enroll in the USMPS and pay its $5.00 annual dues. The USMPS thus was conceived in part to support the study of military science and also to be a political force to promote the Academy.[3]

At the operational level at West Point, all faculty and cadets were automatically members of the USMPS and attended its twice-monthly meetings (except during the Academy's long winter vacation) at Academy Hall. Every member received an engraved diploma with the Society's motto, *Scientia in Bello Pax* ("Science in War is the Guarantee of Peace"). Among the chartering officers were: President, Lieutenant Colonel Jonathan Williams;[4] Recording Secretary, Lieutenant Simon M. Levy; Treasurer, Captain Jared Mansfield; and Keeper of the Cabinet and Librarian, Lieutenant Joseph G. Swift. Williams was the president for the entire existence of the USMPS, and he operated it "as if it were a branch of the Corps of Engineers" [Forman 1945, pp. 273-277]. Even as the Academy struggled during its first years, the USMPS enjoyed some measure of prosperity and by 1807, the Society was considered "a center of scientific activity in America" [Ambrose 1966, p. 32]. "[T]he Society's archives became an international center for the study of

---

[2]During the first year of the Academy's operation, there were nine to twelve cadets, and in September 1802, after no more than three months training at the Academy, Joseph G. Swift and Simon Levy became the USMA's first graduates.

[3]For an account of the anti-Federalist politics involved in the founding of the Academy, see [Crackel 1981].

[4]Williams was promoted from Major to Lieutenant Colonel on July 8, 1802.

military science" [Zuersher 1974, p. 101]. Access to the USMPS library was open to all of its members including cadets. The Society's formal records have been preserved [Anonymous 1802].

Even though Williams had to be absent from West Point often to fulfill his duties as the army's chief engineer and inspector of fortifications, "the academy ... was his darling, his favorite child" [Zuersher 1974, p. 109]. Throughout almost all of his tenure as Superintendent, Williams petitioned the Secretary of War, Congress, and the President to expand the Academy, give it some standards and some academic stature, and provide more teachers and cadets. One of his proposals was to move the Academy to Washington, D.C., and make it the basis of a national university of science and engineering, as was being formed in France at this time. Despite Williams' dreams and pleas, in 1810 he complained that "the military academy, as it now stands, is like a foundling, barely existing among the mountains, and nurtured at a distance out of sight, and almost unknown to its legitimate parents" [Molloy 1975, p. 321]. Matters would get worse before they got better. In 1810, the Secretary of War, William Eustice, decreed that USMA graduates would not be entitled to commissions but rather they would be required to serve as privates in the army. He then displaced the cadets and what was left of the faculty from their quarters and instructional areas by transferring several hundred artillerymen to West Point. Most of the cadets resigned, and in March 1812, after the resignation of the drawing teacher Francis DeMasson, the Academy had no faculty [Zuersher 1974, p. 119; Molloy 1975, p. 330].

Into this void, Congress finally moved. The "Act Making Further Provision for the Corps of Engineers" (April 29, 1812) adopted many of the suggestions Williams had been making for improving the Academy. The faculty was strengthened and expanded. Congress authorized three full professors: in natural and experimental philosophy; in mathematics; and in the art of engineering, each with concomitant army ranks and with appropriate assistants. Provisions were made for teachers of French and drawing. Age and academic standards were set for cadets, and the size of the cadet corps was greatly expanded. Finally, $25,000 was to be spent for physical improvements, including a library [Peters 1845, p. 720].

But the War of 1812 intervened. With tragic bad timing, just three months after Congress acted, Williams resigned on July 24, 1812, in an unrelated conflict. The USMPS disbanded in November 1813. Joseph G. Swift, one of the Academy's first two graduates, succeeded Williams as chief engineer and USMA Superintendent. But Swift was not blessed with Williams' vision of making the Academy into a first class engineering school, rather he was preoccupied with the war and spent most of his time in Washington dabbling in military politics. The notable surveyor Andrew Ellicott was appointed the professor of mathematics on September 1, 1813, and Jared Mansfield returned in 1814, to be the professor of natural and experimental philosophy. For almost all of the period from 1812 to 1817, the acting superintendent was Captain Alden Partridge, who "placed military studies at the head of the Academy's list of priorities" [Molloy 1975, p. 365]. The Academy's academic building, which contained all of the classrooms and the library, was completed in 1815 [Crane and Kieley 1947, p. 38; Boynton 1863]. In order to equip this building and library, Swift dispatched Sylvanus Thayer and William McRee to Europe in the spring of 1815 for:

"An examination of the military establishments, Fortifications, Schools,
Work shops and Libraries in France, Germany and England – partic-
ularly the first and last named nations – to collect Books, Maps and
Instruments." [Molloy 1975, p. 372]

"The Father of the Military Academy," Thayer (1785-1872) came from a line
of Puritan stock which extended back to the seventeenth century in the American
colonies, and included several patriots from the American Revolution. He was a
native of Braintree, Massachusetts. The seeds of educational reform may have
been sown in Thayer's mind during his undergraduate days at Dartmouth College
(1803-1807), where he first encountered George Ticknor, a man who became his
closest and life-long friend. Ticknor enjoyed a long and notable career at Harvard
where he was one of the nation's strongest advocates for progress in American
higher education [Cremin 1980, p. 272]. At Dartmouth, Thayer also took a serious
interest in military history with a special concern for the Napoleonic campaigns.
Even though he stood at the top of his graduating class at Dartmouth, Thayer left
without delivering his commencement address, and within a year had completed
his West Point course and received his commission [Wilson & Fiske 1888; Dupuy
1958; Elliot 1979].

Thayer served along the Canadian frontier and at Norfolk, Virginia, during the
War of 1812. These experiences suggested to him that American military officers
were not as well prepared professionally as their European counterparts [Wilson &
Fiske 1888; *Dictionary of American Biography*, v. 9, p. 410].

When Thayer and McRee sailed for Europe on the ship *Congress*, Napoleon
had just lost to Wellington at Waterloo, there were British troops in Paris, and the
École polytechnique had been closed because it was deemed a breeding ground for
"liberalism and radicalism" [Artz 1966, p. 240; Swift 1890, p. 139]. Called "the most
significant advance in the whole history of higher technical education in Europe"
[Artz 1966, p. 151], the original conception of the mathematician Gaspard Monge
and the chemist Antoine Fourcroy, the École polytechnique was to be a unified
national engineering school. Admission standards were high, and the institution
was to be the gateway to all of the French public service professions – military,
engineering, government and technology [Fourcy 1828, pp. 1-73; Dhombres and
Dhombres 1989, pp. 80-84]. From 1800 to 1816, various intellectual and political
forces removed almost all of the engineering and technology, and when the school
reopened in 1816, Pierre Simon Laplace suggested that it be renamed the "École
mathématique" [Molloy 1975, p. 132]. The École polytechnique that Thayer and
McRee observed may have been more concerned with recruiting French scientists
than with training engineers and public servants. On the other hand, the system
of "répétiteurs," originally designed by Monge and Fourcroy, had been retained
[Fourcy 1828, pp. 335-347; Dhombres and Dhombres 1989, pp. 356-367; 833-838].
This must have made a special impression on Thayer because he adapted many of
its features in his forthcoming reforms of the Academy. The structure of the École
polytechnique "combined the carefully prepared lecture of the German university
professor with the close personal questioning and direction of the English college
tutorial system." Professors were expected to "advance science through their own
research." At the same time, every student exercise was graded by the répétiteurs,
assistants to the professors, and these grades determined class standing, which in

turn dictated the student's position in the public services [Artz 1966, p. 237]. While he was in Paris, Thayer also reportedly met with Adrien Marie Legendre, Pierre Simon Laplace, and Gaspard Monge [Ford 1953, p. 63; Simons 1965, p. 35]. He also encountered polytechnician Claudius Crozet [Ford 1953, p. 64; Simons 1965, p. 36], a protégé of Gaspard Monge and a veteran of Napoleon's Russian campaign. Crozet escaped the Bourbon restoration by coming to America where on October 1, 1816, he was appointed the professor of the art of engineering at the USMA [Anonymous 1904, p. 276]. This finally completed the three major faculty appointments specified in the 1812 reorganization of the Academy. After he came to West Point, Crozet wrote his own *Descriptive Geometry*, a copy of which is in this collection {Crozet 1821}.

At the conclusion of their mission, Thayer and McRee shipped the books that they had bought back to West Point. McRee wrote to Swift:

> "Major Thayer and myself have placed in the hands of W. Carnes, Mercht. of Boston, seven boxes of books, [and] charts marked 'U. S.' and numbered 1, 2, 3, 4, 5, 6, 7, to be shipped on board the *Minerva-Smith* for New York, to your address. These boxes contain between nine & ten hundred volumes besides the charts." [McRee to Swift, Dec. 18, 1816, National Archives, Record Group 77, A11][5]

After short visits to Metz, Brussels, Antwerp, and London, in April 1817, they finally sailed from Le Havre for home aboard the *Maria Theresa* [Ford 1953, pp. 65, 66].

By the Spring of 1817, the antagonisms between Partridge and the rest of the West Point faculty [Swift 1890, p. 167] had so paralyzed the Academy that, after his June inspection, President James Madison decided the institution required a full-time Superintendent. Thayer was chosen over McRee as Partridge's replacement because "he [Thayer] had done the major share of their work in France" [Ambrose 1966, p. 67].

When he returned from Europe, Thayer's philosophy of education, a blend of his Dartmouth years, of many of Jonathan Williams' proposals, and of some of the policies he saw at the École polytechnique, came into focus: "the Thayer of 1817 had little interest in tactics, and placed military engineering in a position of lesser importance to architecture and civil engineering" [Molloy 1975, p. 388-389; Thayer Papers: document of Feb. 1818 entitled "Proposition for the re-organization of the Military Academy."]. With the strong support of Secretary of War, John C. Calhoun, over a period of about five years, Thayer:

(1) established a full, well-defined four-year course of study with a specified calendar which included two examination periods, January and June;

(2) created a Board of Visitors composed of high ranking professional men in and out of the military who heard the cadets' semi-annual examinations and who provided a modest amount of oversight over the curriculum;

(3) eliminated large lectures and organized the academic work into small classes in which every cadet was graded every day;

(4) defined the "merit roll," a largely objective grading system that embraced

---

[5]This type of correspondence is listed under National Archives in the Correspondence section of the References.

TABLE 1. The Succession of West Point Superintendents through 1915

Name	Dates	USMA class	Superintendency
Jonathan Williams	(1750–1815)		1802–1803
			1805–1812
Joseph Swift	(1783–1865)	1802	1812–1814
Alden Partridge	(1785–1854)	1806	1815–1817
Sylvanus Thayer	(1785–1872)	1808	1817–1833
Rene DeRussy	(1790–1865)	1812	1833–1838
Richard Delafield	(1798–1873)	1818	1838–1845
			1856–1861
Henry Brewerton	(1801–1879)	1819	1845–1852
Robert E. Lee	(1807–1870)	1829	1852–1855
John Barnard	(1815–1882)	1833	1855–1856
Pierre Beauregard	(1818–1893)	1838	1861–1861
Alexander Bowman	(1803-1865)	1825	1861–1864
Zealous Tower	(1819-1900)	1841	1864–1864
George Cullum	(1809-1892)	1833	1864–1866
Thomas Pitcher	(1825–1895)	1845	1866–1871
Thomas Ruger	(1833–1907)	1854	1871–1876
John Schofield	(1831–1906)	1853	1876–1881
Oliver Howard	(1830–1909)	1854	1881–1882
Wesley Merritt	(1836–1910)	1860	1882–1887
John Parke	(1827–1900)	1849	1887–1889
John Wilson	(1837–1919)	1860	1889–1893
Oswald Ernst	(1842–1926)	1864	1893–1898
Albert Mills	(1854–1916)	1879	1898–1906
Hugh Scott	(1853–1934)	1876	1906–1910
Thomas Barry	(1855–1919)	1877	1910–1912
Clarence Townsley	(1855–1926)	1881	1912–1916

cadet academic and military grades and, in turn, determined the cadet's army duty after graduation – from engineering at the top to infantry at the bottom [Ambrose 1966, pp. 67–75].

"Perhaps the most singular characteristic of the Thayer System was the emphasis on mathematics, science and engineering . . . using the École polytechnique as a pattern" [Morrison 1986, p. 23]. This system was governed by Thayer's "savants" (the professors) and supported by a generous number of assistants (the répétiteurs) who conducted the many small classes into which the corps of cadets had been organized. Thayer inherited his professors, Ellicott in mathematics, Mansfield in natural philosophy, and Crozet in engineering. It was not until 1831, two years before the end of his tenure as Superintendent, that he finally had his chosen men, all USMA graduates, in these three leading positions. This dynasty produced one of the leading engineering schools in the United States during the ante-bellum period; most of the system's features lasted at least until the end of the century.

2. The Formation and Development of the Library

From the time of the American Revolutionary War, several of the officers stationed at West Point possessed their own significant technical libraries. When Williams arrived in 1802, he brought a considerable private collection, a portion of which he had inherited from his great-uncle, Benjamin Franklin. To help launch the Academy, in July of 1802, Jefferson "authorized" Williams to sell part of his [Williams'] scientific library "at cost" to the Academy's library [Molloy 1975, p. 275]; we have not identified any works in mathematics or mechanics in the current collection as having been owned by Franklin, however.

The "Inventory" of July 11, 1803 prepared by Barron lists 70 works "Rec'd of Colonel Johnathan Williams" [Barron 1803], and is reproduced in Appendix 1. Since the entries of this inventory are brief to the point of being cryptic, ambiguities of identification and provenance abound. Mathematics in the sense of the West Point collection described in this book appears to be well represented by 28 works (40%), and as nearly as we can determine, 20 of these works are in the current collection. However, five of these 20 contain later holographs (e.g., {Euclid 1751}), and there are many attribution questions (e.g., is "Rohaults' Phisics" what we now refer to as {Watts 1776}?).

To purchase at least some books for the Academy's library, Williams had to secure approval from the Secretary of War, Henry Dearborn, who in general was not favorably disposed to President Jefferson's new military Academy. Books were ordered and received from England in the Summer of 1805 and the Spring of 1806, but the booksellers in London were unable to fill all of Williams' orders. Also in 1806, 43 more books (including a copy of Newton's *Optics*) were donated by members of the USMPS.[6] During the Winter of 1806 however, Dearborn rejected one of Williams' major requests for book purchases because "science is advancing so rapidly" [Molloy 1975, pp. 289-290].

Sometime in the Academy's arduous early years, someone made "A list of Books belonging to the Library of the Military Academy." The date on this handwritten document is not clearly discernible, but even though there are a few inconsistencies, our best estimate is that it was composed in 1807 so we will refer to it as [Anonymous 1807?]. A total of 234 works were recorded here, of which 29 appear to qualify as mathematics and mechanics under our current criteria. Unfortunately, eight of the mathematical works from [Barron 1803] do not appear in this list. Again, precise identifications are impossible in most cases, but between 1803 and the compilation of [Anonymous 1807?], approximately ten mathematical works were added to the West Point collection, perhaps most notably {Hutton 1796, 1795} and an unspecified *Cours de mathématiques* which might be any one of six works in the present collection.

From February 1807 until December 1809, Ferdinand Rudolf Hassler served as acting professor of mathematics at the Academy; the sizable scientific library which Hassler brought from his native Switzerland was made available to the cadets. While Hassler was in England ordering instruments and books for the establishment of the U. S. Coast Survey [Cajori 1929, pp. 58-62], he offered this advice concerning the appropriation in the Congressional reform of the Academy in 1812:

> "You want for the establishment indispensably a Library, instruments, philosophical apparatus, etc. To this I would spend the money

---

[6]It is not clear if any distinction was made between the Academy's library and that of the USMPS [Norton 1986, p. 7].

& only the tenth part viz 2500 to making up of the proper rooms for the Academy . . . ." [Norton 1986, p. 7]

During one the Academy's lowest ebbs, from 1810 to 1812, Thayer served as an assistant professor of mathematics, and he produced a curious accounting of at least some of the Academy's library holdings in 1810. Titled "Return of the Books, Instruments, Maps and other public property of the Military Academy" [Thayer 1810], the document does not provide any indication of the place from which such a large collection of works could be "returned," or why they might have been lent. There are 81 "Books" in this list plus approximately 50 items under the heading "Instruments etc.". Thayer added "Remarks" to a few of the entries; for example "Gravesends Philosophy" is described as "damaged," although today {'sGravesande 1784} is in very good to excellent condition. Included are 29 works of mathematics and mechanics, and we note that this list is the earliest evidence that {Mansfield n.d.} is in the West Point library. Thayer closed this document with the declaration, "I certify in honor, that the above is a true statement of *all* [our emphasis] the Books, Instruments, Maps and other public property now at the Military Academy" [Thayer 1810].

In the years following the War of 1812, the Academy was poorly led by Swift and Partridge. Even though Swift was largely absent, in the Summer of 1813, he did purchase $1000 worth of books for the USMA library from a Samuel Campbell of Long Island, whom he had met while overseeing the fortification of New York City during the War. Partridge, as previously mentioned, had very little interest in the Academy's library.

The claim has been made that, while Thayer and McRee were on their European quest, the books purchased for the Academy comprised "virtually every work of importance in mathematics, physics, chemistry, and civil engineering printed in France in the years after 1750" [Molloy 1975, p. 454]. This is a bit of an overstatement with regard to mathematics, but many important volumes were secured. At least as significant as the books themselves was the continuing impulse to develop the Academy's collection. Thayer's commitment to "a large reference library of technical works" [Molloy 1975, p. 454] began with his mission to Europe and continued throughout his superintendency, 1817-1833.

When Thayer arrived as Superintendent, the job of librarian was a collateral duty of Claudius Berard, the teacher of French, and Thayer secured a bonus for Berard for this extra duty [Molloy 1975, p. 454]. An active librarian who served the Academy in this capacity until his death in 1848, Berard produced the first printed "Catalogue" of the USMA library [Anonymous 1822]. Because it listed only about 900 titles, it could not have been complete. The library's second printed "Catalogue" [Anonymous 1830], listing 2,852 books, may also have been incomplete because more than 8,000 volumes were reported to be in the Academy library in 1834 [Norton 1986, p. 15].

When Major Richard Delafield (USMA 1818) became Superintendent in 1838, the library gained a much needed advocate. The damage to the library from the fire that destroyed the Academy's academic building on March 5, 1838, turned out to be minimal because cadets and officers saved almost all of the library's materials

[Berard 1838; Church 1879, p. 25]. The books were carefully housed in the Academy's hotel dining room for three years while a new building was being constructed. In the meantime, on November 12, 1838, Delafield ordered $4,500 worth of books from Paris and $2,500 worth of books from London. He also designed the new building that would house the library. The library would occupy its entire East Wing; the Superintendent's and other offices would be on the first floor of its West Wing, while the second floor would be occupied by lecture and laboratory rooms for natural philosophy; and there would be an observatory, 27 feet in diameter, in the middle of the front of the building. The whole structure measured 160 feet by 78 feet, and when it was finally completed in 1841, it cost over $50,000[7] [Norton 1986, pp. 15, 16].

Delafield's final legacy to the Academy's library was to assign Private André Freis to be the assistant librarian in 1844. A native of Alsace and fluent in French, German and English, Freis learned his librarianship from Berard so well that after he left the Army he retained his position as assistant librarian for a total of fifty years [Norton 1986, p. 17]. The Academy was fortunate to have Freis as the library's shepherd. From 1851 to 1888, there were twelve Academy librarians. All had other primary duties on the faculty; most were junior level assistants; and each was paid a bonus of $120 per year for this service [Norton 1986, pp. 19-26]. Freis thus provided much needed stability to the enterprise. In the late 1840s because of the Mexican War, there were no appropriations for library books [Norton 1986, p. 19]. Yet in 1850, a survey by the Smithsonian Institution noted that the USMA library had grown to 15,000 volumes making it the sixth largest library in the country [Boynton 1863]. Support for the library was erratic, however. Henry Coppee (USMA 1845), who taught French, geography, history and ethics, and was the USMA librarian, complained in 1853 that there was "no permanent fund for the increase of the Library" [Norton 1986, p. 19] and an 1881 inventory listed 28,208 volumes.

By the early 1880s, the library had occupied the same East Wing for forty years and a movement began to renovate the building so that all of its space could be devoted to the library. But by 1893 after no improvements, the library had grown to over 36,000 books, many of which were piled in corners, on tables, and on the floor. The precipitating event was André Freis' death in 1894. A new assistant librarian, Dr. Otto Plate, was appointed, and in January 1895, Congress finally approved the funds for the renovation. The library also benefited from the extended tenure of Peter S. Michie's (USMA 1863) term as librarian, from 1888 to 1901 [Norton 1986, pp. 26-28].

But the work went over budget and stalled, and Michie died in 1901. Fortunately, George W. Goethals (USMA 1880), then an engineering instructor and later commander of the engineers who built the Panama Canal, supervised the completion of the renovation; the library moved into its new quarters in September 1901. By June 1902, it had 46,711 volumes and Samuel E. Tillman (USMA 1869), the professor of chemistry, mineralogy, and geology, served as the "interim librarian." Finally, at this time, the Academy's Superintendent convinced the army that the Academy needed a full-time librarian [Norton 1986, pp. 28, 30].

---

[7]This is equivalent to over $1.2 million in the year 2000. Such a structure would probably cost tens of millions of dollars in the year 2000.

TABLE 2. The Succession of West Point Librarians through 1915

Name	Dates	USMA Class	Other USMA Positions
Claudius Berard	Apr 8, 1816 - May 6, 1848		French Teacher
Henry Coppee	Oct 1, 1851 - May 1, 1855	1845	Ass't French Prof
Absalom Baird	May 1, 1855 - Aug 31, 1859	1849	Ass't Math Prof (Medal of Honor during Civil War)
Oliver Otis Howard	Aug 31, 1859 - Jun 21, 1860	1854	Ass't Math Prof (Medal of Honor during Civil War)
John C. Kelton	Jun 21, 1860 - Apr 24, 1861	1851	Tactics Instr
Herman Biggs	Apr 24, 1861 - Oct 9, 1861;	1856	Ass't Prof
Stephen V. Benet	Oct 9, 1861 - Jan 1, 1864	1849	Gunnery Instr
Edward C. Boynton	Jan 1, 1864 - Sep 1, 1871	1846	USMA Adjutant
Robert Henry Hall	Sep 1, 1871 - Nov 1, 1878	1860	USMA Adjutant
Edgar Wales Bass	Nov 1, 1878 - Jul 1, 1881	1868	Math Prof
Charles E. S. Wood	Jul 1, 1881 - Aug 30, 1882	1874	USMA Adjutant
W. M. Postlethwaite	Sep 7, 1882 - Jan 22, 1885		Chaplain
George B. Davis	Jan 22, 1885 - Aug 28, 1888	1871	Ass't Law Prof
Peter Smith Michie	Aug 28, 1888 - Feb 16, 1901	1863	Natural Phil Prof
Samuel Tillman	Feb 16, 1901 - Jul 1, 1902	1869	Chemistry Prof
Edward S. Holden	Jul 1, 1902 - Mar 16, 1914	1870	Full-time Librarian

When Plate resigned in 1902, Edward S. Holden (USMA 1870) replaced him. Holden had been a student of William Chauvenet at Washington University of St. Louis (B.S. in 1866). In the early 1870s, while an assistant professor of natural and experimental philosophy at the Academy, Holden was influenced by astronomer Henry Draper. Holden's professional career in astronomy formally began in 1873 when he went to the U.S. Naval Observatory to work with Simon Newcomb; he was also the Observatory librarian for two years. Holden's major achievements in astronomy were a biography of William Herschel (1881) and the design and operation of the Lick Observatory of the University of California [*Dictionary of Scientific Biography*; *Dictionary of American Biography*]. When he returned to West Point in 1902, his first job was to assist in the publication of the proceedings of the Academy's Centennial. This task completed, he became the Academy's first full-time librarian. During his relatively short tenure, from 1902 to 1914, Holden accomplished a great deal: he integrated the catalogs of the departmental libraries (libraries still very much in existence today) into the main library's catalog; he began the systematic collection of manuscripts; he instituted the Dewey decimal system; he gave lectures to cadets on library usage; and "for the first time since Thayer's voyage and purchases, an aggressive program of acquisition commenced" [Norton 1986, pp. 30-32]. In 1913, the USMA library contained over 90,000 titles, but still, "it was always a matter of regret to Dr. Holden that the valuable collections of the library were not more freely used ... ." [Norton 1986, p. 32]

The 1841 building served as the Academy's library until the current structure was built starting in 1964. The first archivists in the library were soldiers appointed for that purpose, the most notable being Corporal Sidney Forman, who began in 1943. After the end of World War II, Forman became the first civilian archivist, and later was the USMA librarian. In 1946, the "Thayer Collection" was extracted from the library's circulating collection [Norton 1986, p. 35].

Jonathan Williams' vision for West Point was to create a national engineering school, even if it was not on quite as high a plane as that envisioned by Monge and Fourcroy for the École polytechnique. In the 1820s and 1830s, Sylvanus Thayer and his disciples attempted to fulfill Williams' dream, and in so doing made mathematics and mechanics the cornerstone of the West Point technological education. At the very beginning of the Academy, Williams initiated the mathematics collection with significant works like Isaac Newton's *Universal Arithmetic* [Barron 1803, ms.]. The Parisian mission of Thayer and McRee not only introduced some French educational techniques into the scientific curriculum at West Point and contributed to the French influence in American mathematics in the ante-bellum period [Smith and Ginsburg 1934, pp. 76-83; Parshall and Rowe 1994, pp. 7-8, 12-14], it also turned the focus of the Academy's mathematics and mechanics collection toward France and more generally the European continent and gave this collection a strong historical dimension. After an ambitious start, unfortunately the intellectual drive of the Academy's library was not maintained. Only in recent times has the USMA secured the resources for the preservation, care and appropriate use of its valuable collections.

3. Teaching Mathematics and Mechanics at West Point

Because the United States Military Academy was not burdened with the rigid classical curriculum of eighteenth– and nineteenth–century American colleges, it had the freedom to create itself, to chart its own intellectual course. In particular, the aim of mathematics at West Point in the nineteenth century was characterized by its most prolific expositor, Charles Davies (USMA 1815), as "the union of the French and English systems . . . the exact and beautiful methods of generalization, which distinguish the French school, . . . blended with the practical methods of the English system." Intellectual leadership and practical progress in the growing country, and even abroad, flowed from the contrasting features of West Point mathematics. Mathematics was the basis of the West Point system, and "the graduates of the Military Academy have been sought for whenever science of the highest grade has been needed" {Davies–General 1850, pp. 3,4}. Concomitantly, the Academy's mathematical sciences possessed certain traits which we now associate with the American character and exhibited their own kind of provincialism. Instruction in mechanics and engineering followed a similar path.

3.1. Mathematics. The first mathematics class at West Point was taught on the morning of September 21, 1801 by George Baron. Baron was a product of the Military Academy in Woolwich, England [Molloy 1975, p. 244], and he had been appointed a civilian "teacher of mathematics" or "teacher of arts and sciences" on January 6, 1801, over 14 months before the Military Academy was established by Congress [Zuersher 1974, p. 88]. When he taught, he introduced a "standing slate" on which he wrote with a piece of chalk [Ambrose 1966, pp. 19, 20], and he used the *Course of Mathematics* in two volumes by his Woolwich colleague Charles Hutton. Apparently a contentious person in general, Baron had an infamous and almost violent confrontation with cadet Joseph G. Swift. As a result of other unsavory incidents, he was dismissed on February 11, 1802 [Anonymous 1904, v. 1, pp. 218, 219, 241; Swift 1890, pp. 28-29]. Later that year, he settled in New York City and founded *The Mathematical Correspondent*, the first mathematics journal in the United States. However, within a year or so, his rude and antagonistic management of this enterprise forced him to relinquish control to Robert Adrain [Smith & Ginsburg 1934, pp. 85, 86; Hogan 1976].

As noted, at the founding of the Academy in 1802, Jefferson appointed Jared Mansfield to the new Corps of Engineers as the "acting professor of mathematics" [Anonymous 1904, v. 1, p. 242; *Dictionary of American Biography*]. After a checkered career at Yale, Mansfield had taught at various schools in New Haven and in Philadelphia and had established his reputation as a mathematician with the publication of his *Essays, Mathematical and Physical* {Mansfield n.d.}. The first essay in Mansfield's book, "Use of the Negative Sign in Algebra," represented a contribution to a problem in the foundation of analysis that was sparking significant debate in England at this time [Pycior 1981, pp. 27-31; Pycior 1984, pp. 428-433]. There were also essays on "The Theory of Gunnery" and "Nautical Astronomy." Mansfield's paper on "Fluxionary Analysis" struggled mightily with notions of "quantity"; apparently he was not up to date with the calculus of Euler and Lagrange and unacquainted with their ideas of functions. Baron "derided" Mansfield's *Essays* in the *Mathematical Correspondent*, [Hogan 1976, p. 406].

The second "acting professor of mathematics," William A. Barron had been a classmate of John Quincy Adams and, while a student and tutor at Harvard, a protégé of Samuel Webber, the fourth Hollis professor of mathematics and natural philosophy. On May 6, 1800, Barron was commissioned an officer in the Artillery and Engineers, and shortly after the Academy was formally founded, he was transferred to the new Corps of Engineers as Baron's replacement. [Wilson & Fiske 1888]

The entire academic faculty of the new Academy consisted of Mansfield, Barron, and Superintendent Williams. Classroom instruction was from 8 AM until noon and Williams taught surveying in the afternoons. In mathematics, Mansfield taught algebra and Barron taught "geometrical demonstration" [Anonymous 1904, v. 1, pp. 221, 242]. Barron reportedly had a real flair for teaching.

In the second volume of Hutton's *Course of Mathematics*, the cadets were also exposed to a smattering of what was then called "mechanics" – a somewhat more inclusive and less well defined term than that which we recognize today. The first five editions of this work contained, among others, almost arbitrarily ordered chapters on the laws of motion, hydrostatics, practical gunnery, the air pump, mensuration, fluxions, and miscellaneous "practical" problems.[8] Occasionally, a few of the better cadets were taught some mechanics from Enfield's *Institutes of Natural Philosophy*; Samuel Webber had prepared the American editions of 1802 and 1811. No form of calculus was used in this English work whose sources began with Newton and remained exclusively English. Mechanics only occupied pages 10–80, because the book aimed to lead the student up to a study of descriptive astronomy minus any celestial mechanics {Enfield 1811} [Anonymous 1821]. However, as late as

---

[8]The West Point collection contains copies or partial copies of numerous editions of Hutton's *Course* from the second volume of the first edition {Hutton 1798} through full later London and American editions; many of these books contain student names and notes.

1817, many cadets still graduated from the Academy having had no instruction whatsoever in natural philosophy [Anonymous 1904, v. 1, p. 261].

After less than two years at West Point, Mansfield was promoted to Lieutenant Colonel, appointed surveyor general of the United States, and sent out to Ohio and the Northwest Territory [Mansfield 1897]. When Williams resigned temporarily as Superintendent, from June 20, 1803, to the Spring of 1805, Barron took on the acting superintendency in addition to his teaching. In 1807, Williams charged Barron with "suffering prostitutes to be the company of his quarters..." [Ambrose 1966, p. 32], and rather than endure a court-martial, Barron resigned.

This loss in faculty was relieved at least temporarily with the appointment of Swiss-born Ferdinand Rudolf Hassler as "mathematics teacher in the military academy" in February 1807 {Hassler 1826}. Hassler had the benefit of a scientific education from the University of Bern and experience in geodetics when he immigrated to the United States to be part of a utopian community. When that dream collapsed, through his contacts with members of the American Philosophical Society and the efforts of President Jefferson, Congress established a plan for him to lead a survey of the American coasts. That project also fell through, making Hassler and his large scientific library available at the time of Barron's resignation. In 1807, Hassler began writing his *Elements of Analytic Trigonometry, Plane and Spherical* {Hassler 1826, p. 6}. When this book was finally published by the author himself, it was the first American trigonometry that was not synthetic. (One of the copies in this collection has Hassler's copyright and two letters bound with it.) Contemporary accounts did not rate Hassler very highly as a teacher, however. In 1808, the Secretary of War declared that civilians would not be permitted to teach at the Academy, and so Hassler was forced out. He went on to a controversial career as the founder and first director of the United States Coast Survey and the Superintendent of the Bureau of Weights and Measures, the first scientific governmental agency in the United States [Cajori 1929; Wilson & Fiske 1888].

In 1808 and 1809, Hassler was assisted by then Lieutenant Alden Partridge (USMA 1806), who took charge of mathematics instruction when Hassler departed. When Congress reformed the Academy in the Act of April 29, 1812, it created the formal position of "professor of mathematics" [Molloy 1975, p. 347]. Partridge was the first person to have this official title, holding it from April to September 1813 when he transferred to engineering.

Finally, some stability was established in 1813 with the appointment of Andrew Ellicott, then aged 60, as professor of mathematics. Ellicott came from a prominent family of millers from just west of Baltimore. He had been a Major in the Maryland militia during the American Revolution, he had studied with Robert Patterson of Philadelphia and had made almanacs and briefly taught mathematics in the 1780s. But he earned his reputation as a surveyor. Starting with a group who continued the Mason and Dixon line in 1784, Ellicott surveyed several projects in western Pennsylvania and New York state, the "Territory of Columbia" for George Washington's new capitol, the boundary between the southernmost states and the Spanish possession of Florida, and many others. This work made him one of the leading American mathematical practitioners of the turn of the nineteenth century. For seven years, he taught from Hutton's *Course of Mathematics* at the Academy. Kindly but remote, he was known to the cadets as "Old Infinite Series" and famous for the perfect geometrical constructions that he made with cord and ruler at the blackboard. [*Dictionary of American Biography*; Elliott 1979; Wilson & Fiske 1888].

Despite the congressional reforms in 1812 and despite the presence of a prominent professor (Ellicott), the Academy's instruction in mathematics, along with its general academic performance, failed to rise above the mediocre during the tumultuous years immediately following the War. Then, when Thayer became Superintendent in 1817, the following mathematics, mechanics and engineering courses were gradually instituted within the Academy's four-year course of study:

1st year: algebra, geometry, trigonometry, mensuration, analytical geometry;

2nd year: analytical geometry, descriptive geometry, shades and shadows, calculus, surveying;

3rd year: mechanics; acoustics, optics, astronomy;

4th year: civil engineering; military engineering.

As part of Thayer's reform program, the cadets were divided by academic achievement into two groups; the lower group continued the use of Hutton's *Course of Mathematics* for some time, while the more advanced group was introduced to other texts and Hutton was phased out.

Thayer divided each of Ellicott's classes into four sections, each with its own assistant. Ellicott visited these sections from time to time, but educational progress was hard to achieve. When Claudius Crozet, the new professor of engineering, began teaching descriptive geometry in 1817, he complained of the poor mathematical preparation of his students. He was forced to devote extra time in making up their deficiencies, to the detriment of his engineering courses [Anonymous 1904]. Only forty of the initial 117 cadets who began their studies in 1819 finished Thayer's new academic program. (See also [Church 1879, pp. 41, 49].) Ellicott died in 1820 and was buried at West Point, and Thayer chose Ellicott's son-in-law, Captain David B. Douglass to succeed him in mathematics.

Douglass graduated from Yale and then secured a commission in the Corps of Engineers for which he trained at West Point. He returned to the Academy in the attempted reforms of 1815 as an assistant professor of natural philosophy, and married Ellicott's daughter. After a short term as professor of mathematics, Douglass succeeded Crozet as professor of engineering in 1823 and Davies followed in mathematics [Wilson & Fiske 1888; *Dictionary of American Biography*].

In his capacity as Superintendent, General Joseph Swift had met and recruited Davies to be a cadet while Swift was organizing the army's fortifications up along the St. Lawrence River in 1813. Even though he was almost entirely self educated, Davies easily gained admission to the Academy because the entrance requirements were so low. Under Ellicott, Davies progressed rapidly, graduating on December 11, 1815, and returning to the Academy within a year as Ellicott's choice as an assistant professor of mathematics.

In 1825, two years after Davies had succeeded Douglass, the following mathematics texts were used by first and second year cadets:

1. Lacroix's *Elements of Algebra* (translated by John Farrar) {Lacroix 1825b};

2. Lacroix's *Complément des élémens d'algèbre*
   {Lacroix 1804b} {Lacroix 1817};

3. Legendre's *Geometry* (translated by John Farrar) {Legendre 1825};

4. *Trigonometry* (translated from Lacroix and Bezout by John Farrar)
   {Lacroix 1826};
5. Crozet's *Treatise on Descriptive Geometry and Conic Sections*
   {Crozet 1821};
6. Crozet's *Perspective, Shades and Shadows*;
7. Biot's *Géométrie analytique* {Biot 1826};
8. Lacroix's *Trâité du calcul* {Lacroix 1820b}.

Copies of all of these works and their French originals are in the West Point collection with the exception of Crozet's *Perspective, Shades and Shadows*; since this is not even listed in Crozet's biography [Couper 1936], perhaps it was a set of notes in manuscript. There is some ambiguity concerning the calculus text; probably the cadets studied from Lacroix's *Traité élémentaire du calcul* {Lacroix 1820b} which had been abridged from his three-volume *Traité du calcul* {Lacroix 1800; Lacroix 1810}; [Church 1879, p. 48]. The Farrar translations had been made for the students at Harvard in 1818, 1819 and 1820.

In addition to managing these curricular reforms, very soon Davies began writing his own textbooks. He noted that "In France, Descriptive Geometry is an important element of a scientific education ... indispensable to the Architect and Engineer" {Davies–Descriptive Geometry 1826, Preface}. Not surprisingly, then, his first book was *Descriptive Geometry*. Then, in the eleven short years from 1826 to his resignation from the Academy in 1837, "Old Tush" as Davies was affectionately known to the cadets [Smith 1879, p. 7], wrote/translated/edited/produced thirty-one editions of the following eight books [Karpinski 1980, p. 630]:

Elements of Descriptive Geometry (1826);
Elements of Geometry (1828);
Elements of Surveying (1830);
A Treatise on Shades and Shadows (1832);
The Common School Arithmetic (1833);
Elements of Algebra ... Bourdon (1835);
Elements of Analytic Geometry (1836);
Elements of the Differential and Integral Calculus (1836).

In his spare time, Davies studied law and was admitted to the bar!

Exhausted, Davies resigned his West Point professorship in May 1837 and went to Europe to recuperate. On his return, he moved to Hartford, Connecticut, where he formed a partnership with Alfred S. Barnes to publish and promote his sequence of mathematics textbooks. But Davies very soon found his contacts with the business world distracting and distasteful, and the partnership was dissolved in favor of a royalty agreement which still paid him generously. Even though Davies only lived in Hartford until 1841, he managed to write/compile five new textbooks (and numerous revisions) there. Except for the years 1841 to 1845, when he was treasurer of the Academy, Davies spent the rest of his life either teaching and writing or just living on the income from his textbooks while producing more revisions and more books. The Davies textbooks were assembled into "The West Point Course" and formed the foundation of A. S. Barnes and Company's commercial success. Several of these texts continued to be edited and revised by other authors for twenty or more years after Davies' death [Davies 1895 pp. 68-73; Wilson & Fiske 1888; Elliott 1979]. Davies wrote some valuable mathematical works other than his textbooks. To mention just two: his *Logic and Utility of Mathematics* {Davies–General

1850} was the first American book devoted to methods of teaching mathematics; the *Mathematical Dictionary* {Davies and Peck 1855} (written with his son-in-law William G. Peck (USMA 1844), later a mathematics professor at Columbia College) was a revealing statement of the state of the art of American mathematics at mid-century. Even though Davies' first exposure to "higher" mathematics came from Hutton's *Course of Mathematics*, his textbooks manifest the French influence in American mathematics and mathematics education. Davies wrote practical works for the widespread education needed by, and promised to, a new and growing America.

Albert Church (USMA 1828) came to West Point as a cadet in the summer of 1824, instead of entering Yale and preparing for a law career. He stood number one in his graduating class, remained at the Academy as an assistant professor of mathematics under Davies, and in fact, except for a year or so in the early 1830s, spent his whole professional life at West Point [Church 1879]. A patient, yet uninspiring teacher and a textbook writer, Church intended his books as improvements over the Davies texts used by West Point cadets, but he never tried to redo Davies entire series. By 1860, and under Church's guidance, mathematical instruction at the Academy had been subdivided into eight separate courses: algebra; geometry; trigonometry; mensuration; descriptive geometry; analytical geometry; calculus; and surveying. Each of the four classes was divided into sections of ten to fifteen cadets, and a large contingent of assistant professors, "under the general supervision of the professor," conducted these daily sessions. As Church himself explained:

> "Every member of each section is, if possible, required to daily explain, at the blackboard or wall slate, one or more propositions of the lesson given out on the previous day, and is thoroughly examined by questions on a portion or all of it." [Anonymous 1904, v. 1, p. 247].

When Church died in 1878, he was succeeded by one of his former students, Edgar W. Bass (USMA 1868). Bass served in the Corps of Engineers and his academic preparation consisted of two terms as an assistant in natural philosophy as well as a tour as an assistant astronomer on the 1875 United States expedition to New Zealand to observe the transit of Venus. In his second year as professor, Bass introduced the method of least squares into the curriculum, and in 1880, a portion of the advanced sections in algebra were devoted to a partial course in determinants. Bass believed that the definitions in Church's calculus text were unclear; he conceived of calculus as "the science of rates" [Anonymous ca. 1990]. For twelve years, Bass supplemented the Academy's calculus course with pamphlets that he published himself and issued to the cadets. Finally, he assembled these into his own textbook {Bass 1896}, and after 53 years, Church's *Calculus* was replaced [Anonymous 1904, v. 2, p. 180]. As a teacher, Bass was characterized as "strict" but "no man was ever really more helpful to his pupils, no man fairer or more just" [Anonymous ca. 1990]. When his eyesight failed in 1898, Bass retired.

Wright P. Edgerton (USMA 1874) served in the artillery for eight years, as an assistant in mathematics for ten more years, and finally as the first associate professor of mathematics during the last five years of Bass' tenure as professor. After this extended apprenticeship, he succeeded Bass and was only professor of mathematics for the six years from 1898 to 1904. In 1900, he retired Davies' *Bourdon's Algebra* and reorganized the curriculum so that the cadets would no longer take algebra one year and geometry the next but rather combined their algebra with their geometry into one year, studying them on alternate days. Edgerton converted most of the semi-annual examinations in mathematics from oral to written tests in 1900, and in 1902 surveying was transferred to engineering. Edgerton died in 1904 [Anonymous 1904, v. 1, pp. 249-250].

Continuity in mathematics at West Point was preserved when Charles P. Echols (USMA 1891) was appointed Edgerton's successor. Echols had worked his way up in mathematics at the Academy from instructor (1895-1897), to assistant professor (1897-1898), to associate professor (1898-1904). "In an effort to analyze and critique different teaching techniques and environments in Europe, Lieutenant Colonel Echols visited a variety of schools in Europe from July 1905 until July 1906" [Anonymous ca. 1990]. Echols' service as professor of mathematics at USMA (1904-1931) was second in length only to that of Albert Church, and he was fondly remembered as a rigorous and idiosyncratic teacher [Anonymous ca. 1990].

3.2. Mechanics. The reform Act of 1812 transformed the informal and infrequent instruction in mechanics taken from Hutton and from Enfield into the department of natural and experimental philosophy. Jared Mansfield returned to West Point in the spring of 1814 as the first professor of this department, and in the summer of 1816, mechanics was made a part of the announced curriculum. As was the case in mathematics, under Thayer's reforms, classes in natural philosophy were split by ability into two sections; the higher section used the *Mechanics* by Olinthus Gregory, Hutton's successor at Woolwich {Gregory 1815}. When Gregory's text was attempted in the lower section, it was found to be too difficult and so, in 1824, Bridge's *Mechanics* was used instead [Anonymous 1904, pp. 261-264]; see also {Bridge 1814}.

Even though Jared Mansfield was "beloved and respected by the cadets" [Anonymous 1904, p. 264], he was a thorn in Thayer's side. In 1828, he was finally forced to resign due to old age, and the result was a contest for his position between Edward H. Courtenay (USMA 1821), one of his assistants, and David Douglass, then professor of engineering. Courtenay had entered the Academy at age fourteen during Thayer's first year as Superintendent, and had graduated a year ahead of time number one in his class, thus becoming one of Thayer's protégés. He won the appointment to Mansfield's professorship [Molloy 1975, pp. 402, 403]. Courtenay spent most of his short tenure in this position preparing a translation of the *Mechanics* of Jean Louis Boucharlat. From its second edition, Boucharlat's work boasted that it "seeks to encompass all of the essential theories of modern Mechanics ... introduced into this science by Lagrange and Laplace .... " {Boucharlat 1827, p. v}. In this program, mechanics was divided into four branches, statics, dynamics, hydrostatics, and hydrodynamics. Courtenay tailored Boucharlat to the special needs of his West Point cadets by adding a few explanations, some of which he borrowed from Gregory and other sources, and omitting a couple of sections {Courtenay 1833, p. 3}. After six years in his post, Courtenay resigned, taught mathematics briefly at the University of Pennsylvania, and spent several years as an engineer. In 1842, he succeeded J. J. Sylvester as professor of mathematics at the University of Virginia, and while there composed his massive two volume *Calculus* {Courtenay 1855}; [Wilson & Fiske 1888; Elliott 1979].

TABLE 3. The Succession of West Point Mathematics "Professors" through 1915

Professor	Position/Notes	Term
George Baron (1769–?)	Acting Professor of Math (USMA founded 16 Mar 1802)	6 Jan 1801 - 11 Feb 1802
Jared A. Mansfield (1759–1830; Yale 1777)	Acting Professor of Math	3 May 1802 – 14 Nov 1803
William A. Barron (1769-1825; Harvard 1787)	Acting Professor of Math (Forced to resign)	6 Jul 1802 – 14 Feb 1807
Ferdinand R. Hassler (1770–1843)	Acting Professor of Math	14 Feb 1807 – 31 Dec 1809
Alden Partridge (1785–1854; USMA 1806)	Acting Professor of Math Professor of Math	Dec 1809 – Apr 1813 Apr 1813 - 1 Sep 1813
Andrew Ellicott (1754–1820)	Professor of Math	1 Sep 1813 – 29 Aug 1820
David B. Douglass (1790–1849; Yale 1813)	Professor of Math Married Ellicott's daughter	29 Aug 1820 – 1 May 1823
Charles Davies (1798–1876; USMA 1815)	Professor of Math Married Mansfield's daughter	1 May 1823 – 31 May 1837
Albert E. Church (1807–1878; USMA 1828)	Acting Professor of Math Professor of Math	1 Jun 1837 – 13 Mar 1838 13 Mar 1838 – 30 Mar 1878
Edgar W. Bass (1843–1918; USMA 1868)	Professor of Math Librarian (1878–1881)	17 Apr 1878 – 7 Oct 1898
Wright P. Edgerton (1852–1904; USMA 1874)	Professor of Math	7 Oct 1898 – 25 Jun 1904
Charles P. Echols (1867–1940; USMA 1891)	Professor of Math	29 Jun 1904 – 30 Sep 1931

William H. C. Bartlett (USMA 1826) also came to West Point with very little formal education and yet was able to graduate number one in his class. When Bartlett succeeded Courtenay as professor of natural and experimental philosophy, the cadets had just begun using Courtenay's version of Boucharlat's *Mechanics* and Bartlett continued its use until 1850. Bartlett became a notable American astronomer during this time. From July 1 to November 20, 1840, he traveled in Europe inspecting observatories and facilities for making instruments. On his return, he directed the construction and outfitting of the West Point observatory (1843), located in the front of the new library building. Bartlett used this observatory to produce papers on comets and optics, and his most famous research involved the partial solar eclipse of 1854 for which he "was the first [in America] to obtain quantitative results from astronomical photography" [Holden 1911, p. 181]. When

TABLE 4. The Succession of West Point Natural Philosophy (Mechanics) "Professors" through 1915

Professor	Notes	Term
Jared A. Mansfield (1759–1830; Yale 1777)	Professor of Natural & Experimental Philosophy (appointed Oct 7, 1812)	Apr 10, 1814–Aug 31, 1828
Edward H. Courtenay (1803-1853; USMA 1821)	Acting Professor of Natural & Experimental Philosophy	Sep 1, 1828 –Feb 16, 1829
	Professor of Natural & Experimental Philosophy	Feb 16, 1829–Dec 31, 1834
William H. C. Bartlett (1804-1893; USMA 1826)	Acting Professor of Natural & Experimental Philosophy	Nov 22, 1836–Apr 20, 1836
	Professor of Natural & Experimental Philosophy	Apr 20, 1834–Feb 14, 1871
Peter S. Michie (1838-1901; USMA 1863)	Professor of Natural & Experimental Philosophy Librarian (1888–1901)	Feb 14, 1871–Feb 16, 1901
William B. Gordon (1853-1938; USMA 1877)	Professor of Natural & Experimental Philosophy	Feb 16, 1901–May 7, 1917

Bartlett introduced his text, *Elements of Natural Philosophy. I.–Mechanics* in 1850 (later title: *Synthetic Mechanics*), the Academy's Academic Board protested that the work did not use calculus and therefore was not up to West Point standards [Holden 1911, p. 182]. This prompted Bartlett to publish what became his most important work, his *Elements of Analytical Mechanics*. In its preface, he made his philosophical position clear:

> "All physical phenomena are but the necessary results of a perpetual conflict of equal and opposing forces, and the mathematical formula expressive of the laws of this conflict must involve the whole doctrine of Mechanics." {Bartlett 1853, p. iii}

This philosophy continued to appear in his later editions, in the form:

> "That formula [is] no other than the simple analytical expression of what is now generally called the law of conservation of energy . . . ." {Bartlett 1874, p. v}

That is, Bartlett aimed to derive all of the properties of the mechanics of solids and of fluids "deductively" from the law of conservation of energy. He envisioned mechanics, which "begins with the simplest elements of physics," as "the groundwork of the whole" of science and technology [Holden 1911, p. 181]. In addition to being a well respected teacher, Bartlett was one of the original incorporators, in 1863, of the National Academy of Sciences. In 1871, "after (more than) forty years of continuous service," Bartlett was induced to retire from the army in order to begin a second and remuneratively more successful career as an actuary in New York City [Wilson & Fiske 1888; Elliott 1979].

Peter S. Michie (USMA 1863), Bartlett's successor, was a native of Scotland. He was well prepared for the Academy at the Woodward High School in Cincinnati, Ohio, and graduated number two in his West Point class. After distinguished service in the American Civil War in the final campaigns around Richmond and Appomattox, Michie returned to West Point as an assistant professor and then went to Europe in 1870 to study ways in which iron could be used to strengthen sea-coast defenses. Michie continued using Bartlett's *Mechanics* for fifteen years before publishing his own *Analytical Mechanics* (1886) which was intended to "cover all of the important principles of Mechanics which form the basis of that scientific knowledge now required by the military profession" {Michie 1887, p. iii}. Michie's text, derived from sources (Siméon Denis Poisson (1781-1840), Edward John Routh (1831-1917), Maurice Levy (1838-1910)) other than Bartlett, was at first restricted to the mechanics of solids. Fluid mechanics was added by 1890, but otherwise the book was reprinted and used at West Point well into the first decade of the twentieth century [Wilson & Fiske 1888; *Dictionary of American Biography*]. As noted, Michie was also the part-time librarian who presided over much of the renovation of the library building during the years 1895 to his death in 1901.

3.3. Curriculum Development and Leadership. This discussion of mathematics and mechanics professors, textbooks, and curricula, underscores the uniqueness of the USMA as an institution in nineteenth-century American life. The Academy was not a college; in fact, baccalaureate degrees were not awarded to graduating cadets until 1933 [Ambrose 1966, p. 290]. Moreover, unlike almost all American colleges at the start of the nineteenth century, the USMA was secular, and it has been national in its intent, outlook and programs from the very beginning. On the other hand, of the college, the seminary, the academy, the lyceum, etc., the college of nineteenth-century America is still the closest corporate body to which we can compare the Academy.

After the turn of the nineteenth century in American colleges, mathematics was alloyed with other disciplines (most often, natural philosophy) and instruction in the sciences was not well differentiated within the classical curriculum. Rote recitation of computational rules also characterized college arithmetic and algebra, and also in geometry, rote memory from one of the English Euclids (Robert Simson, Thomas Simpson, or John Playfair, for instance) was the usual practice [Karpinski 1980]. Some additional mathematical study – selected from trigonometry, navigation and surveying, mensuration, conic sections, fluxions, etc. – might be irregularly offered to some students at some institutions. College mathematics at this time had two basic purposes: to discipline the mind; and to a lesser degree to be a tool for commerce. Slowly, as the first decades of the new century unfolded, there was a "redirection of mathematical sights from Britain to Continental Europe" [Parshall and Rowe 1994, p. 15], mathematics gradually gained some self-definition, and the nation's leading colleges (Harvard, Yale, University of Virginia, etc.) began to offer various elective programs in "higher" mathematics [Cremin 1980, pp. 111, 270-280, 403-409; Parshall and Rowe 1994, pp. 3-4, 7-8, 12-20, 50].

Instruction in mechanics was also given in some of America's early nineteenth-century colleges, but textbooks in mechanics were scarce. Enfield's *Institutes of Natural Philosophy* was in widespread use, but it (first edition, 1785) was a badly spliced together compilation based almost entirely on sources from the first half of the eighteenth century [Anonymous 1821]. Indeed, the format of the *Institutes* was copied from Newton's *Principia* – Propositions followed by Scholia – which made it ideal for rote recitations. An early issue of *The American Journal of Science* damned Enfield in a review, implying that he was "... one who takes on himself the responsibility of adding to the number of books without adding to the amount of knowledge in the world" [Anonymous 1821, 3, p. 127].

At the USMA, Jonathan Williams' attempts to overcome these legacies and forge a new kind of institute of higher and especially technical learning were, in spite of their shortcomings, truly heroic against the standard of the American college. Moreover, around the time Thayer assumed command at West Point " ... Americans became interested in the pursuit of science .... This intensification of activity and the consequent availability of nationally distributed media made possible for the first time the development of an 'American scientific community' " [Daniels 1968, pp. 3,4]. The main goal of Thayer's program was to train engineers: "Surveying in the early nineteenth century was considered to be a genuinely scientific occupation because of its connection with astronomy ... the West Point-trained surveyor was far more than an engineer. Under the leadership of Thayer, there was assembled at the Military Academy some of the best scientific talent in the country ...." [Daniels 1971, p. 180].

As we have seen, Thayer drew his faculty from the cream of the Academy's graduates. The assistants were young graduates who, with only a few exceptions, held these positions for only a year or two or three. A small number of assistants were novitiates for the professorships. As it turned out, three of Thayer's professors dominated the Academy for over thirty years through the middle third of the century: Church; Bartlett; and Dennis H. Mahan, the professor of engineering from 1832 to 1871. Through their long tenures on the Academic Board, they "controlled every phase of the academic operation at West Point" [Morrison 1986, p. 43]. Over a span of many years, almost every Saturday, Church, Bartlett and Mahan rowed across the Hudson River to Cold Spring to play whist and have dinner [Holden 1911, p. 188]; one imagines that the real business of the Academy was transacted across the game or the dinner table. A certain insularity and academic stagnation became inevitable [Morrisson 1986, p. 59]. "The Academy continued to teach the same subjects, often with the same professors, from the same textbooks" for too many years [Ambrose 1966, p. 193]; for instance, Church's *Descriptive Geometry* was used up into the 1930s! Even though Thayer had certainly been inspired by many features of the École polytechnique, as the century wore on, the paths of the two institutions separated.[9] The Academy's professors "were not as keen to follow the progress of their sciences as they certainly would have been had they lived in an intellectual atmosphere like that of Paris" [Holden 1911, pp. 185, 186].

---

[9]The West Point Collection contains some of Cauchy's pathbreaking works in analysis, {Cauchy 1826} {Cauchy 1827}, but the Academy's calculus courses did not lead the way in introducing Cauchy's calculus in the United States.

At West Point, by the late 1820s, mathematics served as the language of science and engineering. This attitude quickly grew at other American colleges as well, as one of the strains on the classical curriculum. New institutions such as Rensselaer were established outside of the college family. George Ticknor organized Harvard's curriculum into departments and students were permitted a few choices, and when the University of Virginia was opened in 1825, a system of electives was built in. However, "to lay the foundation of a superior education," the uniform college curriculum of the classics, mathematics, and the natural sciences was defended in the famous Yale Report of 1828 [Cremin 1980, p. 272]. "Later, when these sciences [mathematics, chemistry, physics, and engineering, as well as laboratory instruction] entered the curriculum of the liberal arts colleges, the substance and methods developed at West Point and Rensselaer became the models" [Cremin 1980, p. 403].

West Point's example also informed the first "college-affiliated schools of technology," the Lawrence Scientific School at Harvard and the Sheffield Scientific School at Yale. Henry L. Eustis (USMA 1842) was recruited from the Academy to be Lawrence's first engineering professor and went on to become Lawrence's second dean, while William A. Norton (USMA 1831) was the first engineering professor at the Sheffield School. Paradoxically, as the nation's established colleges looked to West Point for leadership in the new technological education, the Academy was no longer the pre-eminent scientific and engineering school in the country and the faculty was not so much a collection of distinguished scholars as a group of very efficient teachers of the Thayer system [Ambrose 1966, pp. 197, 198, 204]. Relative to the rest of American higher education in general and progress in mathematics and science in particular, the Academy's trajectory was actually in decline in the second half of the century. "West Point had stressed science and civil engineering so early as to be in effect the nation's first engineering college ... But West Point itself passed its zenith in engineering during the 1830s ... ." By 1860, according to the Superintendent's report to Congress, "the Military Academy was 'no longer a school of engineering' but 'an institution for the purposes of national defense' " [Bruce 1987, pp. 160-164].

4. The West Point Collection

As noted, William A. Barron revealed the early wealth of mathematics and mechanics materials at the Academy in his "Inventory" of 1803 [Barron 1803]. Almost all of the mathematics and mechanics volumes were in English, whereas most of the works in fortification and engineering were in French. Perhaps the most notable work in [Barron 1803] was "Barrow's Euclid." Since [Anonymous 1807?] and [Thayer 1810] also recorded a "Barrow's Euclid," we might be tempted to think that this is the same volume as {Euclid 1751}; however, the book in the current collection has Partridge's holograph, so it remains unclear.

The two new professors resulting from the Congressional reforms following the War of 1812 also contributed significantly to the West Point collection. We have identified more than a dozen works, from a very broad spectrum, which belonged to Andrew Ellicott. Several are astronomical works, from {Halley 1752} to an early textbook {Keill 1730}. Manifesting Ellicott's international scientific reputation, there are also two presentation copies, {Biot 1802} (presented by the National Institute of France) and {Delambre 1806} (presented by the author). None of Ellicott's books appear to be closely connected to his teaching duties at the Academy, however.

In the spring of 1815, Jared Mansfield complained that the library had "fewer scientific books than 13 years ago" [Mansfield to Swift, March 28, 1815, National Archives, Record Group 77, C81]. During his first tenure at West Point, Mansfield had apparently given some volumes to the library, but no records and no books with his autograph or other identification remain. As the Academy struggled in the aftermath of the War and just before the departure of Thayer and McRee for Paris, Mansfield called for "appropriations of monies for the purpose of Library & Apparatus," estimating that at least $3000 should be spent for books at "Booksellers in Europe" [Mansfield to Swift, March 28, 1815, National Archives, Record Group 77, C81]. While Thayer and McRee were in Paris, Mansfield reported to the Superintendent that "The Library ... is miserably deficient in those [books] of Science," especially "those great works in science which are the Admiration of Europe" [Mansfield to Swift, April 26, 1816, National Archives, Record Group 77, C84]. By the spring of 1817, Mansfield had assembled "a Catalogue of books which I have purchased" and which included books for his classes, such as Martin's *Philosophy*, and about fifty additional works in mathematics and mechanics for the library. All of these works were in English and some were purchased from the collection of Samuel Webber, formerly Hollis Professor at Harvard. Notable among Mansfield's acquisitions were {Newton 1745}, {Bossut 1803}, {Stone 1730}, and several of Charles Hutton's works which were not textbooks such as {Hutton 1796} [Mansfield to Swift, May 10, 1817, National Archives, Record Group 77, C91].

The next significant addition to the collection owed to the efforts of Thayer and McRee in Paris in 1816. Of the books that they had shipped back, no more than 200 were purchased as bound; more than 800 were obtained unbound and "bound afterwards, according to instructions" [McRee to Swift, Dec. 18, 1816, National Archives, Record Group 77, A11]. Thayer and McRee engaged the three Paris firms Kilian, Tessier, and Picquet to purchase these books and to do the binding [McRee to Swift, Jan. 8, 1817, Thayer Papers]. The design of the bindings had several important features:

(1) the plates were to have been pasted on a blank sheet, "uniformly placed at the end of the volume ... and unfolding entirely out of the book: by which means you read with the whole plate constantly before you ... ." [McRee to Swift, Dec. 18, 1816, National Archives, Record Group 77, A11]; and

(2) each casing would be of full leather with gold trim and print on the spine as well as a decorative border around the edge of both the front and back covers, and the distinctive gold stamping

U. S.
Military Academy
West - Point

would appear on the front [Anonymous 1975]. (This is what we have called the *Thayer binding*. See Figure 25 for a photograph of a copy of the binding.) McRee complained that "this plan has not been an economical one," because it did not conform to the practices of the Parisian bookbinders, but he asserted that the arrangement of the plates was "fully worth the additional expense" [McRee to Swift, Dec. 18, 1816, National Archives, Record Group 77, A11]. Unfortunately, McRee observed, and current observations agree, that there was still "a good deal of incongruity in the binding" [McRee to Swift, Jan. 8, 1817, Thayer Papers]. Somewhere between sixty and one hundred books were not included in this shipment because their bindings were not complete.

From the accounting of Claudius Berard, confirmed by Superintendent Swift, there were "1142 vols. of text & plates – not including atlasses, & loose charts, nor the 287 vols. of the Encyclopedia" [Berard 1817] which were deposited in the Academy library by September 1817. Of course, the great majority of these works were of military interest, from the *Campagnes de Bonaparte en Italie* to the classic Vauban *Défense et attaque des places, et traité des mines*. Also, there were many volumes on engineering, such as Gaspard Monge's *Arte de fabriquer les canons* [Monge 1794]. In mathematics, Thayer and McRee acquired some pioneering works in descriptive geometry such as {Frezier 1737} and Monge's *Géométrie descriptive* (today the West Point collection has a later edition, {Monge 1820}), single copies of numerous textbooks like {Legendre 1813} and Garnier's translation of {Euler 1807}, and some of the earliest work in pedagogy especially {Lacroix 1816a}. Significantly, there were also a good number of higher level scholarly works such as {Montucla 1798} and Peyrard's {Euclid 1814a}, the first edition of Euclid's *Elements* to be based on a copy of a manuscript more ancient than those of Theon of Alexandria (4th century AD) [Heath 1956, vol. 1, pp. 46, 103]. Lagrange's *Théorie des fonctions analytique* {Lagrange 1797} was not included, however Wronski's *Réfutation* {Wronski 1812} was, along with Labby's translations of {Euler 1796} and {Carnot 1813}.

In addition to the first eleven volumes of the *Journal de l'École Polytechnique*, Thayer and McRee purchased in all 56 works in mathematics, of which 45 still remain in the collection. The mathematics volumes they sent back to the Academy, all French works, are not quite the same as what the library today calls "The Thayer Collection"; for instance {Keith 1820} has the distinctive Thayer binding but is obviously a later addition.

During Thayer's reforms of the Academy between 1817 and 1822, a number of mathematics textbooks were added to the library, from the rather old {Sturm 1709} and {Camus 1750} to the very much up-to-date {Dupin 1813}, {Hachette 1818} and {Leslie 1811 or Leslie 1820a}. In the printed "Catalogue ... , August, 1822" [Anonymous 1822], presumably prepared by Claudius Berard, the West Point library was reported to have contained a total of 941 volumes,[10] of which 160 were in "Mathematics and Navigation" and "Natural Philosophy". We note that 109 (68%) of these "mathematics" (in the wide sense of the term) works were in French; the whole collection, as tallied in this "Catalogue", was almost exactly 60% French at

---

[10]Despite its authority, this accounting is surely incomplete in view of the 1142 volumes acquired by Thayer and McRee, which of course were added to whatever remained of the library assembled by Williams and Swift.

this time. Almost all of these mathematics and mechanics works came from the late eighteenth and early nineteenth centuries.

Thayer played a strong role in expanding the Academy's collection throughout his superintendency, retaining an account with Kilian in Paris and keeping up with catalogs from Didot and Bachelier, the most prominent scientific publishers in Paris at that time [Thayer to O'Connor, May 21, 1824, Thayer Papers]. Through the "Book Fund of the Military Academy," great numbers of textbooks were purchased from Paris, from London, and in the United States, together with many volumes that added stature to the West Point library.

The Academy began a serious effort to build a collection in classical Greek mathematics. Early translations of some of the extant portions of Apollonius' *Conics* were obvious choices: {Apollonius 1655} and {Apollonius 1661}. The "most remarkable of the extant Books" of the *Conics* was Book V, which "deals with normals to conics regarded as maximum and minimum straight lines drawn from particular points to the curve." This Book is a "veritable geometrical tour de force" [Heath 1981, vol. 2, p. 159]. It was restored in a controversial work by Vincentio Viviani {Hutton 1796, vol. 2, pp. 660-661}, and {Viviani 1659} was added to the West Point collection in the 1820s. We are surprised to find one of Apollonius' minor works: "The two books of *Plane Locii* are at once the least recoverable and most to be missed of the five lost minor works of Apollonius in the 'Domain of Analysis' " [Jones 1986, vol. 2, p. 539]. Pappas gave "a pretty full account of the contents" [Heath 1981, vol. 2, p. 185] of this work in Book VII of *The Collection* (a later edition of Commandino's translation is {Pappas 1660}). Based on Pappas' account, reconstructions were made by Pierre Fermat (published in 1679), Frans Van Schooten (1656), and Fr. Gasper Schott (1658). Finally, the most complete reconstruction, {Apollonius 1749}, was also "accompanied by an excellent study of the methods of ancient geometry" by Robert Simson [ver Eecke 1933, pp. lxxii-lxxv; Jones 1986, vol. 2, pp. 539-546 and vol. 1, pp. 104-112].

The intellectual level of the mathematics and mechanics collection in the West Point library was considerably elevated during the decade of the 1820s with the addition of what we call *scholarly books*, i.e., works like the translations of some of the Greek classics that were not textbooks and not in any direct way supportive of textbooks or of contemporary Academy courses. We call attention to {Copernicus 1543} and {Alhazen 1572}, and some noteworthy collected works: {Bernoulli, Jacques 1744}; {Bernoulli, Johann 1742}; {Cardano 1570}; {Fermat 1679}; {Newton 1779}; {Stevin 1634}; {Viète 1646}; and {Wallis 1695}. In more up-to-date scholarly books, three of Euler's works in calculus were added, {Euler 1744}, {Euler 1768}, and {Euler 1790}. When a copy of the *Disquisitiones arithmeticae* was purchased, it was the French translation {Gauss 1807}, the first translation of this work into any other language.

Another measure of the ascent in quality of the West Point mathematics and mechanics collection during this time was the acquisition of full runs of the *Journal de l'École Polytechnique*, of Gergonne's *Annales de Mathématiques Pures et Appliquées*, and of the *Transactions of the Royal Society*. The first of these journals also allowed the Academy's professors to keep up with the French curricular reforms of the decade. We do not include a complete listing of the Academy's journal holdings in this catalog.

In the late summer and fall of 1825, the Military Academy Book Fund paid Kilian $1095.40 for 48 different titles among which were sixty copies of Biot's *Essai de géométrie analytique*, and then on February 10, 1826, {D'Alembert 1784} and the fifth volume of {Laplace 1799} were purchased through the same account [MacKay to Kilian, July 1, 1825; Sept. 10, 1825; Nov. 14, 1825; Feb. 9, 1826, USMA Papers][11]. In 1822, Garnier's translation of the *Algèbre* {Euler 1807} was in the West Point library, and so the expanded English translation {Euler 1822} was added to support Academy courses. Twelve textbooks of Sylvestre Lacroix, eleven of Thomas Simpson, seven of William Emerson, and others account for a good deal of the quantitative gain of this collection during the 1820s.

The Academy purchased many copies of Hassler's *Trigonometry, Geometry,* and *Arithmetic* textbooks directly from the author. During 1824 and 1825, Thayer negotiated the purchase of a sizable portion of Hassler's scientific library for the Academy, judging "that the books were of a rare and valuable kind & are wanted in order to render our library tolerably complete in the Departments of Mathematics, Philosophy & Astronomy" [Thayer to Swift, Dec. 27, 1824; Dec. 30, 1824; Jan. 26, 1825 Thayer Papers] [Thayer to Macomb, Sept. 14, 1825, National Archives, Record Group 94, 613]. Douglass estimated the value of this part of Hassler's library at $1352 [Douglass to Thayer April 26, 1825, National Archives, Record Group 94]. Thayer directed other purchases of books in the United States [Thayer to Brown, Jan. 31, 1826, Thayer Papers][12], and proposed comparison of the West Point scientific collection with that of Nathaniel Bowditch, known to have had an outstanding private library [Thayer to Ticknor, Oct. 12, 1826, Thayer Papers].

The second printed "Catalogue" of the USMA library [Anonymous 1830], again presumably prepared by Berard, showed the dramatic growth both in quantity and in quality of the Academy's mathematics and mechanics collection. The library had grown to a total of 2,852 volumes, and in just "Mathematics" (excluding "Natural Philosophy-Navigation") there were now 366 titles. For an American institution, European works were quite prominent, 124 (34%) of the strictly "Mathematics" works were French and 18 (5%) were German. But it was the increase in quality of the Academy's mathematics and mechanics collection that was most striking. Considerably more than half of the most important pre-nineteenth-century works which we have listed in this bibliography had been secured for the Academy by 1830. In addition to acquiring many of its most valuable individual volumes, this collection took on its unique character during Thayer's term as Superintendent.

As noted, the fire of 1838 lightly damaged fifty mathematics and mechanics books, according to Berard's report [Berard 1838]. Perhaps the poor condition in which we found the valuable and multi-volumed {Maseres 1791} resulted from this fire; from this distance this is a hard judgment to make. Berard also reported twenty-one mathematics and mechanics works missing after the fire. In comparing his report with our bibliography, we see that volume 11 of Gergonne's *Annales* was either found or replaced and we can say the same for at least a half dozen other works.

---

[11]This type of correspondence is listed under USMA Papers in the Correspondence section of the References.

[12]This type of correspondence is listed under Thayer Papers in the Correspondence section of the References.

When Thayer left West Point in 1833, the collection's growth slowed and by 1840 there were 405 strictly "Mathematics" works listed [Anonymous 1840]. The most scholarly volumes added were {Maseres 1791} and the anthology {Maseres 1795}. The Academy kept abreast of mathematical developments in this country by subscribing to Robert Adrain's *Mathematical Diary* and to Charles Gill's *Mathematical Miscellany*. Although it may not have been apparent at the time, the most valuable additions to this collection during the 1830s were probably the *Exercices* of {Cauchy 1826}, the *Nouveaux exercices* {Cauchy 1835}, and the manuscript "Cours d'analyse" {Cauchy 1827}.

Today, there are four categories of works in the West Point collection that deserve special attention. First, the primary materials at the Academy for the study of the development of calculus extend back to the seventeenth-century origins of the subject. We noted {Fermat 1679} above and there is also a copy of the book that introduced analytic geometry to the community of scientists, {Descartes 1649}. Special mention must be made of René de Sluse's copy of {Saint Vincent 1647}. In addition to several editions of the *Principia*, there are {Newton 1706}; {Newton 1736}; and {Newton 1745}. Correspondence between Leibniz and Johann Bernoulli is contained in {Leibniz and Bernoulli 1745}, and the English side of the Newton-Leibniz dispute is presented in {Anonymous 1725b}. From the eighteenth century, there are many works bearing on the development of and the controversies surrounding fluxions and calculus; perhaps the most interesting are {Maclaurin 1742} and {Maclaurin 1749}. The struggles for reform at the end of the century and in the early decades of the nineteenth century are well represented; for instance, see {Carnot 1813} and perhaps especially {Cauchy 1827}.

Part of the legacy the United States Military Academy owes to the École polytechnique is the substantial collection of descriptive geometry books, both texts and collections of plates. There are even several valuable predecessors to the works of Gaspard Monge such as {Frezier 1737}, from what might be called the prehistory of descriptive geometry. Many of the French, English, and American texts, published throughout the nineteenth century, are included. Especially noteworthy are the several works of Théodore Olivier and, of course, many of the editions of Church's texts.

For one who wishes to study the history of logarithms, and especially their tables, an obvious place to start is {Napier 1614}. "The first extensive work on logarithms of the decimal system" [Henderson 1926, pp. 41, 55, 63-64] was that of Henry Briggs, and three of his earliest works on this subject are {Briggs 1624a}, {Briggs 1624b}, and {Briggs 1633}. The number of other seventeenth- and eighteenth-century books on logarithms is small, but these gaps are offset by the massive {Maseres 1791} which contains republications of many treasures (e.g., Kepler's *Chiliades*). Also note that {Hutton 1811} contains Hutton's "Large and Original" history of logarithms. The tables {Babbage 1827} were printed on yellow paper as part of an experiment in which Charles Babbage made a study of typography, layout, design, and color in his effort to improve the ease and usefulness of such works. Of special West Point interest, {Hassler 1830} was reviewed by Gauss and then translated into several other languages.

In supporting the courses established during and following Thayer's curricular reforms, West Point accumulated an extensive collection of some of the earliest higher mathematics textbooks. The textbook was a genre which gradually emerged in France and England in the eighteenth century and the first years of the nineteenth century. In the United States, West Point authors played a leading role as the nineteenth century unfolded.

There were several *Cours de mathématique*'s during the French "ancien régime," and one of the earliest was {Ozanam 1697} (See Figure 8 in Appendix 2). Usually these were multi-volumed works that contained a potpourri of mathematics and related science applicable to the artillery and other military uses and to various facets of cartography, construction and the beginnings of engineering. The most popular such *Cours* was that of {Bezout 1777}, first published in 1764 for the French Navy and then revised and used well into the nineteenth century [Dhombres 1985].

The process of transforming Isaac Newton's *Principia* and his other advances in physics into textbook form was begun by J. T. Desaguliers and W. J. van 'sGravesande. The latter attended Desagulier's public lectures (which featured experiments) in London in 1715, then used them as the basis for his own course at the University of Leiden in 1717. 'sGravesande turned his lectures into a book, of which {'sGravesande 1784} is a later edition. It was translated into English {Desaguliers 1719} and became a well-known textbook [Taylor 1966, pp. 6, 7, 121, 122]. Another independent English lecturer who produced a number of "generally useful" early higher mathematics textbooks, some of which are at West Point, was the "optician and globe maker" Benjamin Martin {Hutton 1796}.

The establishment in 1741 of the Royal Military Academy at Woolwich provided a base for several English textbook authors who are represented in this collection. Comparable to the French *Cours de mathématique* was {Muller 1757}. There are several textbooks of Thomas Simpson's, including a first edition of his *Fluxions* {Simpson 1734}; some of his texts retained their popularity into the first two decades of the nineteenth century. We have mentioned some of the textbooks of Charles Hutton and his student Olinthus Gregory above.

During the Revolutionary period (1789 to 1802) in France, "a whole system of scientific education was established" beginning with the Central Schools (later the Lycées) and extending to the École polytechnique and the Écoles d'application, which were specialized engineering and military schools [Dhombres 1985, p. 125; Dhombres and Dhombres 1989, pp. 596-604]. "Uniformization of mathematical education" [Dhombres 1993, pp. 4-11] was the goal and this led to a "golden age" (1797 to 1819) of publication of mathematics books, especially textbooks [Dhombres 1985, pp. 135–136]. Mathematics was beginning to separate itself from other sciences, and some authors like J. G. Garnier were translating and adapting older classics to classroom use; for example {Clairaut 1801} and {Euler 1807}. By the time of the Empire, with the endorsement of the state [Dhombres 1985, p. 105], Lacroix and others subdivided the *Cours de mathématique* into many separately titled works (arithmetic, algebra, geometry, trigonometry, etc.) intended to be used as separate textbooks at different stages in a student's development; there are many editions of Lacroix's texts in this collection. Many of the reforms in French mathematical education were described in {Lacroix 1816a}, first published in 1813 [Dhombres and Dhombres 1989, pp. 252-256; 356-367; 417-421].

The English reform in university mathematics, especially in calculus, was stimulated by {Woodhouse 1803} and actually began with {Lacroix 1816b}. However, by this time, thanks to Thayer's and McRee's expedition, the major influences on mathematics at West Point were French. So this collection has only a small number of nineteenth-century English textbooks, although {De Morgan 1842} was the first calculus book written in English and based on limits [Howson 1982, p. 82].

From the periods of the Empire and the Restoration, the following French textbooks were translated and/or edited for American students by Academy faculty and/or graduates: {Legendre 1813} became {Davies–Geometry and Trigonometry 1834}; {Bourdon 1823} and {Bourdon 1828} were first {Ross 1831} and then {Davies–Algebra 1835}; {Boucharlat 1826} and {Boucharlat 1830} were the major influences on {Davies–Analytical Geometry 1836}; {Biot 1826} became {Biot 1840}; and {Boucharlat 1827} led to {Courtenay 1833}.

Throughout the nineteenth century and into the twentieth, the West Point library continued to secure numerous French higher mathematics textbooks, for just two examples {Bertrand 1864} and {Poincaré 1912}. It is disappointing that, with the presence of many editions and copies of nineteenth-century mathematics and mechanics textbooks by Academy authors, there are no complete sets. Furthermore, with only a small number of exceptions (e.g., {Pike 1788}), there are very few early American mathematics textbooks by non-West Point authors.

The "birth of the advanced [mathematics] textbook" is difficult to define exactly [Dhombres 1985, p. 98] [Dhombres 1989, p. 28]. However, within some limitations, the West Point collection provides a rich source for this study of our heritage.

From textbooks to works of analysis to some of the farthest reaches of what we can call "mathematics," the USMA has both provided for and been the recipient of a fortuitous conjunction of historical forces. The collection of primary materials in the history of mathematics at West Point is – and is not – an "American" collection. Only 26% of the works are American, whereas 33% are French and 27% are English. On the other hand, the founding of the Academy and the development of the new country, from the post-Revolutionary era to the settling and creation of the new western states, were quintessentially American impulses. West Point was perfectly positioned to house what is now a unique resource. At the beginning of the nineteenth century in the United States, mathematics slowly began to gain a certain amount of freedom from the constraints of its ancillary roles in the classical education of gentlemen, and the Academy was very much a leader in this transformation. As a result, during the first half of the century, the nation produced a number of mathematical practitioners whose main occupations were to support the growth of technology, engineering, and science (of which the art of war turned out to be just a part). The collection's architects, Jonathan Williams, Sylvanus Thayer, and their successors were among the first persons in the United States to recognize the centrality of mathematics to science and engineering. They were in a position to exercise this insight, and we are the benefactors of their foresight and labors.

5. Organization of the Catalog

Scholarship is a community undertaking wherein each person's contribution builds on that of predecessors, and one's work is accomplished for the benefit of those to follow. A bibliography is one of the foundations of this whole enterprise. Our main goal is to provide as complete and accurate a record possible of the mathematics collection at the United States Military Academy.

According to {Hutton, vol. 2, 1795, p. 81}, the word "mathematics" originated with the Greek "$\mu\alpha\theta\eta\sigma\iota\varsigma$", progressed to the Latin "mathesis", and originally signified "discipline or science in general." As Charles Davies, West Point's most famous mathematician, explained it, the Greek root of the word "mathematics" meant simply "to learn" {Davies and Peck, 1855, p. 356}. History would thus appear to dictate that our view of mathematics for this collection should be broad. In addition to including all the works which we would consider mathematics today, we have also embraced all works of mechanics which we have found. Astronomy, surveying, cartography, optics, acoustics, navigation, hydrostatics, and even natural philosophy books have been examined with a generous eye, and those with a significant mathematical content have been included also. Some difficult judgments have ensued. For instance, as might be expected, the West Point library contains many works on fortification such as [Ozanam 1727] and [Tartaglia 1588], and works on military engineering such as [Stevin 1617] and [Monge 1794]. Some of these books contain a certain amount of geometry and some mechanics; however we were compelled to exclude those works whose primary purpose was something other than "mathesis."

The oldest work in our catalog is {Regiomontanus 1496}, and the newest is {Darboux 1917}, thus establishing our boundaries in time. We decided not to cut the collection off at the turn of the twentieth century because we encountered a number of late nineteenth-century works, especially textbooks, whose various volumes and editions ran into the early years of the twentieth century; {Chomé 1898b} is just one example. Since the collection is so heavily French, World War I seemed like a more natural boundary. The listings in this bibliography are alphabetical by author.

There are 798 authors and 1195 book listings in our catalog, which represent 1340 different works, when all volumes and editions of the books are included. Since there are multiple copies of many books, we estimate over 1500 books are included in the cataloged collection. Of the 1340 works listed, Table 5 summarizes the breakdowns by century, language and place of publication.

All of the works should be in one of the following three locations: USMA library Special Collections; USMA library circulating collection; USMA Department of Mathematical Sciences library in Thayer Hall. Our goal has been to record every mathematics work (in the above sense) at West Point and to do so with accuracy and completeness. Inevitably, some mistakes will have slipped by us. We hope they are so minor as to be obviously correctable and that they will not detract from the main objectives of our work.

The details of our bibliographic style originate with *The Chicago Manual of Style*, 13th ed., form B. However, to provide as much information as possible about

TABLE 5. The Century, Language, and Place of Publication for the 1340 volumes in the catalog.

Century	15th	16th	17th	18th	19th	20th	no date
	1	11	53	290	862	98	25

Language	English	French	Latin	German	Spanish	Italian	Greek
	710	478	92	54	3	2	1

Place of Publica-tion	France	Eng-land	US	Ger-many	Scot-land Ireland	Italy	Belgium Holland	Other	Unk
	444	360	349	85	25	9	15	32	21

the works in this collection, we have evolved some variations. In broad terms but with two exceptions to be explained below, each of our entries is organized into the following major elements:

A. The author's name (for the author's first listing);
B. The body of the bibliographical information;
    1. Date of publication;
    2. Title material;
    3. Publication data;
    4. Pagination;
C. Two categories of annotations;
D. Cross references.

Element A.

The entries in this bibliography are alphabetized by the author's last name. In the case of works by multiple authors, we use the last name of the first author to determine the work's placement in our bibliography.

The first work by an author is headed by the author's name in bold-face, last name first. When known, birth and death dates are included, and when appropriate, the author's USMA class date is also given. We use as much of the most common spelling of the author's name as we know, and whenever possible, the *Dictionary of Scientific Biography* (DSB) has been our guide to the spelling of non-English names.

When an author has two or more works in the bibliography, we list them in the chronological order of publication. Then the author's name is omitted for succeeding works. The entries of the two West Point authors, Albert E. Church (1807-1878) and Charles Davies (1798-1876), are the only exceptions to this ordering, and the organization of their works will be described below.

For works by multiple authors, all of the authors and their birth and death dates, etc. are included. When a work was edited, we made the following distinction: in the rare occasions when the editor functioned as a compiler and thus was the primary author, the entry has been listed under the editor's name; in most cases, however, the editor functioned as a reviser and therefore we placed his name after the work's title in the body of the entry and merely added a cross reference under the editor's name. The same distinction and practice was employed for translators.

There are 19 anonymous entries for which there are no authors, editors, etc., specified on the title page of the work itself.

Element B.

1. The publication date is also bold-faced. Additional works by the same author are listed chronologically. When two or more of an author's works have the same publication date, we append lower case letters, e.g., 1811a first, then 1811b, etc. Different editions of a work are given separate entries in our bibliography, according to their publication dates.

With multi-volume works whose publication dates are not all the same, we list these dates, separated by semi-colons, in the order of the volumes.

When there is no date on the title page but if through other means we know the publication date, we include it enclosed in square brackets. In some instances, when we think the copyright date may be of help, we include it, as for example, "c.1811." When we have no reliable information about the publication and/or the copyright date(s), we simply write "n.d.".

2. The title of each work is italicized. Each title is given in full, with all subtitles, in the language of the work and in the spelling actually appearing on the title page. We have preserved the existence (and in the case of some of the French works, the absence) of the accent marks on the title page. Capitalization follows the rules of the language of the work. In a small number of cases, we have made minor changes to the punctuation for clarity.

In some multi-volume works, there are variations, some substantial, in the titles of the individual volumes. In these cases, we give all of the different titles in full.

If the author's name is embedded in the title, we usually do not include it; the major exceptions are most of the works in Latin and the few Greek works. We omit the biographical information often found underneath an author's name.

For editions of a work after the first, we reproduce the work's description of itself, for example, "Deuxième édition, corrigée et augmentée".

Next, we list any translator(s) and/or editor(s).

We specify the number of volumes with Hindu-Arabic numerals.

Finally, if a work is a member of a series and if that fact is noted on the title page, we reproduce this information.

3. The publication facts begin with the name of the city in the orthographic form and in the grammatical case in which it is given on the title page. When clarity would be served, we add the English name of the city in square brackets. When there are several cities listed and it is clear one of them is primary, that is the only one we give.

The publisher's name is given as it appears on the title page (e.g., "Firmin Didot"). When several publishers from the same city are given on the title page,

TABLE 6. Numbers of entries in different subjects for West Point authors Church and Davies.

Albert Ensign Church	entries
Trigonometry	1
Analytical Geometry	11
Descriptive Geometry	14
Calculus	8

(these entries date from 1842 to 1902)

Charles Davies	entries
Arithmetic	4
Algebra	11
Geometry and Trigonometry	16
Analytical Geometry	5
Linear Perspective	5
Descriptive Geometry	10
Surveying	12
Calculus	3
General Mathematics	1

(these entries date from 1826 to 1890)

we list only the first. On the other hand, when a work has clearly been published in more than one city by different publishers, we specify these facts. This required that we make some judgments. In particular, especially for some English and some American volumes, additional firms listed on the title page were, we believed, not publishers but rather booksellers, and so usually they were omitted.

In some works, in place of a publisher, the work is described as "Printed by" or "Printed for" some person or business. In these cases, we have repeated this information along with the first named firm, but otherwise we have not included the name(s) of printers when they appear together with that of the publisher(s).

In some multi-volume works, the publishers, like the publication dates and some of the titles, varied from volume to volume and we have recorded these variations from the title pages.

4. The pagination begins with the notation "pp." The most common arrangement begins with the front matter, followed by the body of the text, then appendices, and finally the number of plates. The front matter and the appendices are almost always numbered in Roman numerals and the body of the work in Hindu-Arabic numerals. Pagination information is presented in the order in which it appears in the copy of the book(s) we have examined, and distinct sets of paginations within each volume are separated by a "+" sign. For each set of unnumbered pages, the reported pagination is enclosed in square brackets.

Element C.

We have provided two kinds of annotations.

1. Annotations denoted by a diamond (◊) provide general information about the book or work. For instance, an unusual frontispiece or a portrait; use of the book at the Academy as a textbook; errata pages; a dedication or a special introduction; etc.

2. Annotations with the small circle (○) indicate our observations of the particular volume in the West Point collection. For example, multiple copies; holographs and/or marginal notes; the Thayer binding; major defects in the physical book such as missing pages or plates; etc.

Element D.

Following an author's last entry, we direct the reader to other works in the West Point collection in which the author had a major role but which we have listed under another author. We have provided cross references for all secondary authors, translators and editors. Most editors and translators will have only cross references to listings under the names of primary authors.

The works of Albert Ensign Church (USMA, 1828) and Charles Davies (USMA, 1815) are almost entirely textbooks. There are a good many of these texts, and in addition, almost all of them appeared in many, many editions spread out over not just years but decades. Our goal is to render an accurate accounting of these various works, many in multiple copies.

First, as noted at the start of the entries for each of these authors, we have gathered the Church and the Davies works into classes by subject. See Table 6 for the breakdown of these subjects. Each subject class is ordered chronologically.

For each entry, the body of the bibliographical information and the two kinds of annotations are arranged exactly as in the rest of this bibliography. The annotations for each of these entries are especially noteworthy. Not only are there multiple copies of many of these books, some contain valuable handwritten notes. With some of the Davies books especially, there are variations in the titles, spin-offs and transformations of books, and there are revisers and secondary authors; the Davies textbooks were enthusiastically marketed through much of the nineteenth century. As generous as the West Point collection is in the Church and Davies textbooks, comparison with [Karpinski 1940] shows that many editions and even titles prior to 1850 are unfortunately not included.

At the end of each of these classes, we have given appropriate cross references to other related works in this bibliography, e.g., to Bourdon from Davies (*Algebra*).

There are over 140 works listed in the catalog that were used as textbooks at the USMA. Over 15 of these volumes were published at West Point and many titles contained phrases to the effect "for use by cadets at West Point." While Davies and Church were the most prolific of the West Point authors, there are 31 authors in the catalog who either graduated from or taught at West Point. This list is provided in Table 7.

TABLE 7. List of West Point authors who have listings in the catalog.

Name	USMA Class	USMA Position(s)
Alvord, Benjamin (1813-1884)	1833	Asst Professor, Mathematics 1837–1839
Barnard, John (1815-1882)	1833	Superintendent 1855-1856
Bartlett, William H. Chambers (1804-1893)	1826	Professor, Nat & Exp Philosophy 1834–1871
Bass, Edgar (1843-1918)	1868	Professor, Mathematics 1878–1898
Blakely, George (1870-1965)	1892	Asst Professor, Mathematics 1898-1901
Church, Albert Ensign (1807-1878)	1828	Professor, Mathematics 1837–1878
Courtenay, Edward Henry (1803-1853)	1821	Professor, Nat & Exp Philosophy 1829–1834
Crozet, Claudius (1790-1863)	—	Professor, Engineering 1817–1823
Cullum, George W. (1809-1892)	1833	Professor, Engineering 1848–1855 Superintendent 1864–1866
Davies, Charles (1798-1876)	1815	Professor, Mathematics 1823–1837
Gordon, William B. (1853-1938)	1877	Professor, Nat & Exp Philosphy 1901–1917
Hackley, Charles (1809-1861)	1829	Asst Professor, Mathematics 1829–1833
Hagadorn, Charles (1866-1918)	1889	Asst Professor, Drawing 1896–1897
Hardy, Arthur S. (1847-1930)	1869	Artillery Instructor 1869
Hassler, Ferdinand (1770-1843)	—	Professor, Mathematics 1807–1809
Knight, J. G. D. (1846-1919)	1868	Asst Professor, Mathematics 1874–1881
Lee, Thomas Jefferson (1808-1891)	1830	cadet
Ludlow, Henry H. (1854-1926)	1876	Asst Professor, Mathematics 1879–1883
Mahan, Dennis Hart (1802-1871)	1824	Professor, Engineering 1830–1871
Mansfield, Jared (1759-1830)	—	Professor, Mathematics 1802–1803 Professor, Nat & Exp Philosphy 1814–1828
Michie, Peter S. (1838-1901)	1863	Professor, Nat & Exp Philosphy 1871–1901
Peck, William Guy (1820-1892)	1844	Asst Professor, Mathematics 1846–1855
Quinby, Issac Ferdinand (1821-1891)	1843	Asst Professor, Mathematics 1845–1847
Robinson, James W. (1827-1918)	1852	cadet
Ross, Edward C. (1801-1851)	1821	Asst Professor, Mathematics 1824–1833
Smith, Edmund Dickerson (1854-1900)	1879	Asst Professor, Mathematics 1884–1895
Smith, Francis H. (1811-1890)	1833	cadet
Smith, Richard S. (1813-1877)	1834	Asst Professor, Drawing 1840–1855
Steese, James Gordon (1882-1958)	1907	cadet
Wendell, Abraham (1796- 1817)	1815	cadet
Young, Gordon R. (1891-1963)	1913	cadet

6. References

6.1 Correspondence

National Archives. Record Group 77. Records of the Office of the Chief of Engineers. Military Academy Papers, 1813–1818. Letters from Box 1 have prefix A; those from Box 3 have prefix C.

National Archives. Record Group 94. Records of the Adjutant General's Office, 1780's-1917. Correspondence relating to the Military Academy. Entry 212, Box 7. The letter number is given.

Thayer Papers. *The West Point Thayer Papers, 1808–1872*, edited by Cindy Adams, et al. West Point, NY: Association of Graduates, 1965. 11 volumes. This transcription of letters from many sources is arranged by date. [Copy in Special Collections, USMA.]

USMA Papers, 1824–1829. Box marked "U.S. Military Academy papers, 1824–1829. Library. Book and stationary dealers. Book publishers and book sellers." File: Kilian.

6.2 Other References

Ambrose, Stephen E. 1966. *Duty, Honor, Country: A History of West Point*. Baltimore: The Johns Hopkins Press.

Anonymous 1802, "The United States Military Philosophical Society, Manuscript Minutes and Records, Membership Lists, Correspondence and Papers Written for the Society, 1802-1813," 4 vols. New York Historical Society.

Anonymous 1807?, "A List of Books belonging to the Library of the Military Academy." Handwritten manuscript. West Point, NY: United States Military Academy.

Anonymous 1821, "Remarks on Dr. Enfield's Institutes of Natural Philosophy – Third American Edition – 1820," *American Journal of Science*, 3, 125-157.

Anonymous 1822. *Catalogue of Books in the Library of the Military Academy*. Newburgh, NY: Ward M. Gazlay.

Anonymous 1830. *Catalogue of the Library of the U. S. Military Academy at West Point*. New York: J. Desnoues.

Anonymous 1851, "Elementary works on physical science," *North American Review*, 72, 358-395.

Anonymous 1853. *Catalogue of the Library of the U. S. Military Academy, West Point, N. Y. Exhibiting its Condition at the Close of the Year 1852*. New York: John F. Trow.

Anonymous 1904 (reprint 1969). *The Centennial of the United States Military Academy at West Point, New York, 1802–1902*. 2 Volumes. New York: Greenwood Press.

Anonymous 1975, "The Thayer Collection. Its History, Significance and an Account of the Restoration Project," *USMA Library Bulletin No. 13*, West Point, NY: United States Military Academy.

Anonymous ca. 1990. "History of the USMA Mathematics Department." Typewritten manuscript.

Artz, Frederick B. 1966. *The Development of Technical Education in France, 1500–1850.* Cambridge, MA: The Society for the History of Technology and the MIT Press.

Ball, W. W. R. 1889. *A History of the Study of Mathematics at Cambridge.* Cambridge: Cambridge University Press.

Barron, William A. 1803. "Inventory of Books, Maps and Charts, belonging to the Military Academy at West Point." Handwritten manuscript. Manuscripts Department, Lilly Library, Indiana University, Bloomington, IN.

Berard, Claudius 1817, "Catalogue of Books contained in ten cases, lately received from France, and sent to the Library of the Military Academy at W–Point by General J. G. Swift." Handwritten manuscript, dated Sept. 15, 1817. USMA, Special Collections. Copied from the National Archives, Record Group 217.

Berard, Claudius 1838, "A Statement of the damage and loss sustained by the Library of the U. S. Military Academy at the late fire." Handwritten manuscript. Special Collections, USMA Library, West Point, NY.

Boynton, Edward C. 1863 (reprint 1970). *History of West Point, and its Military Importance During the American Revolution: and the Origin and Progress of the United States Military Academy.* Freeport, NY: Books for Libraries Press.

Bruce, Robert V. 1987. *The Launching of Modern American Science, 1846-1876.* New York: Alfred A. Knopf.

Cajori, Florian 1929 (reprint 1980). *The Chequered Career of Ferdinand Rudolph Hassler.* New York: Arno Press.

Church, Albert E. 1879, "Personal Reminiscences of the Military Academy, from 1824 to 1831," A Paper read to the U. S. Military Service Institute, West Point, March 28, 1878. West Point, NY: U.S.M.A. Press.

Couper, William 1936. *Claudius Crozet: Scholar, Educator, Engineer (1789-1864).* Charlottesville, VA: The Historical Publishing Co.

Crane, John and Kieley, James F. 1947. *West Point: The Key to America.* New York: Whittlesey House, McGraw-Hill.

Crackel, Theodore Joseph 1981, "The founding of West Point: Jefferson and the politics of security," *Armed Forces and Society*, 7, 529-543.

Cullum, George W. 1891. *Biographical Register of the Officers and Graduates of the United States Military Academy at West Point, N. Y. from its Establishment, in 1802, to 1890 with the Early History of the United States Military Academy.* Boston and New York: Houghton, Mifflin and Company.

Cremin, Lawrence A. 1980. *American Education: The National Experience, 1783-1876*. New York: Harper & Row.

Daniels, George H. 1968. *American Science in the Age of Jackson*. New York: Columbia University Press.

Daniels, George H. 1971. *Science in American Society: A Social History*. New York: Alfred A. Knopf.

Daniels, George H. 1972. *Nineteenth–Century American Science: A Reappraisal*. Evanston, IL: Northwestern University Press.

Davies, Henry Eugene 1895. *Davies Memoir: A Genealogical and Biographical Monograph on the Family and Descendents of John Davies of Litchfield, Connecticut*. Privately Printed.

Dhombres, Jean 1985, "French mathematical textbooks from Bezout to Cauchy," *Historia Scientiarum*, 28, 91-137.

Dhombres, Jean 1989, "Grandeurs et servitudes de la vie mathématique en France: le passage du XVIIIeme siecle au XIXeme siecle," in *Les Amateurs de Science et de Technique, Cite des Sciences et de l'Industrie, Cahiers d'Histoire & de Philosophie des Sciences*, nouvelle serie, No. 27, 17-44.

Dhombres, Jean 1993, "Les repères d'une culture mathématique vers 1800: le témoignage de deux listes de livres," *Revista di storia della Scienze*, (2) No. 1, 1-83.

Dhombres, Nicole and Dhombres, Jean 1989. *Naissance d'un nouveau pouvoir: Sciences et savants en France 1793-1824*. Paris: Éditions Payot.

Dictionary of American Biography 1928-1958, under the auspices of the American Council of Learned Societies. Edited by Allen Johnson. New York: Charles Scribner's Sons. 22 volumes.

Dictionary of National Biography 1885+. Edited by Leslie Stephen and Sidney Lee. London: Smith, Elder. 22 volumes with supplements.

Dictionary of Scientific Biography 1981, published under the auspices of the American Council of Learned Societies. Edited by Charles Coulston Gillispie. New York: Charles Scribner's Sons. 18 volumes.

Dupuy, R. Ernest 1958. *Sylvanus Thayer, Father of Technology in the United States*. West Point, NY: The Association of Graduates, United States Military Academy.

Elliott, Clark A. 1979. *Biographical Dictionary of American Science: The Seventeenth Through the Nineteenth Centuries*. Westport, CT: Greenwood Press.

Ford, Norman Robert 1953. *Thayer of West Point*. St. Johnsbury, VT: Thayer Book Press.

Forman, Sidney 1945, "The United States Military Philosophical Society, 1802-1813," *William and Mary Quarterly*, 2, 273-285.

Fourcy, Ambroise 1828. *Historie de l'École polytechnique*. Paris: Republished by Belin (1987) with an introduction by Jean Dhombres.

Guralnick, Stanley M. 1975. *Science and the Ante-Bellum American College.* Philadelphia: The American Philosophical Society.

Heath, Thomas 1956. *The Thirteen Books of Euclid's Elements*, Second Edition Revised with Additions, 3 Volumes. New York: Dover Publications.

Heath, Thomas 1981. *A History of Greek Mathematics*, 2 Volumes. New York: Dover Publications.

Henderson, James 1926. *Bibliotheca Tabularum Mathematicarum, Being a Descriptive Catalogue of Mathematical Tables. Part I, Logarithmic Tables.* Cambridge: Cambridge University Press.

Hogan, Edward R. 1976, "George Baron and the Mathematical Correspondent," *Historia Mathematica*, 3, 403-415.

Holden, Edward S. 1911, "Biographical Memoir of William H. C. Bartlett, 1804-1893," *National Academy of Sciences Biographical Memoirs*, Volume 7, 173-193.

Howson, Geoffrey 1982. *A History of Mathematics Education in England.* Cambridge: Cambridge University Press.

Jones, Alexander (Editor) 1986. *Pappas of Alexandria, Book 7 of the "Collection"*, 2 Parts. New York: Springer–Verlag.

Karpinski, Louis C. 1980. *Bibliography of Mathematical Works Printed in America Through 1850*; and Supplement and Second Supplement. New York: Arno Press.

Langins, Janis 1990, "The Ecole Polytechnique and the French Revolution: merit, militarization, and mathematics," *Llull*, 13, 91-105.

Mansfield, Edward D. 1897 (reprint 1970). *Personal Memories Social, Political, and Literary with Sketches of Many Noted People, 1803-1843.* Freeport, NY: Books for Libraries Press.

Molloy, Peter Michael 1975. "Technical Education and the Young Republic: West Point as America's Ecole Polytechnique, 1802-1833." Ph.D. dissertation. Providence, RI: Brown University.

Monge, Gaspard 1794. *Description de l'art de fabriquer les canons, faite en execution de l'arrete du Comitu de Salut public, du 18 pluviose de l'an 2 de la Republique francaise, une et indivisible.* Paris: Impr. du Comite de Salut public.

Morrisson, James L. Jr. 1986. *The Best School in the World: West Point, the Pre-Civil War Years, 1833–1866.* Kent, OH: Kent State University Press.

Ozanam, Jacques 1727. *A Treatise of Fortification, ... The Second Edition.* Translated by J. T. Desagulieurs. London: Printed by J. Jackson.

Parshall, Karen Hunger and Rowe, David E. 1994. *The Emergence of the American Mathematical Research Community, 1876-1900: J. J. Sylvester, Felix Klein, and E. H. Moore.* Providence: American Mathematical Society.

Pycior, Helena M. 1981. "George Peacock and the British origins of symbolical algebra," *Historia Mathematica*, 8, 23-45.

Pycior, Helena M. 1984. "Internalism, externalism, and beyond: 19th-Century British algebra," *Historia Mathematica*, 11, 424-441.

Simons, William E. 1965. *Liberal Education in the Service Academies*. New York: Published for the Institute of Higher Education, Teachers College, Columbia University.

Smith, David Eugene and Ginsburg, Jekuthiel 1934. *A History of Mathematics in America Before 1900*. (The Carus Mathematical Monographs, No. 5) Chicago: The Mathematical Association of America and Open Court Publishing Co.

Smith, Francis H. 1879. "West Point Fifty Years Ago," An address delivered before the Association of Graduates of the U. S. Military Academy, West Point, at the annual reunion, June 12, 1879. New York: D. Van Nostrand.

Street, William B. 1993. "The Military Influence on American Engineering Education," *Cornell Engineering Quarterly*, Winter, 3–10.

Stevin, Simon 1617. *Castramentatio dar is Legermeting*. Rotterdam: Jan van Waesgerghe.

Swift, Joseph Gardner 1890. *The Memoirs of Gen. Joseph Gardner Swift, First Graduate of the United States Military Academy, West Point, Chief Engineer U. S. A. from 1812 to 1818. 1800-1865*. [Worcester, MA:] Privately Printed.

Tartaglia, Nicholas 1588. *Three Bookes of Colloquies Concerning the Arte of Shooting in Great and Small Peeces of Artillerie, ...* Translated by Cyprian Lucar. London: John Harrison.

Taylor, E. G. R. 1966. *The Mathematical Practioners of Hanoverian England, 1714-1840*. Cambridge: Cambridge University Press.

Thayer, Sylvanus 1810, "Return of the Books, Instruments, Maps and other public property at the Military Academy." Handwritten manuscript. West Point, Feb. 17, 1810. Jonathan Williams Papers, Manuscript Deapartment, Lilly Library, Indiana University, Bloomington, IN.

Ver Eecke, Paul 1933. *Pappas D'Alexandrie. La Collection Mathématique. Oeuvre Traduite pour la Premiere Fois du Grec en Francais, avec une Introduction et des Notes*. 2 Tomes. Paris et Bruges: Desclee de Brouwer et Cie.

Wilson, James Grant and Fiske, John (Editors) 1888. *Appleton's Cyclopaedia of American Biography*. New York: D. Appleton and Co.

Zuersher, Dorothy 1974. "Benjamin Franklin, Jonathan Williams and the United States Military Academy." Ed.D. dissertation. Greensboro, NC: University of North Carolina at Greensboro.

CHAPTER 2

Catalog of the West Point Collection

Abakanowicz, Bruno Abdank (1852–1900).

1886. *Les intégraphes la courbe intégrale et ses applications. Étude sur un nouveau système d'intégrateurs mécaniques.* Paris: Gauthier–Villars. pp. x + 156.

○ The library has two copies.

○ One copy has a hand written letter to the author pasted on pages 84–85.

Abel, Niels Henrik (1802–1829). See Peslouan 1906.

Adams, George, Senior (1704–1773).

1810. *A Treatise, Describing the Construction and Explaining the Use of New Celestial and Terrestrial Globes; Designed to Illustrate, in the Most Easy and Natural Manner, the Phenomena of the Earth and Heavens, and to Shew the Correspondence of the Two Spheres; with a Great Variety of Astronomical and Geographical Problems.* Thirtieth edition in which a comprehensive view of the solar system is given; and the use of the globes is farther shewn, in the explanation of spherical triangles. London: Dudley Adams. pp. xxiv + 242 + 14 plates.

◇ Frontispiece of a new Terrestrial Globe of D. Adams. Dedicated to the King. First edition was 1766. Author is father of George Adams.

Adams, George (1750–1795).

1803. *Geometrical and Graphical Essays, Containing, a General Description of the Mathematical Instruments Used in Geometry, Civil and Military Surveying, Levelling and Perspective; with Many New Practical Problems.* Third edition corrected and enlarged by William Jones. London: W. Glendinning. pp. 16 + 518 + 34 copper plates + 14 +2.

◇ See Figure 22 for a photograph of the frontispiece, which is the "Great Theodolite."

◇ First edition was 1791. Author was son of George Adams, Senior.

◇ Used as a textbook at West Point beginning in 1803.

Adams, John (fl. 1758–1796).

1796. *The Mathematician's Companion, or a Table of Logarithms from 1 to 10,860. Also, the Logarithmic Sines, Tangents, and Secants, Advantageously Disposed to Every Minute of the Quadrant, and the Sines and Tangents to 180 Degrees; with Useful Additions, by which These Logarithms are Easily Extended 108,600; and the Sines and Tangents Rendered much more Extensive and Convenient than Heretofore in Trigonometry, Astronomy, Gunnery, and Other Mathematical Calculations, to which is Prefixed, a Short Introduction to Decimal and Logarithmic Arithmetic.*

London: Printed for Mount and Davidson. pp. 22 + unnumbered pages of tables.
◇ Errata follow the title page.

Adhémar, Joseph Alphonse (1797–1862).

1832a. *Cours de mathématiques a l'usage de l'ingénieur civil; géométrie descriptive.* Paris: Carilian–Goeury. pp. vi + 7–216.
◇ The text refers to plates, which are found in Adhémar 1832b.

[1832b]. *Cours de mathématiques a l'usage de l'ingénieur civil; géométrie descriptive.* Paris: Carilian–Goeury. pp. 52 plates.
◇ Plates for Adhémar 1832a.

1836a. *Traité de perspective a l'usage des artistes.* Paris: Bachelier; Carilian–Goeury; the author. pp. viii + 5–250.
◇ The 316 figures on 68 plates are in Adhémar 1836b.

1836b. *Traité de perspective a l'usage des artistes.* Paris: Bachelier; Carilian–Goeury; the author. 62 plates.
◇ Plates for Adhémar 1836a.

Adrain, Robert (1775–1843). See Hutton 1812b; Hutton 1818; Hutton 1822; Hutton 1825; and Ryan 1828.

Agnesi, Maria Gaetana (1718–1799).

1801. *Analytical Institutions, in Four Books: Originally Written in Italian.* 2 vols. Translated by John Colson. Edited by John Hellins. London: Printed by Wilks and Taylor. pp. xxvii + xlvi + 251; 371.
◇ Errata for both volumes on pp. 365–366 of volume 2. There is a short biography of the author on pp. xiii–xvi of volume 1. Dedicated to Maria Teresa of Austria, Empress of Germany, Queen of Hungary, Bohemia, etc.

Airy, George Biddell (1801–1892).

1826. *Mathematical Tracts on Physical Astronomy, the Figure of the Earth, Precession and Nutation, and the Calculus of Variations. Designed for the Use of Students in the University.* First edition. Cambridge: Printed by J. Smith. pp. viii + 180.

1831. *Mathematical Tracts on the Lunar and Planetary Theories, the Figure of the Earth, Precession and Nutation, the Calculus of Variations, and the Undulatory Theory of Optics. Designed for the Use of Students in the University.* Second edition. Cambridge: Printed by John Smith. pp. v + 410 + 5 plates.

d'Alembert, Jean Le Rond, (1717–1783).

1743. *Traité de dynamique, dans lequel les loix de l'equilibre & du mouvement des corps sont réduites au plus petit nombre possible, & démontrées d'une maniére nouvelle, & où l'on donne un principe général pour trouver le mouvement de plusieurs corps qui agissent les uns sur les autres, d'une maniére quelconque.* Paris: David l'aîné. pp. xviii + 4 plates + 186 + [2].

◇ Errata on the second page following p. 186. Plates pasted on fold out sheets.

○ Two copies of this book are in the library.

1749. *Recherches sur la précession des equinoxes et sur la nutation de l'axe de la terre, dans le systême Newtonien.* Paris: David l'aîné. pp. xl + 184 + 4 plates.

◇ Errata on p. 184. Dedicated to the Marquis Lomellini.

1754; 1754; 1756. *Recherches sur differens points importans du systême du monde.* 3 vols. Paris: David. pp. lxviii + 260 + 1 plate; vi + 290 + 3 plates; xlvi + 263 + 1plate.

◇ Errata for volume 1 are on p. lxvii; errata for all three volumes are on pp. 261-263 of volume 3.

d'Alembert, Jean Le Rond (1717–1783); Bossut, l'Abbé Charles (1730–1814); Lalande, Joseph–Jerome Lefrançais de (1732–1807); and Condorcet, Marie–Jean–Antoine–Nicolas Caritat, Marquis de (1743–1794).
1784; 1785; 1789. *Encyclopédie méthodique. Mathématiques.* 3 volumes plus a fourth volume of plates. Paris: Panckoucke; Liège: Plomteux. pp. cxiv + 721; 787; xxvii + 56 + 184; 98 plates.

◇ Topics arranged alphabetically: Volume 1 contains A–E; volume 2 contains F–R; volume 3 contains a list of the authors of the entries of this encyclopedia (p. xxviii), a table and review of the number of volumes of text and of plates which make up the encyclopedia, (pp. 1–56), S–Z, pp. 1–174, additions and corrections for the astronomy part pp. 175–176, a supplement to the whole dictionary pp. 177–181, and a table of lectures pp. 182–184. The volume of plates has no title page.

d'Alembert, Jean Le Rond. See also Gaudin 1825.

Alhazen (965–1039).
1572. *Opticae thesaurus Alhazeni Arabis libri septem, nunc primùm editi. Eiusdem liber de crepusculis & nubium ascensionibus. Item Vitellonis Thuringopoloni libri X. Omnes instaurati, figuris illustrati & aucti, adiectis etiam in Alhazenum commentarijs, a Federico Risnero.* Basileae [Basel]: Per Episcopios. pp. [vi] + 288 + [vii] + 474.

◇ "Frontispiece" on verso of title page.

○ Signatures on title page: "Jacobi Fabricus" and "Le Comte de Plelo." There are a few worm holes in the margin near the end.

Allaize, M.; Puissant, Louis (1769–1843); Billy; and Boudrot.
1813. *Cours de mathématiques, a l'usage des Écoles impériales militaires; Rédigé par ordre de M. le général de division Bellavène, Commandant–directeur des études de l'Ecole spéciale militaire de Saint–Cyr.* Paris: Magimel. pp. xii + 608 + 13 plates.

◇ Dedicated to Prince Alexander. Errata follow p. 608. The authors' names appear on p. [vi]. Abridged version of Bézout 1777–1800.

○ Thayer binding.

Allman, George Johnston (1824–1904).
1889. *Greek Geometry from Thales to Euclid.* Dublin University Press Series, Dublin: Hodges, Figgis, & Co. pp. xii + 237 + 2. Errata on p. viii.
◇ Frontispiece contains portrait of Archytas.

Alvord, Benjamin (1813–1884; USMA 1833).
1855. *The Tangencies of Circles and of Spheres.* Smithsonian Contributions to Knowledge Series, Washington, D.C.: The Smithsonian Institution. pp. 16 + 9 plates.
○ This copy was owned by Professor Albert E. Church and presented by him to the USMA library.

Amiot, A. (?–1865).
1853. *Leçons nouvelles de géométrie descriptive.* Paris: Guiraudet et Jouaust. pp. iii + 190 + 21 plates.
◇ Errata follow p. iii. An extra copy of pp. 145–146 follows p. 190. The plates carry a separate title page, "Planches," so this may have originally been two volumes.

Angel, Henry.
1880a. *Practical Plane Geometry and Projection, for Science Classes, Schools, and Colleges, Adapted to meet the Requirements of the Higher Stages in the Higher Stages in the Science and Art Department Syllabus. Volume 1, Text.* London: William Collins, Sons, and Co. pp. i + 359.
○ Signed by Frank E. Harris inside front cover.

1880b. *Practical Plane Geometry and Projection, for Science Classes, Schools, and Colleges, Adapted to meet the Requirements of the Higher Stages in the Higher Stages in the Science and Art Department Syllabus. Volume 2, Plates.* London: William Collins, Sons, and Company. 81 plates.
○ Copy signed by Frank E. Harris.

Anger de La Loriais, Isadore.
1846. *Essai sur les fonctions elliptiques.* Paris: Carilian–Goeury et V^{or} Dalmont. pp. 35.
◇ The Introduction (pp. 1–6) gives a review of the contributions to elliptic functions of Maclaurin, d'Alembert, Fagnano, Euler, Lagrange, Landen, and Legendre.

Anonymous.
1809. *Recueil de problêmes rèsolus par des considerations purement géométriques.* Paris: Courcier. pp. 244 + 4 plates.
◇ Errata follow p. 244.

Anonymous.
1812. *Tables de multiplication, a l'usage de MM. les géomètres, de MM. les ingéneurs–vérificateurs du cadastre, pour le calcul de la surface des plans, et de MM. les directeurs des contributions, pour la conversion des mesures anciennes en*

mesures nouvelles, et pour l'application des tarifs défintifs au classement des propriétés foncières; Approuvées par son excellence le Ministre des Finances. Seconde Edition. Paris: Imprimerie de Valade. pp. 8 + 500 pages of tables.

◇ Errata follow p. 500.

○ Thayer binding.

Anonymous.

1821. *Cambridge Problems: Being a Collection of the Printed Questions proposed to the Candidates for the Degree of Bachelor of Arts at the General Examinations from 1801 to 1820 Inclusive.* Cambridge: Printed by J. Smith for J. Deighton and Sons. pp. [vi] + 425.

◇ There is a list of moderators from 1800 to 1820 on pp. [v]–[vi].

Anonymous.

1833. *Supplement to Chap. 17th Sganzin and Supplement to Chap. 22 Sganzin.* Lithographed. pp. 61 + page hand written notes + 2 plates + 16.

○ G. W. Cullum's copy signed in 1833.

Anonymous.

1848-1849. *Ecole polytechnique cours de physique.* 2 vols. Paris: [no publisher]. pp. 12 + 580; 312.

◇ Lithographed.

◇ Many illustrations embedded in the text. Handwritten table of contents (incomplete) tipped in at the end of volume 2.

Anonymous.

1894. *Résumé du cours d'algèbre supérieure.* Ecole Militaire (Belgium). pp. 159.

◇ Lithographed.

○ This copy has several War Department stamps and went through the military attache LT Harris in 1894. Cover has a letter written to Prof. Edgerton, USMA.

Anonymous.

1900. *Examples from C. Smith's Treatise on Algebra.* West Point: Press of U. S. Military Academy. pp. 136.

Anonymous.

1901. *Examples from Murray's Integral Calculus.* West Point: Press of U. S. Military Academy. pp. 48.

◇ See also Murray, Daniel Alexander.

Anonymous.

1915. *Exercises in Descriptive Geometry.* West Point, NY: USMA Printing Office. pp. 19.

◇ This book was composed by members of the Department of Mathematics, USMA.

Anonymous.
n.d.
  ◇   An anthology of papers by various authors assembled and bound together by an unknown person.
  ◇   See also Babbage 1815; Babbage 1819; Brinkley 1807; Horner 1819; Milner 1778; Milner n.d.; Playfair 1778; Walker 1801; Woodhouse 1801.

Anonymous.
n.d. *Elements of Algebra.* pp. iii + 398.
  ◇   Missing the title page and front pages.
  ○   Stamped "Miller, A.M." on cover. Signed by Alexander Macomb Miller, 1861, Washington City, D.C. Also signed by W. Gadsby. Notes throughout.

Anonymous.
n.d. *Èléments d'algèbre avec de nombreaux exercises par les frères des Ècoles chrétiennes.*   Neuvième édition. Cours de mathématiques élémentaires. Tours: Alfred Mame & Fils; Paris: Ch. Poussielgue.

Anonymous.
n.d. *Epures de géométrie déscriptive.*   pp. 30 plates + 8 plates + 19 plates + 42 plates.
  ◇   No title page.

Anonymous.
n.d.
  ◇   French manuscript containing four parts: "Géométry," 132 numbered pages; "Traitté de la stereometrie et de la trigonomtrie", 49 numbered pages; "Traitté de l'art de fortifier", 197 numbered pages; and "L'Usage du compas de proportion", 39 numbered pages. Publication appears to be late 18th century.

Anonymous.
n.d. *The Ladies' Diary: or Woman's Almanack.* London: Printed for the company of stationers. pp. 576.
  ◇   This book is composed exclusively of copies of the *Ladies Diary* for the years 1812, 1813, . . . , 1823. There were 48 pages in the *Ladies Diary* each year; thus, this volume has 576 pages total.

Anonymous.
n.d. *Lancaster Eliptic Tables.* pp. [i] + 15.
  ◇   Hand inked copy. Binding indicates 18th century.

Anonymous.
n.d. *Notes to Accompany Michies [sic] Mechanics.* West Point, NY: Department of Philosophy, U. S. Military Academy. pp. 132.
  ○   Three copies in the library.
  ○   One copy given by Lieut. Jas. G. Steese.

Anonymous.

n.d. "The Practical Application of the Method of Least Squares." Annapolis, Md. pp. 18.

◇ No Title page. This is a handwritten pamphlet. Compiled by members of the Department of Mathematics, United States Naval Academy.

Anonymous.

n.d. *Subjects in Mechanics.* pp. 18.

◇ Pamphlet. Outline and brief questions keyed to an unidentified textbook.

Anthémius of Tralles (fl ca. A.D. 550)

1777. *Fragment d'un ouvrage Grec d'Anthémius sur des paradoxes de mécanique. Revu & corrigé sur quatre manuscrits avec une traduction francoise & des notes.* Translated by M. Dupuy. [Paris]: [L'imprimerie royale]. pp. xiii + 41 + 1 plate.

◇ Pages 0–21 have the Greek on the left page and the French translation facing it on the right page. Notes for the translation are on pp. 21–26. There is the "Observation sur le premier problème, qui consiste à faire tomber les rayons solaires constamment à un point donné, à toute heure & en toute saison" (Anonymous) on pp. 27–41.

Antomari, X. (1855–1902).

1897. *Cours de géométrie descriptive a l'usage des candidats a l'École polytechnique, a l'École normale supérieure, aux Écoles centrale des arts et manufactures, des ponts et chaussées et des mines de Paris et de Saint-Étienne.* Paris: Nony & Cie. pp. 572.

1904. *Traité de géométrie descriptive a l'usage des élèves des classes de mathématiques des aspirants au baccalauréat et des candidats a l'Institut agronomique et a l'École navale.* Troisième édition. Paris: Vuibert et Nony. pp. 160.

Apollonius of Perga (Second half of third century B.C. to early second century B.C.).

1655. *Apollonii Pergaei conicorum libri IV. Cum commentariis R. P. Claudii Richardi.* Antverpiae [Antwerp]: Apud Hieronymun & Ioannem Bapt. Verdussen. pp. 15 plates + [lxxii] + 398 + 13 plates.

◇ The title page contains a portrait of D'Guliel. Raym. de Moncada. Dedicated to Moncada. Errata follow p. 398. The plates are bound at the beginning of the volume.

1661. *Apollonii Pergaei conicorum lib. V. VI. VII. paraphraste Abalphato Asphahanensi nunc primùm editi. Additus in calce Archimedis assumptorum liber, ex codicibus arabicis m.ss.* Florentine [Florence]: Ex typographia Iosephi Cocchini. pp. [xxxvi] + 415.

1749. *Apollonii Pergaei locorum planorum libri II. Restituti a Roberto Simson.* Glasguae [Glascow]: Excudebant Rob. et And. Foulis. pp. xviii + 233.

◇ Restored by Robert Simson (1687–1768). Errata follow p. 233. The volume contains Book I (pp. 1–115) and Book II (pp. 117–233) of Apollonius' *Plane Loci.*

1764. *The Two Books of Apollonius Pergaeus, concerning Tangencies, as They have been Restored by Franciscus Vieta and Marinus Ghetaldus. With a Supplement.* Translated by John Lawson (1723–1779). Cambridge: T. Fletcher and F. Hodson. pp. vii + 17 + 6 plates.

⋄ The Supplement (pp. 14–17) is written by John Lawson. Title page signed "John King's, 4, 1st July, 1918." This work is also known by the title "On Contacts."

1795. *Apollonii de tactionibus quae supersunt, ac maxime lemmata Pappi in hos libros Graece nunc primum edita e codicibus mscptis, cum vietae librorum Apollonii restitutione, adjectis observationibus, computationibus, ac problematis Apolloniani historia a Joanne Guilielmo Camerer.* Gothae [Gotha]: Apud Car. Grillielmi Ettinger. pp. 112 + 66 + 3 plates (54 figures).

⋄ François Viète's "restitution" of Book II of the *Tactionibus* constitutes the 66–page section. Corrigenda on pp. 65–66.

Apollonius of Perga. See also Thévenot 1693.

Appell, Paul (1855–1930).
1905. *Éléments d'analyse mathématique.* Paris: Gauthier–Villars. pp. ix + 421.

Appell, Paul (1855–1930) and E. Lacour (1854–?).
1897. *Principes de la théorie des fonctions elliptiques et applications.* Paris: Gauthier–Villars et Fils. pp. ix + 421.
⋄ Function tables at end of book.

Archimedes (ca. 287 BC – 212 BC).
1807. *Oeuvres d'Archimède, traduites littéralement, avec un commentaire, par F. Peyrard; Suivies d'un mémoire du traducteur, sur un nouveau miroir ardent, et d'un autre mémoire de M. Delambre, sur l'arithmétique des Grecs.* Paris: Chez François Buisson. pp. xlviii + 601 + 2 plates and 1 fold–out table.
⋄ Frontispiece is an engraving of a bust of Archimedes. Errata follow p. 601. Dedicated to the King.
∘ Thayer Binding.

1897. *The Works of Archimedes Edited in Modern Notation with Introductory Chapters.* Edited by T. L. Heath. Cambridge: University Press. pp. clxxxvi + 326.

Archimedes. See also Apollonius 1661; Euclid 1751; Euclid 1803; Wallis 1699.

Argand, Jean Robert (1768–1822).
1881a. *Imaginary Quantities: Their Geometrical Intrepretation, Translated from the French of M. Argand.* Translated by Arthur Hardy. Preface by J. Houel. New York: Van Nostrand. pp. xvi + 17–135.
⋄ Reprinted from Van Nostrand's Magazine. Pages 83–135 contain "Notes on the Geometrical Interpretation of Imaginary Quantities" by Hardy.
∘ Two copies are present in the library.

Aristarchus of Samos (ca. 310–230). See Heath 1913.

Atkinson, James, Senior.
1790. *Epitome of the Whole Art of Navigation or, a Short, Easy, and Methodical Way to become a Complete Navigator and Astronomer, Containing an Introduction to Decimal and Logarithmic Arithmetic, Practical Geometry, Trigonometry Plane and Spheric, Geometrically, and Logarithmically, with Their uses in Navigation, viz. In Plane, Mercator's and Middle Latitude Sailing, Geography, and Nautical Astronomy; Illustrated with Charts and Diagrams. With an Introduction to the Lunar Method of Determining the Longitude at Sea. The Gregorian or New Calendar, Description and Use of the Plane Chart, Mercator's Chart, also of Hadley's Octant and Sextant. A Table of the Latitude and Longitude of Places; Tables of Latitude and Departure to every Quarter Point and Degree of the Compass to 300 Miles Distance; a Table of Meridional Parts, Solar Tables, Natural Sines, &c. &c. also, a Table of 10,860 Logarithms, and Logarithmic Sines, Tangents and Secants.* Much improved and enlarged from the best authors on these subjects by John Adams. London: Printed for Mount and Davidson. pp. vi + 219 + unnumbered pages of tables + iv + 3 plates.

Atwood, George (1746–1807).
1784. *A Treatise on the Rectilinear Motion and Rotation of Bodies; with a Description of the Original Experiment Relative to the Subject.* Cambridge: J. Archdeacon; Printed for J. & J. Merrill. pp. [4] + xv + 1 page errata + 436 + 1 page errata + 7 plates.

Azemar, L. P. V. M. and Garnier, J. G. (1766–1840)
1809. *Trisection de l'angle, par L. P. V. M. Azemar, suivie de recherches analytiques sur le même sujet, par J. G. Garnier.* Paris: Courcier. pp. viii + 118 + 3 plates.
◇ Errata on p. viii. Pages 93–118 are by Garnier.

Babbage, Charles (1792–1871).
1815. "An Essay Towards the Calculus of Functions."
◇ A paper, not a book, bound together with other papers in Anonymous n.d.

1819. "On Some New Methods of Investigating the Sums of Several Classes of Infinite Series."
◇ A paper, not a book, bound together with other papers in Anonymous n.d.
○ Title page inscribed "Wildig, Caius College."

1820. *Examples of the Solutions of Functional Equations.* London: No publisher listed. pp. [iii] + 42 + 1 plate.
○ Two copies are present in the library. One copy bound with Herschel 1820.

1827. *Table of the Logarithms of the Natural Numbers, from 1 to 108,000.* Stereotyped. London: Printed for J. Mawman. pp. xx + unnumbered pages of tables.

◇   Printed on yellow paper. Dedicated to Lieutenant–Colonel Colby, Royal Engineers.

o   Three copies are present in the library.

Bachet, de Méziriac, Claude–Gaspar (1581–1638). See Diophantus 1670.

Bailly, Jean Sylvain (1736–1793).
1779; 1779; 1782. *Histoire de l'astronomie moderne depuis la fondation de l'école d'Alexandrie, jusqu'a l'époque de MDCCXXX.* 3 vols. Paris: les Freres de Bure. pp. xvi + 728 + 13 plates; 751 + 5 plates; 415.

o   The title of volume 3 specifies to the epoch of 1782. Errata in volume 1 on p. 728. Glossary in volume 2, pp. 733–747. Volume 3 has an index, pp. 345–414.

o   Thayer binding.

1781. *Histoire de l'astronomie ancienne, depuis son origine jusqu'à l'établissement de l'ècole d'Alexandrie.* Seconde édition. Paris: De Bure fils aîné. pp. xxiv + 527 + 3 plates.

o   Thayer binding.

Baire, René (1874–1932).
1907. *Leçons sur les théories générales de l'analyse. Tome I. Principes fondamentaux. – variables réelles.* Cous d'analyse de la faculté des sciences de Dijon. Paris: Gauthier–Villars. pp. x + 347.

Baker, Arthur Latham (1853–1934).
1890. *Elliptic Functions. An Elementary Textbook for Students of Mathematics.* New York: John Wiley. pp. i + 118.

Ball, W. W. Rouse (1850–1925).
1889. *A History of the Study of Mathematics at Cambridge.* Cambridge: University Press. pp. xvii + 264.

o   The library has two copies.

1893. *An Essay on Newton's "Principia."* London: Macmillan and Co. pp. x + 175.

◇   Eight pages of advertisements for other books by Rouse Ball follow p. 175.

Banneker, Benjamin (1731–1806). See Norris 1854.

Bardet de Villeneuve, P. P. A.
1740. *Traité de la geometrie pratique, a l'usage des officiers: qui enseigne toutes les opérations les plus nécessaires, tant sur le papier que sur le terrain.* La Haye [The Hague]: Jean van Duren. pp. 195 + [9] + 17 plates.

◇   Frontispiece is labeled "La Science Militaire" and depicts a battle scene. See Figure 15 for a photograph.

o   The name of John Rutherford, Esq. of Edgerstown is listed on the inside of the front cover. Donated in memory of MAJ Donald J. MacLachlan [USMA 1907].

Bardin, Etienne Alexandre (1774–1840).
1837. *Notes et croquis de géométrie descriptive.* 2me. édition, revue, corrigée et augmentée. Paris: L. Mathias (Augustin). pp. 28 plates (each one is two pages).

Barker, Arthur Henry.
1902. *Graphical Calculus. With an Introduction by John Goodman.* Second edition. London: Longmans, Green, and Co. pp. xi + 195 + 1 plate.

Barlow, Peter (1776–1862).
1814a. *A New Mathematical and Philosophical Dictionary; Comprising an Explanation of the Terms and Principles of Pure and Mixed Mathematics, and such Branches of Natural Philosophy as are Susceptible of Mathematical Investigation. With Historical Sketches of the Rise, Progress and Present State of the Several Departments of these Sciences, and an Account of the Discoveries and Writings of the Most Celebrated Authors, Both Ancient and Modern.* London: G. and S. Robinson, et. al. pp. vii + approximately 1000 unnumbered pages + 13 plates.
◇ Alphabetically arranged. Errata at the end of the text and before the plates.

1814b. *New Mathematical Tables Containing the Factors, Squares, Cubes, Square Roots, Cube Roots, Reciprocals, and Hyperbolic Logarithms, of all Numbers from 1 to 10,000; Tables of Powers and Prime Numbers; an Extensive Table of Formulae, or General Synopsis of the Most Important Particulars Relating to the Doctrines of Equations, Series, Fluxions, Fluents, etc. etc. etc..* London: Printed for G. and S. Robinson. pp. lxi + 336.
◇ This is the first edition of this book.

Barnard, John Gross (1815–1882; USMA 1833).
1867. *Analysis of Rotary Motion, as Applied to the Gyroscope.* Pamphlet reprinted from Barnard's *American Journal of Education,* June 1867. pp. 537–560.
◇ Barnard served as Superintendent of the Academy (1855–1856).
○ Presented by Major Barnard to Lieutenant F. E. Prime.

Barrow, Isaac (1630–1677).
1734. *The Usefulness of Mathematical Learning Explained and Demonstrated: Being Mathematical Lectures Read in the Publick Schools at the University of Cambridge. By Issac Barrow. To which is Prefixed the Oratorical Preface of Our Learned Author, Spoke before the University on his being Elected Lucasian Professor of the Mathematics.* Translated by the Rev. John Kirkby. London: Printed for Stephen Austen. pp. xxxii + 440 + 1 plate + 8.
◇ Portrait of Barrow on frontispiece.

1735. *Geometrical Lectures: Explaining the Generation, Nature and Properties of Curve Lines. Read in the University of Cambridge, by Issac Barrow, D.D. Professor of Mathematicks, and Master of Trinity–College, etc. Translated from the Latin Edition, Revised, Corrected and Amended by the late Sir Isaac Newton. By Edmund Stone, F.R.S.* London: Printed for Stephen Austen. pp. vi + 309 (pages near the end of one copy are incorrectly numbered).

⬦ Newton is thanked for his help, p. iv.
○ The library has two copies.

Barrow, Isaac. See also Collins 1725; Euclid 1751.

Bartlett, Dana Prescott (1860–1936).
1915. *General Principles of the Method of Least Squares, with Applications.* Third edition. Boston: The Author. pp. [vi] + 142 + xi.
⬦ This book was used as a text from 1933 until 1941 at USMA.
○ The library has two copies.
○ One is signed [Col H. J.] "Katz" (USMA 1936). Printed slip pasted in at p. 11.
○ "Lesson sheet" for 3rd class mathematics, USMA 1934, glued in copy.

Bartlett, William Holms Chambers (1804–1893; USMA 1826).
1851. *Elements of Natural Philosophy. I. Mechanics.* Second edition. New York: A. S. Barnes & Company. pp. 632.
⬦ Bartlett served as Department Head of Natural and Experimental Philosophy, USMA (1836–1871).
○ The library has two copies.

1853. *Elements of Analytic Mechanics.* New York: A. S. Barnes & Company. pp. vii + 445.
⬦ Analytic mechanics utilized calculus.
○ Some notes in the margins.

1855. *Elements of Analytic Mechanics.* Third edition, revised and corrected. New York: A. S. Barnes & Company. pp. viii + 448.
⬦ Used as a textbook at USMA, 1853–1887.

1858. *Elements of Analytic Mechanics.* Fifth edition, revised, corrected, and enlarged. New York: A. S. Barnes & Company. pp. viii + 508.
⬦ Dedicated to COL Sylvannus Thayer.

1860a. *Elements of Analytic Mechanics.* Seventh edition, revised, corrected, and enlarged. New York: A. S. Barnes & Burr. pp. viii + 508.
⬦ Used as a textbook at USMA, 1876–1882. Dedicated to COL Sylvannus Thayer.

1860b. *Elements of Synthetic Mechanics.* Sixth edition, revised and corrected. New York: A. S. Barnes & Burr. pp. 632.
⬦ Synthetic mechanics does not utilize calculus. Used as a textbook at USMA.

1865. *Elements of Analytic Mechanics.* Eighth edition, revised, corrected, and enlarged. New York: Barnes & Burr. pp. viii + 510.
⬦ Dedicated to COL Sylvannus Thayer.
○ The library has two copies. One copy is a gift of Norman R. Dilley.

1874. *Elements of Analytic Mechanics.* Ninth edition, revised and corrected. New York: A. S. Barnes and Company. pp. vii + 530.
⬦ Dedicated to COL Sylvannus Thayer. Contains the preface to the Second edition (1858) and the preface to the Ninth edition (1874).

Barton, Samuel Marx (1859–1926).
1899. *An Elementary Treatise on the Theory of Equations.* Boston: D. C. Heath & Co. pp. xii + 199 + 10 pages of advertisements.

Bass, Edgar Wales (1843–1918; USMA 1868).
1887a; 1889. *Introduction to the Differential Calculus,* Parts I, II, and III. West Point: United States Military Academy Press. pp. iii + 220; 221–319.
⋄ Bass served as Department Head, USMA Department of Mathematics from 1878 to 1898.
⋄ This book was used as a text at USMA from 1888 until 1895.
○ Part I rebound in 1952. One part signed by Pratt, Professor of Math at USMA. Notes in blue and red throughout the text.

1887b. *Introduction to the Differential Calculus.* West Point: United States Military Academy Press. pp. 95.
○ Two copies present in the library.
○ Corrections pasted into both of the books.
○ One copy is signed "Cdt P. S. Michie with Compliments of Edgar W. Bass, Dec 4, 1887" inside front cover. Same copy signed "copyright 1887 by Edgar W. Bass" on page 2.

1889. *Differential Calculus.* Part I. West Point: U. S. Military Academy Press. pp. 93–220.
⋄ Missing pages 1–92.
○ There are six copies of this book in the library.
○ Notes made throughout the books with the notations "blue–lower sections. red–upper sections."

1895. *Differential Calculus.* Part II. pp. 1 + 222–319.
⋄ The title page is missing.

1901a. *Elements of Differential Calculus.* Second edition, first thousand. New York: John Wiley and Sons. pp. viii + 356 + 16 pages of tables of books published by Wiley.
⋄ This book was used as a text from 1896 until 1907 at USMA. Preface contains a list of authors that Bass consulted in preparation of this book.
○ The library has two copies.
○ One copy has some red and blue highlighting throughout.

1901b. *Examples from Bass' Calculus.* West Point: Press of U. S. Military Academy. pp. [viii] + 19.
⋄ Two title pages, one from 1901 and one from 1902.

Bass, Edgar W. See also Ludlow and Bass 1890; Ludlow and Bass 1893; Ludlow and Bass 1896; Ludlow and Bass 1899; Ludlow and Bass 1900.

Beaune, Florimond de (1601–1652). See Descartes 1649.

Beckett, Joseph (fl. 1804).
1822. *Elements and Practice of Mensuration and Land Surveying: Adapted Both to Public and Private Instruction: with an Appendix, Containing Rules for Measuring Haystacks, Marl-pits and Canals.* London: W. Taylor. pp. xv + 342 + 4 plates.

Beebe, [William] (1851-1917).
n.d. *Four-Place Tables.* pp. [22].
◇ No title page.

Belidor (1697–1761).
1757. *Nouveau cours de mathématique, a l'usage de l'artillerie et du genie, oú l'on applique les parties les plus utiles de cette science à la théorie & à la pratique des différens sujets qui peuvent avoir rapport à la guerre.* Nouvelle edition, corrigée & considérablement augmentée. Paris: Noyon. pp. xxxii + 656 + 34 plates.
○ Bookplate of Ludovicus Comes Lanckoronski inside the front cover. According to the inscription on the verso of the title page, this book was given to Louis Lanckoronski by Pozzo di Borgo on 7 September 1804. The stamp of Charles Lanckoronski is on the title page.

Beman, Wooster Woodruff (1850–1922). See Fink 1903 and Klein 1897.

Berard, Joseph Balthasar (1763–?).
1810. *Statique des voutes, contenant l'essai d'une nouvelle théorie de la poussée, et un appedice sur les anses de panier.* Paris: Didot. pp. xx + 160 + 4 plates.

Bergery, Claude Lucien (1787–1863).
1825. *Géométrie appliquée à l'industrie, à l'usage des artistes et des ouvriers. Sommaire des leçons publiques données dans l'hôtel de ville de Metz.*   Metz: Lamort; A. J. Kilian. pp. [xxiv] + [335] + [13 plates].

Berniolle, Paul.
1904. *Cours de géométrie descriptive conforme au programme du 11 août 1902 pour la préparation à l'École militaire de Saint-Cyr.*   Paris: Henry Paulin et Cie. pp. 297 + 2.

Bernoulli, Jakob [Jacques] (1654–1705).
1713. *Jacobi Bernoulli, profess. Basil. & utriusque Societ. Reg. Scientiar. Gall. & Pruss. Sodal. mathematici celeberrimi, ars conjectandi, opus posthumum. Accedit tractatus de seriebus infinitis, et epistola Gallicè scripta de ludo pilae reticularis.* Basileae [Basel]: Impensis Thurnisiorum, Fratrum. pp. [iv] + 306 +35 + 1 plate.
◇ On pp. 241–306 is the "Tractatus de seriebus infinitis earumque summa finita, et usu in quadraturis spatiorum & rectificationibus currarum." Following p. 306, there is the "Lettre à un Amy, sur les Parties du Jeu de Paume." The preface, pp. iii–iv, is by Nicolas Bernoulli, the son of the author.

1744. *Jacobi Bernoulli, Basíleensis, opera.* 2 vols. Genevae [Geneva]: Sumptibus Haeredium Cramer & Fratrum Philibert. pp. viii + 48 + 663 + 28 plates; 667–1139 + 10 plates.

1801. *L'art de conjecturer, traduit du Latin de Jacques Bernoulli; avec des obser-vations, éclaircissemens et additions par L. G. F. Vastel, Premiere partie.* Caen: G. Le Roy. pp. ii + vi + 180 + 8 page catalogue.
⋄ The translator's "Observations, etc." are on pp. 101–180. This translates only the first part of Bernoulli's *Ars Conjectandi* on pp. 7–99. It contains a treatise by C. Huygens, "De la maière de raisonnes dans les jeux de hasard," together with remarks by Bernoulli.

Bernoulli, Jakob. See also Maseres 1795.

Bernoulli, Johann [Jean] (1667–1748).
1742. *Johann Bernoulli, M. D. Matheseos Professoris, Regiarum Societatum Paris-iensis, Londinensis, Petropolitanae, Berolinensis, Socii & c. Opera omnia, tam antea sparsim edita, quam hactenus inedita.* Each of the four volumes carries a separate subtitle, as follows: (1) *Tomus primus, quo continentur ea quae ab anno 1690 ad annum 1713 prodierunt.* (2) *Tomus secundus, quo continentur ea quae ab anno 1714 ad annum 1726 prodierunt.* (3) *Tomus tertius, quo continentur ea quae ab anno 1717 ad hanc usque diem prodierunt. Accedunt lectiones mathematicae de calculo integralium, in usum ilustr. March. Hospitalii conscriptae.* (4) *Tomus quar-tus, quo continentur* ANEKΔOTA. Lausannae & Genevae [Lausanne & Geneva]: Sumptibus Marci–Michaelis Bousquet & Sociorum. pp. xxiv + 563 + 23 plates interspersed; 620 + plates 24–40; 563 + plates to 76; 588 + plates to 91.
⋄ Portraits of Frederick III and Johann Bernoulli in Volume 1.

Bernoulli, Johann. See also Leibniz and Bernoulli 1745.

Bernoulli, Johann II (1744–1807). See Leonhard Euler 1822.

Bernoulli, Nicolas (1687–1759). See Bernoulli, Jakob 1713.

Bertaux–Levillain, C.
1847. *Éléments de géométrie descriptive, contenant les matières exigées pour l'admission aux écoles polytechnique, militaire, navale et forestière, et renfermant en outre des considérations générales sur les plans tangents, avec quelques applica-tions aux surfaces conques, cylindriques et sphériques.* Paris: Carilian–Goeury et Vor. Dalmont. pp. xv + 180 + 27 plates.
⋄ The plates contain a separate title page, "Atlas."

Bertrand, Joseph Louis François (1822–1900).
1855. *Traité d'algèbre.* Deuxième édition conforme aux derniers programmes officiels de l'enseignement dans les lycées. Paris: L. Hachette et Cie. pp. 495.

1864; 1870. *Traité de calcul différentiel et de calcul intégral.* 2 vols. Paris: Gauthier–Villars. pp. vii + xliv + 780; xii + 725.

◇ Volume 1 is differential calculus and volume 2 is integral calculus. Volume 1 marked received July 16, 1880. The preface in volume I (pp. i–xliv) is an essay on the history of calculus. There is an index.

○ The library has two copies.

1889. *Cacul des probabilités.* Paris: Gauthier–Villars et Fils. pp. lvii + 332.

○ The library has two copies. In one copy a librarian has written his first name as "Jules."

Betrand, Joseph Louis François. See also Lagrange 1853.

Bessiere, Jacques–Felix.
1833. *L'arithmétique elémentaire, traitée simplement, ou exposition des elemens de la science des nombres, suivant la marche, reguliere de l'intelligence.* 2 vols. Paris: Bachelier, Augustins. pp. [ii] + xvi + 1 + 4–78 + 3 plates; [iv] + 350.

◇ Both volumes bound together.

Bézout, Étienne (1730–1783).
1777; 1771; 1812; 1799; 1799; 1800. *Cours de mathématiques, a l'usage des gardes du pavillon et de la marine.* 6 vols. The title given above appears on volumes 1, 2, 6; volume 3 is titled *Cours de mathématiques, a l'usage de la marine et de l'artillerie*; volumes 4 and 5 are titled *Cours de mathématiques, a l'usage des gardes du pavillon et de la marine et des èleves de l'École polytechnique ... Nouvelle edition, revue et considérablement augmentée par le citoyen Garnier* (except Garnier's name is missing on Volume 5). These six volumes have various subtitles. Volume 1 is subtitled *Premiere partie. Éléments d'arithmétique.* Volume 2 is subtitled *Seconde partie. Contenant les eléments de géometrie, la trigonométric rectiligne & la trigonomeé sphérique.* Volume 3 is subtitled *Troisiéme partie, contenant l'algébre et l'application de l'algébre a la géométrie, avec des notes explicatives par A.-A.-L. Reynaud.* Volume 4 is subtitled *Quatrieme partie. Contenant les principes généraux de la méchanique, précédés des principes de calcul qui servent d'introduction aux sciences physico–mathématiques.* Volume 5 is subtitled *Suite de la quatrieme partie. Contenant l'application des principes généraux de la méchanique, á différens cas de mouvement et d'equilibre.* Volume 6 is subtitled *Sixieme et derniere partie, contenant le traité de navigation. Nouvelle edition, augmentée de deux tables des logarithmes des nombres; et ceux des sinus, cosinus, tangentes et cotangentes, beaucoup plus exactes et plus étendues que les anciennes.* These six volumes also have different publishers. Paris: J. B. G. Musier fils for volumes 1 and 2; Courcier for volumes 3, 4, 5, and 6. pp. xvi + 253; viii + 356 + 7 plates; viii + 322 + 216 + 4 plates; viii + 366 + 5 plates; iv + 402 + 10 plates; xvi + 288 + tables + 10 plates (bound in incorrect order).

◇ Allaize et al. 1813 is an abridgement.

◇ Dedicated to the Duke of Choiseul.

○ The accession numbers indicate that this set was acquired at three different times.

1779. *Théorie générale des équations algébriques.* Paris: Ph.–D. Pierres. pp. [iv] + xxviii + 471.

◇ Dedicated to DeSartine.

Bézout, Étienne. See also Lacroix 1820c; Lacroix 1826.

Bhascara [Bhaskaracarya] (1114–?). See Colebrook 1817.

Bicquilley, C. F. de.

1788. *Die Rechnung des Wahrscheinlichen. Aus dem Französischen des Herrn C. F. de Biquilley übersetzt und mit Anmerkungen versehen von M. Christ. Friedr. Rüdiger. Nebst einer Kupfertafel.* Leipzig: Engelhard Benjamin Schwickert. pp. viii + 314 + 4 plates.

Biot, Jean Baptiste (1774–1862).

1802. *Relation d'un voyage fait dans le département de l'Orne, pour constater la réalité d'un météore observé à l'Aigle le 6 floréal an 11.* Paris: Baudouin. pp. 47.

◇ Frontispiece is a map of the region of Aigle. Floréal was the eighth month of the calendar year established by the first French republic.

○ This copy presented to Andrew Ellicott by the National Institute of France.

1813; 1814. *Mémoires de la classe des sciences mathématiques et physiques. Mémoire sur une nouvelle application de la théorie des oscillations de la lumiére. Mémoire sur les propriétés physiques que les molécules lumineuses acquièrent en traversant les cristaux doués de la double réfruction.* pp. 38 + 1 plate.

◇ No title page. This appears to be two papers bound together. The paper on the oscillation theory of light is on pp. 1–30.

1826. *Essai de géométrie analytique, appliquée aux courbes et aux surfaces du second ordre; Ouvrage destiné à l'enseignement public, par arrêté de la Commission de l'instruction publique, en date du 22 février 1817.* Septième édition. Paris: Bachelier (successeur de Mme. Ve. Courcier). pp. x + 447 + 9 plates.

◇ This book was used as a text at USMA until 1833.

○ The library has two copies of this volume. One copy has a small amount of writing in the margins.

1840. *An Elementary Treatise on Analytical Geometry: Translated from the French of J. B. Biot for the Use of the Cadets of the Virginia Military Institute, at Lexington, Va.; and Adpated to the Present State of Mathematical Instruction in the Colleges of the United States.* Translated by Francis H. Smith. New York and London: Wiley and Putnam. pp. xii + 212.

◇ Francis Henney Smith (1811–1890, USMA 1833) was a professor at the Virginia Military Institute.

Biot, Jean Baptiste (1774–1862). See also Davies (Analytical Geometry) 1836.

Blakely, George (1870–1965; USMA 1892).
1899. *Exercises in Descriptive Geometry.* West Point: United States Military Academy Printing Office. pp. 23.

Bland, Miles (1786–1868).
1816. *Algebraical Problems, Producing Simple and Quadratic Equations, with Their Solutions. Designed as an Introduction to the Higher Branches of Analytics.* Second edition with additions. Cambridge: Printed by J. Smith. pp. [vi] + 367.
⋄ This is a collection of worked problems.

1824. *Algebraical Problems, Producing Simple and Quadratic Equations, with Their Solutions. Designed as an Introduction to the Higher Branches of Analytics.* Fourth edition. Cambridge: Printed by J. Smith. pp. [viii] + 383.

1828. *Mechanical Problems, Adapted to the Course of Reading Pursued in the University of Cambridge: Collected and Arranged for the Use of Students.* London: Printed by R. Gilbert. pp. viii + 184.
⋄ Dedicated to the Rev. William Whewell. Errata on p. vii. No table of contents.

1842. *Geometrical Problems Deducible from the First Six Books of Euclid, Arranged and Solved: to Which is Added, an Appendix Containing the Elements of Plane Trigonometry, for the Use of Younger Students.* Fourth edition. London: Whittaker & Co. pp. xlviii + 288.

Blum, Isaac Auguste (1812–1877).
1844; 1845. *Cours complet de mathématiques, a l'usage des aspirants a toutes les écoles du gouvernement. Renfermant les connaissances exigées pour l'admission aux écoles polytechnique, normale, navale, militaire de Saint–Cyr, forestière, des arts et manufactures et des beaux–arts.* 2 vols. Paris: Carilian–Goeury et Vor Dalmont. pp. xxvii + 480; xliii + 477 + 24 plates.
⋄ Volume 1 contains arithmetic and elementary algebra; volume 2 contains elementary geometry, rectilinear trigonometry, spherical trigonometry, and the elements of descriptive geometry. Errata are contained in volume 2, pp. xli–xliii.

Böcklern, Georg Andr.
1690. *Arithmetica nova militaris. Das ist: Neues arithmetisches Kriegs–Manual handelt von der {Gemeinen, Zehentheiligen, Sechszigtheiligen} Arithmetic, sempt beygefügter Rabdologia Nepperiana. Oder Künstlicher Stäblein Rechnung; Allen Kriegs Officirern und Ingenieurs, nützlich und dienlich/mit vielen schönen Exemplis erkläret.* Nürnberg: Joh. Andreas Endters. pp. [xxxiv] + 695 + 36 plates.
⋄ Frontispiece with the "queen" of arithmetic raised up by the steps of notation, addition, subtraction, multiplication, and division.

Bode, Johann Elert (1747–1826). See Ptolemy 1795.

Böger, Rudolf (1854–?).
1900. *Ebene Geometrie der Lage.* Sammlung Schubert Series VII. Leipzig: G. J. Göschensche Verlagshandlung. pp. [x] + 289.

Bohnert, F. (1862–?)
1902. *Elementare Stereometrie.*   Sammlung Schubert Series IV. Leipzig: G. J. Göschensche Verlagshandlung. pp. [vii] + 183.

Boisaymé, Jean Marie Joseph Aime du.
1811. *De la courbe que décrit un chien en courant après son maitre.* Paris: Didot. pp. 21 + 1 plate.
◇ The curve in question is the tractrix.
○ Some notes in the margins.

Bois–Bertrand, E. D.
1810; 1811. *Cours d'algèbre, à l'usage des aspirants à l'École polytechnique.*   2 vols. Paris: Didot; A. J. Kilian. pp. xii + 375; 422.
◇ Errata for volume I follows p. 375.

[Boissiere, Claude de] (1554–1608).
1705. *Elemens de geometrie de monseigneur le duc de Bourgogne.*   Paris: Jean Boudot. pp. [xxiv] + 190 + 30.
◇ Dedicated to the Duke of Bourgogne.
◇ The last 30 pages contain "Problems of Arithmetic and Geometry."
○ There are many notes in the margin in French, and there are three pages in which the margins fold out and reveal several figures. These drawings have been very carefully made by a previous owner of the book.

Boivin, Jean (1665–1726). See Thévenot 1693.

Bonnycastle, John (1750–1821).
1818a. *A Treatise on Plane and Spherical Trigonometry: with Their Most Useful Practical Applications.* Third edition, corrected and improved. London: Printed for Cadell and Davies; et al. pp. xxxiv + 438.
◇ Attractive and appropriate engraving is found on the title page. See Figure 23 for a photograph of the title page of this volume.
◇ The introduction, pp. [v]–xxx, contains a history of trigonometry.

1818b. *Elements of Geometry; Containing the Principal Propositions in the First Six, and the Eleventh and Twelfth Books of Euclid. With Critical Notes, and an Appendix, Containing Various Particulars Relating to the Higher Parts of the Science.* Sixth edition. London: Printed for J. Richardson; et al. pp. 1 + 431.
◇ Errata follow p. 1. Pages [408]–431 contain "A Synopsis of All the Data for the Construction of Triangles, from which Geometrical Solutions have Hitherto been in Print. With References to the Authors, where those Solutions are to be Found," by John Lawson, first published in 1783.

1820. *A Treatise On Algebra, in Practice and Theory, with Notes and Illustrations; Containing a Variety of Particulars relating to the Discoveries and Improvements that have been made in this Branch of Analysis.* Second edition, revised and greatly improved. London: Printed for J. Nunn, et al. pp. xxxvi + 411.
◇ The library has volume I only. Errata on pp. xxxv–xxxvi.

1840. *Introduction to Mensuration and Practical Geometry. To Which are Added a Treatise on Guaging [sic]: and also the Most Important Problems in Mechanics, by James Ryan.* Philadelphia: Kimber & Sharpless. pp. [xii] + 288.
⋄   Copyright 1833.

Bonnycastle, John. See also Bossut 1803; Hutton 1812a.

Boole, George (1815–1864).
1865. *Treatise on Differential Equations. Supplementary Volume.* Cambridge and London: Macmillan and Company. pp. xii + 235.
⋄   "The present volume contains all that Professor Boole wrote for the purpose of enlarging his Treatise on Differential Equations." (From the preface by I. Todhunter.) There is a list of Boole's published papers and books, pp. viii–xi.

1872. *A Treatise on the Calculus of Finite Differences.* Edited by J. F. Moulton. Second edition. London: Macmillan & Co. pp. xii + 336 + 32.

1877. *A Treatise on Differential Equations.* Fourth edition. London: Macmillan & Co. pp. xv + 496 + 1 plate.
⋄   The book contains [53 pp.] of a catalog of educational books published by Macmillan & Co.

Borda, Jean Charles de (1733–1799). See Delambre 1800.

Borel, Émile (1871–1956).
1904. *Algèbre, second cycle.* Deuxième édition. Cours de mathématiques rédigé conformément aux nouveaux programmes (31 Mai 1902). Paris: Armand Colin. pp. viii + 379.

1909. *Elements de la théorie des probabilites.* Paris: Librairie Scientifique Hermann et Fils. pp. viii + 191.

1914. *Le hasard.* Deuxième édition. Paris: Félix Alcan. pp. iv + 312.

Borel, Émile. See also Niewenglowski 1894.

Bossut, Charles (1730–1814).
1808; 1800; 1802. *Cours de mathématiques.* 3 vols. Nouvelle édition revue et augmentée. Paris: Firmin Didot. pp. xxiv + 455; xvi + 421 + 16 plates; xxiv + 444 + 13 plates.
⋄   Errata for volume 1 follows p. 455, for volume 2 on p. xvi, and for volume 3 on p. 444. Volume 1 contains arithmetic and algebra, and volume 2 contains geometry and the application of algebra to geometry. The first and third volumes are announced as new editions. The second volume contains a letter from Carnot to Bossut providing several new aspects of trigonometry.
○   Thayer binding.

1803. *A General History of Mathematics from the Earliest Times to the Middle of the Eighteenth Century, Translated from the French of John* (sic) *Bossut, to which* *is affixed a Chronological Table of the Most Eminent Mathematicians.* Translated by John Bonnycastle. London: Printed for J. Johnson. pp. xxvi + 540.

⬦ A two–page table of "eminent mathematicians" assembled by John Bonnycastle follows p. 540.

Bossut, Charles (1730–1814), and Viallet, Guillaume (?–1772).

1764. *Recherches sur la construction la plus avantageuse des digues: ouvrage qui a remporté le prix quadruple proposé par l'Académie royale des sciences, inscriptions & belles–lettres de Toulouse, pour l'année 1762.* Paris: Charles–Antoine Jombert. pp. 64 + 7 plates.

Boucharlat, Jean Louis (1775–1848).

1810. *Théorie des courbes et des surfaces du second ordre, précédée des principes fondamentaux de la géométrie analytique.* Seconde édition. Paris: F. Bechet. pp. xx + 344 + 6 plates.

⬦ Dedicated to Senator Count Lagrange.
○ Copy has a few notes in the margins.

1826. *Élémens de calcul différentiel et de calcul intégral.* Troisième édition, revue et augmentée. Paris: Bachelier, successor de Mme Ve Courcier. pp. xii + 410.

⬦ This book was the basis of Charles Davies' calculus textbook.
⬦ This particular book (in French) was used as a text at USMA from 1821 until 1833.

1827. *Élémens de mécanique.* Seconde édition, considérablement augmentée. Paris: Bachelier. pp. xx + 472 + 10 plates.

⬦ This book was used as a textbook at USMA from 1833 to 1850.
○ The library has two copies.
○ One copy has tag inside cover, "A. J. Kilian." This copy was given to USMA by the heirs of COL Jas. Duncan.

1830. *Élémens de calcul differentiel et de calcul integral.* Quatrième édition, augmentée considerablement. Paris: Bachelier. pp. xvi + 540 + 5 plates.

○ Copy contains numerous student notes.

Boucharlat, Jean Louis (1775–1848). See also Davies (Calculus).

Boucher, Maurice.

1903. *Essai sur l'hyperspace. Le temps, la matière et l'énergie.* Paris: Félix Alcan. pp. 204.

Bourdon, Louis Pierre Marie (1779–1854).

1821. *Élémens d'arithmétique.* Paris: Mme Ve Courcier. pp. xvi + 382.
⬦ Errata follow p. 382.

1823. *Élémens d'algèbre, ouvrage adopté par l'université.* Troisième édition. Paris: Bachelier, successeur de Mme Ve Courcier. pp. xxiv + 670.

1828. *Élémens d'algèbre; ouvrage adopté par l'université.*   Cinquième édition. Paris: Bachelier (successeur de Mme Ve Courcier); Brussels: Librairie Parisienne. pp. xiv + 692.

1831a. *Elements of Algebra, Translated from the French of M. Bourdon, for the Use of the Cadets of the U. S. Military Academy.* Translated by Edward C. Ross. New York: E. B. Clayton. pp. vii + 389.
  ◇ Errata follow p. 389.

1831b. *Elements of Algebra, Translated from the French of M. Bourdon, for the Use of the Cadets of the U. S. Military Academy.* Volume 2 (only) of 2 volumes. Translated by Edward C. Ross. New York: E. B. Clayton. pp. iv + 233–389.

1837. *Application de l'algébre a la géométrie.*   Quatrième édition. Paris: Bachelier. pp. xvi + 606 + 15 plates.
  ◇ Errata on p. xvi.
  ○ The library has two copies.

1843. *Éléments d'algèbre, ouvrage adopté par l'université.*   Neuvième édition. Paris: Bachelier. pp. xvi + 690.

Bourdon, Louis Pierre Marie. See also Ross 1831; Davies (Algebra); Davies (Analytical Geometry) 1836.

Bourlet, Carlo (1866–1913).
1896a. *Leçons d'analyse.*   Paris: Armand Colin & Cie.

1896b. *Leçons d'algèbre élémentaire.*   Cours complet de mathématiques élémentaire. Publié sous la direction de M. Darboux. Paris: Armand Colin & Cie. pp. xii + 548.

Boutroux, Pierre (1880–1922).
1908. *Leçons sur les fonctions définies par les équations différentielles du premier ordre professées au Collège de France.*   Collection de monographies sur la théorie des fonctions. Publiée sous la direction de M. Émile Borel. Paris: Gauthier–Villars. pp. [ii] + 190 + 64 blank pages.
  ◇ Contains a note by Paul Painlevé.

1914; 1919. *Les principes de l'analyse mathématique exposé historique et critique.* 2 vols. Paris: A. Hermann & Fils. pp. xi + 547; 512.
  ◇ Errata for both volumes are in volume 2, pp. 508–510.

Bowditch, Nathaniel (1773–1838). See Bowditch 1840; Peirce 1840; Pickering 1838.

Bowditch, Nathaniel Ingersoll (1805–1861).
1840. *Memoir of Nathaniel Bowditch.* Second edition. Boston: Charles C. Little and James Brown. pp. 172.
  ◇ The author is Nathaniel Bowditch's son. This memoir was bound together with

volume four of Nathaniel Bowditch's translation of Laplace's *Mécanique Céleste*. There are portraits of Nathaniel Bowditch and his wife Mary Bowditch. The Appendix (pp. 159–172) contains seven notes, a list of persons and institutions to whom Bowditch gave copies of his translation of Laplace, and a copy of a letter defending Bowditch's role in Dr. Kirkland's resignation from Harvard.

○ Presentation copy to Dr. William H. Vanburen. A letter from J. Ingersoll Bowditch (Jan. 3, 1863) to Dr. Vanburen is in the volume.

Brahe, Tycho (1546–1601). See Flamsteed 1725; Kepler 1627.

Brahmagupta (598–665). See Colebrook 1817.

Brakenridge [Braikenridge], William (1701-1762).
1733. *Exercitatio geometrica de descriptione linearum curvarum.* Londini [London]: Ric. Hett et Joh. Nourse. pp. viii + 70.
◇ Dedicated to John Comiti. Errata on p. 70.

Brancker, Thomas (1633–1676). See Maseres 1795.

Branford, Benchara.
1908. *A Study of Mathematical Education Including the Teaching of Arithmetic.* Oxford: Clarendon Press, pp. xii+392+16 pages of advertising.

Breckenridge, William Edwin (1869–?).
n.d. *The Polyphase Slide Rule. A Self Teaching Manual with Tables of Setting, Equivalents and Gauge Points.* New York: Keuffel & Esser Co. pp. [ii] + 83 + [2].
○ Volume was a gift of COL Leydecker.

Bremiker, Karl (1804–1877).
1852. *Logarithmorum VI decimalium nova tabula Berolinensis et numerorum vulgarium ab 1 usque ad 100000 et functionum trigonometricarum ad decades minutorum secundorum.* Berolini [Berlin]: Prostat in libraria Friderici Nicolai. pp. 82 + unnumbered pages of tables.
◇ Dedicated to J. F. Encke. Errata follow the last page of the tables.

Bremiker, Karl. See also Vega 1857; Vega 1859; Vega 1883; Crelle 1880.

Bresson, Charles.
1842. *Traité élémentaire de mécanique appliquée aux sciences physiques, et aux arts.* Paris: Bachelier. pp. xxi+664.
◇ Three volumes in one. The 18 plates are bound separately in an "atlas.".

Brewster, David (1781–1868).
1814. *[James] Ferguson's Lectures on Select Subjects in Mechanics, Hydrostatis, Hydraulics, Pnenmatics, Optics, Geography, Astronomy, and Dialling.* 3 vols. A

new edition, corrected and enlarged with notes and an appendix, adapted to the present state of the arts and sciences. Second American edition, carefully revised and corrected by Robert Patterson. Philadelphia: M. Carey. pp. xlvi + 356; vii + 500; [iv] + 48 plates.

◊   There is a "short account of the life" of James Ferguson in vol. 1, pp. xv–xlvi. Volume 2 concludes with a general index, pp. 485–500.

Brewster, David. See also Davies (Geometry and Trigonometry); Leonhard Euler 1835.

Brianchon, Charles Julien (1785–1864).
1818. *Application de la théorie des transversales, cours d'opérations géométriques sur le terrain, fait a l'École d'artillerie de la garde royale, en Mars 1818.* Premier cahier. Paris: Bachelier. pp. [iv] + 62 + 1 plate.

Bridge, Bewick (1767–1833).
1814a. *A Treatise on Mechanics: Intended as an Introduction to the Study of Natural Philosophy.* 2 vols in one book. Broxbourne: Watts. pp. [x] + 369 + 228.
○   Hand written notes inside the back cover.

1814b. *A Treatise on Mechanics: Intended as an Introduction to the Study of Natural Philosophy.* 2 vols in one book. London: Printed for T. Cadell and W. Davies, etc. pp. [x] + 362 + 228.
◊   Errata on p. [x].

1817. *A Treatise on the Construction Properties and Analogies of the Three Conic Sections.* Second edition with alterations and corrections. London: Printed by R. Watts. pp. [vi] + 137.
◊   Followed by Bridge 1818 in the same binding.

1818. *A Compendious Treatise on the Elements of Plane Trigonometry: with the Method of Constructing Trigonometrical Tables.* Second edition. London: Printed by R. Watts. pp. [v] + 77.
◊   Bound after Bridge 1817.

1821. *A Compendious Treatise on the Theory and Solution of Cubic and Biquadratic Equations, and of Equations of the Higher Orders.* London: Printed by R. Watts. pp. [v] + 227.
◊   Errata on p. [v]. Four other books by the same author are listed on p. [229].

Bridges, William.
1718. *An Essay to Facilitate Vulgar Fractions; after a New Method, and to Make Arithmetical Operations Very Concise: Containing, Many New and Excellent Rules on Vulgar Fractions and Mix'd Numbers; some Critical Observations on Numbers; Multiplication and Division of Money, Weight and Measure; with these following Practical Rules, viz. I. Fellowship, II. Barter, III. Loss and Gain, IV. Interest, V. Rebate or Discount, VI. Exchange. To which is Annexed ... ious Practical and, Lastly, Miscellaneous Questions, with Their Answers.* London: Printed for A. Bettesworth. pp. [xii] + 153 + [15].

Briggs, Henry (1561–1630).

1624a. *Logarithmicall Arithmetike. Or Tables of Logarithmes for Absolvte Nvmbers from an Unite to 100000; as also for Sines, Tang[s], an Secantes for every Minute of a Quad[t], with a Plaine Description of Their Use in Arithmetike, Geometrie, Geographie, Astronomie, Navigation, &c. These Numbers were First Invented by the Most Excellent Iohn Neper Baron of Marchiston, and the Same Were Transformed, and the Foundation and Use of Them Illustrated with his Approbation by Henry Briggs, Sir Henry Savils Professor of Geometry in the University of Oxford. The Uses Whereof were Written in Latin by the Author Himself, & Since his Death Published in English by Diverse of his Friends According to his Mind, for the Benefit of Such as Understand not the Latin Tongue. Deus nobis usuram vitae dedit, et ingenii, tanquam pecuniae, nulla praestituta die.* London: Printed by George Miller. pp. [iii] + 54 + several hundred unnumbered pages of tables.

◇ Errata on p. [iii].

○ The title page has been lost and replaced long ago by a very neatly hand–lettered page. The initials "M.E." have been carved into the bottom of the pages.

1624b. *Arithmetica logarithmica sive logarithmorum chiliades triginta, pro numeris naturali serie crescentibus ab unitate ad 20,000: et a 90,000 ad 100,000. Quorum ope multa persiciuntur arithmetica problemata et geometrica. Hos numeros primus invenit clarissimus vir Iohannes Neperus Baro Merchistonij: eos autem ex eiusdem sententia mutavit, eorumque ortum et usum illustravit Henricus Briggius, in celeberrima Academia Oxoniensi geometriae professor Salivianus. Deus nobis usuram vitae dedit, et ingenii, tanquam pecuniae, nulla praestituta die.* Londini [London]: Excudebat Gulielmus Iones. pp. [vii] + 88 + unnumbered pages of tables.

◇ Errata on p. [vii]. After the last page (it contains the logarithm of 100,000 and is marked "finis" at the foot), there are 10 slightly smaller pages containing the logarithms of 100001 to 101000.

○ A description of the book has been pasted on the flyleaf by the USMA Library.

1633. *Trigonometria Britannica: sive de doctrina triangulorum libri duo. Quorum prior continet constructionem canonis sinuum tangentium & secantium, una cum logarithmis sinuum & tangentium ad gradus & graduum centesimus & ad minuta & secunda centesimis respondentia: a clarissimo doctissimo integerrimoque viro domino Henrico Briggio geometriae in celeberrima academia Oxoniensi professore Saviliano dignissimo, paulo ante inopinatam ipsius e terris emigrationem compositus. Posterior vero usum sive applicationem canonis in resolutione triangulorum tam planorum quam sphaericorum e geometricis fundamentis petita, calculo facillimo, eximiisque compendiis exhibet: ab Henrico Gellibrand astronomiae in collegio Greshamensi apud Londmenses professore constructus.* Govdae [Gouda]: Excudebat Petrus Rammasenius. pp. [vi] + 110 + approximately 200 pages of tables.

◇ Errata on p. [vi]. The feature of this table is that each degree is subdivided into one–hundreths rather than minutes and seconds. The text of Book I explaining the construction of the table and the table itself were composed by Henry Briggs, ca. 1600. Book II describing some uses of trigonometry was written by Henry Gellibrand. The volume was "published by the direction of Gellibrand in 1633, it having been printed at Gouda under the care of Vlacq,..." [Hutton 1812a, vol. 1, pp. 388–390].

Briggs, Henry. See also Sherwin 1761 and Vlacq 1633.

Brinkley, John (1763–1835).
1801. *On Determining Innumerable Portions of a Sphere, the Solidities and Spherical Superficies of which Portions are at the Same Time Algebraically Assignable.* Dublin: George Bonham. pp. 15 + 1 plate.
◇ This is a paper from the Transactions of the Royal Irish Academy, read November 2, 1801, which has been bound.
◇ This deals with Viviani's Florentine enigma of 1692.

1807. "An Investigation of the General Term of Important Series in the Inverse Method of Finite Differences."
◇ A paper bound together with other papers in Anonymous n.d.

Brisse, Ch. (1843–1898).
1895. *Cours de géométrie descriptive a l'usage des élèves de l'enseignement secondaire moderne.* Paris: Gauthier–Villars et Fils. pp. xx + 144 + 116.
◇ Two parts bound together.

Brisson, Barnabe (1777–1828). See Heather 1851; Monge 1820; Monge 1827.

Brown, John.
1826. *Mathematical Tables; Containing the Logarithms of Numbers, Logarithmic Sines, Tangents, and Secants, Natural Sines, Traverse Table, and Various Tables Useful in Business; to which are Prefixed, the Construction and Use of the Tables, Plane and Spherical Trigonometry, with Their Applications. For the Use of Schools.* Fourth edition, improved and enlarged by J. Wallace. Stereotyped. Edinburgh: Printed for Stirling and Kenney, and John Fairbairn. pp. [iv] + 180 + 93.

Brownell, W. R. See Carnot 1832.

Burali–Forti, Cesare (1861–1931) and Marcolongo, Roberto (1862–1943).
1910. *Éléments de calcul vectoriel avec de nombreuses applications à la géométrie, à la mécanique et à la physique mathématique.* Édition française traduite de l'Italien et augmentée d'un supplément par S. Lattès. Paris: A. Hermann et Fils. pp. vii + 229.
◇ There are "Historical and Critical Notes" on pp. 213–223.

Burckhardt, Johann Karl (1773–1825).
1814; 1816. *Table des diviseurs pour tous les nombres du deuxième million, ou plus exactement, depuis 1020000 à 2028000, avec les nombres premiers qui s'y trouvent.* Bound together with *Table des diviseurs pour tous les nombres du troisième million, ou plus exactement, depuis 2028000 à 3036000, avec les nombres premiers qui s'y trouvent.* Paris: Ve Courcier. pp. viii + 112; [iv] + 112.
○ Thayer binding.

Burg, Adam (1797–1882).

1847. *Traité du dessin géométrique ou exposition complète de l'art du dessin linéaire, de la construction des ombres et du lavis, a l'usage des Industriels, des Savants et de ceux qui veulent s'instruire sans le secours de maîtres et Spécialement destiné pour l'enseignement dans les Écoles royales d'Artillerie prussienne.* 2 vols. 2me édition, complétement refondue et augmentée, traduit de l'allemand par le Dr. Regnier. Paris: J. Corréard, pp. vii + 9–352; 30 plates.

⋄ Volume 1 is the text and volume 2 is the "Atlas", i.e., the plates.

Burg, Adam. See also Delambre 1806.

Bürja, Abel (1752–1816).

1786. *Der selbstlernende Algebrist, oder deutliche Anweisung zur ganzen Rechenkunst, worunter sowohl die Arithmetik und gemeine Algebra, als auch die Differenzial–und Integral–Rechnung begriffen ist.* 2 vols. Berlin und Libau: Lagarde und Friedrich. pp. xviii + 320; 332.

⋄ Frontispiece is a classical calculator. Errata follow p. 332 of volume 2.

1788. *Erleichterter Unterricht in der höheren Messkunst, oder deutliche Anweisung zur Geometrie der krummen Linien.* 2 vols. Berlin und Libau: Lagarde und Friedrich. pp. xxxiiii + 382; 388.

⋄ Volume 1 has an interesting frontispiece. See Figure 20 for a photograph.
⋄ Errata following p. 388 of volume 2.

1791. *Grundlehren der Dynamik oder desjenigen Theiles der Mechanik welcher von den festen Körpern im Zustande der Bewegung handelt.* Berlin: F. Lagarde. pp. [x] + 416 + 2 pages of errata + nachricht.

⋄ Signatures uncut. Contains interesting frontispiece.

Burnside, William Snow (1839–1920) and Panton, Arthur William (?–1906).

1892. *The Theory of Equations: with an Introduction to the Theory of Binary Algebraic Forms.* Dublin University Press Series. Third edition. Dublin: Hodges, Figgis, & Co.; London: Longmans, Green, & Co. pp. xvi + 496.

Büsch, Johann Georg (1728–1800).

1790a. *Versuch einer Mathematik zum Nuzzen und Vergnügen des bürgerlichen Lebens, welcher das Nuzbarste aus der abstrakten Mathematik und eine praktische Mechanik enthält.* Dritte und sehr vermehrte Ausgabe. Hamburg: Benjamin Gottlob Hoffmann. pp. viii + 295 + 9 plates + 344 + 9 plates.

1790b. *Versuch einer Mathematik zum Nuzen und Vergnügen des bürgerlichen Lebens, Zweiter Theile, welcher die Hydrostatik, Aerometrie und Hydraulik enthält.* Hamburg: Benjamin Gottlob Hoffmann. pp. viii + 464 + 7 plates.

Byerly, William Elwood (1849–1935).

1889 [1888]. *Elements of the Integral Calculus, with a Key to the Solution of Differential Equations, and a Short Table of Integrals.* Second edition, revised and enlarged. Boston: Ginn and Company. pp. xvi + 339 + 65–76 appendix + 1–32 table of integrals.

◇ This book is bound with *A Short Table of Integrals* compiled by Benjamin O. Peirce [32 pp.] published in 1889.

1893a. *Chauvenet's Treatise on Elementary Geometry.* Revised and abridged. Philadelphia: J. B. Lippincott. pp. 8 + 314.

1893b. *An Elementary Treatise on Fourier Series and Spherical, Cylindrical, and Ellipsoidal Harmonics, with Applications to Problems in Mathematical Physics.* Boston: Ginn and Company. pp. ix + 287.

Cagnoli, Antoine (1743–1816).

1808. *Trigonométrie, rectiligne et sphérique, traduite de l'Italien par N. M. Chompré.* Seconde édition considérablement augmentée. Paris: Courcier; Lyon: Maire; Marseille: Jean Mossy; Geneva: Paschaud. pp. xvi + 504 + 9 tables + 7 plates.

◇ A list of new articles included in the second edition is on p. xv. Errata are on p. xvi.

◇ Translator is Nicolas Maurice Chompré (1750–1825).

Cajori, Florian (1859-1930).

1890. *The Teaching and History of Mathematics in the United States.* Washington, D.C.: Government Printing Office. pp. 400.

◇ This was Bureau of Education Circular of Information No. 3, 1890. This volume contains information about the mathematics department at West Point.

1894. *A History of Mathematics.* New York: MacMillan & Co. pp. xiv + 422 + 22 pages of publications.

◇ First printing of the first edition. "Books of Reference" (101 items, pp. ix–xiv) includes "books, pamphlets, and articles" used by Cajori for this work.

◇ c. 1893.

Callet, Jean François (1744–1798).

1795 (Printed 1825); (Printed 1827); (Printed 1829). *Tables portatives de logarithmes contenant les logarithmes des nombres, depuis 1 jusqu'à 108,000; les logarithmes des sinus et tangentes, de seconde en seconde pour les cinq premiers degrés, de dix en dix secondes pour tous les degrés du quart de cercle; et, suivant la nouvelle division centésimale, de dix–millième en dix–millième, précédées d'un discours préliminaire sur l'explication, l'usage et la sommation des logarithmes, et sur leur application à l'astronomie, à la navigation, à la géométrie–pratique, et aux calculs d'intérêts; suivies de nouvelles tables plus approchées, et de plusieurs autres utiles à la recherche de longitudes en mer, etc.* Édition stéréotype. Paris: Firmin Didot. pp. vi + 7–118 + plate + several hundred unnumbered pages of tables.

◇ This is an expansion of the *Tables portatives de logarithmes* of Callet and W. Gardiner (1783). The 1795 tables were "the first set of tables to be stereotyped".

○ The library's copy of the 1825 book belonged to Cadet Theodore Franks (admitted 1824, but did not graduate). A later owner has signed that copy.

Camerer, Johann Wilhelm von (1763–1847). See Apollonius of Perga 1795.

Campbell, John Edward (1862–1924).
1903. *Introductory Treatise on Lie's Theory of Finite Continuous Transformation Groups.* Oxford: Clarendon Press. pp. xx + 416.

Cambridge University. See Anonymous 1821.

Camus, Charles Etienne Louis (1699–1768).
1750. *Cours de mathématique. Seconde partie. Élémens de géometrie théorique et pratique.* Paris: Durand. pp. vi + 568 + 25 plates.
◇ Errata on p. 536. The text ends on p. 536. A list of theorems, which serves as a table of contents begins on p. 537, but ends abruptly on p. 568 (the last page cited is p. 493, so several leaves are missing).
○ The library has volume 2 only.

Cardano, Girolamo (1501–1576).
1570. *Hieronymi Cardani mediolanensis, civisque bononiensis, philosophi, medice et mathematici clarisimi, opus novum de proportionibus numerorum, motuum, ponderum, sonarum, aliarumque rerum mensurandarum, non solum geometrico more stabilitum, sed etiam uarijs experimentis & observationibus rerum in natura, solerti demonstratione illustratum, ad multiplices usus accommodatum, & in V libros digestum. Praeterea artis magnae, sive de regulis algebraicis, liber unus, abstrausissimus & inexhaustus plane totius arithmeticae thesaurus, ab authore recens multis in locis recognitus & auctus. Item de aliza regula liber, hoc est, algebraicae logisticae suae, numeros recondita numerandi subtilitate, secundum geometricas quantitates inquirentis, necessaria coronis, nunc demum in lucem edita. Opus physicis & mathematicis imprimis utile & necessarium.* Basileae [Basel]: [Ex officina Henricpetrina] Cum Caes. Maiest. Gratia & Priuilegio. pp. [xv] + 271 + 173 + [viii] + 111.
◇ The Ars Magna is dated 1645 on the reverse of the title page.

Carhart, Daniel (1839–1926).
1888. *A Treatise on Plane Surveying.* Boston: Ginn and Company. pp. xvii + 411 + 87.
○ Volume is rebound.

Carnot, Lazare Nicolas Marguerite (1753–1823).
1797. *Oeuvres mathématiques du citoyen Carnot.* Basle [Basel]: J. Decker. pp. xvi + 208 + 1 plate.
◇ This volume contains Carnot's important "Reflexions sur la metaphysique du calcul infinitesimal" (pp. 125–204).
◇ Portrait of Carnot is the frontispiece.

1801. *De la corrélation des figures de géométrie.* Paris: Duprat. pp. viii + 188 + 4 plates.
○ Thayer binding (but the border is floral).

1803a. *Géométrie de position.*   Paris: J. B. M. Duprat. pp. xxxix + 489 + 15 plates.

◊ Errata follow p. xxxix.

○ Thayer binding.

1803b. *Principes fondamentaux de l'équilibre et du mouvement.* Paris: Deterville. pp. xxii + 262 + 1 errata + 2 plates.

○ "U. S. Military Academy West–Point" gold and red Thayer binding.

1806. *Mémoire sur la Relation qui existe entre les distances respectives de cinq points quelconques pris dans l'espace; suivi d'un essai sur la théorie des transversales.*   Paris: Courcier. pp. 111 +3 plates.

◊ Corrections and observations follow p. 111.

○ Thayer binding.

1813. *Réflexions sur la métaphysique du calcul infinitésimal.*   Seconde édition. Paris: Ve Courcier. pp. vi + 252 + 1 plate.

○ Thayer binding.

1832. *Reflexions on the Metaphysical Principles of the Infinitesimal Analysis.* Translated by W. R. Brownell. Oxford: Printed for J. H. Parker, et al. pp. vii + 132.

Carnot, Lazare Nicolas Marguerite. See also Bossut 1800; 1802; 1808.

Carnoy, Joseph (1841–1906).
1877. *Cours de géométrie analytique. Géométrie de l'espace.*   Seconde édition. Louvain: D. Aug. Peeters–Ruelens; Paris: Gauthier–Villars. pp. xiii + 516.

○ On p. 209 there are red checks in the margin.

1880. *Cours de géométrie analytique. Géométrie plane.*   Troisième édition. Louvain: Aug. Peeters; Paris: Gauthier–Villars. pp. x + 419.

◊ The spine of the binding suggests that this is volume 1 and Carnoy 1877 is volume 2 of the same work. However, note the different editions and publication dates.

1892. *Cours d'algèbre superieure. Principes de la théorie des déterminants; théorie des équations; introduction a la théorie des formes algébriques.*   Louvain: A. Uystpruyst; Paris: Gauthier–Villars & Fils. pp. xi + 537.

◊ Errata follow p. xi.

Carr, George Shoobridge (1837–?).
1886. *A Synopsis of Elementary Results in Pure Mathematics: Containing Propositions, Formulae, and Methods of Analysis, with Abridged Demonstrations. Supplemented by an Index to the Papers on Pure Mathematics which are to be Found in the Principal Journals and Transactions of Learned Societies, Both English and Foreign, of the Present Century.* London: Francis Hodgson. pp. 935 + 20 plates.

◊ Errata face the copyright page.

○ Two copies are in the library.

Carroll, Lewis (1832–1898). See under Dodgson, Charles Lutwidge.

Carvallo, Jules.
1853. *Étude sur la stabilité des voûtes.* Paris: Carilian–Goeury et Vor Dalmont. pp. 79 + 3 fold out tables + 2 plates.
◇ Extracted from the *Annales des ponts et chaussées,* 1853.

Casey, John (1820-1891).
1888. *A Sequel to the First Six Books of the Elements of Euclid, Containing An Easy Introduction to Modern Geometry, With Numerous Examples.* Dublin University Press Series. Fifth edition, revised and enlarged. Dublin: Hodges, Figgis, & Co.; London: Longmans, Green, & Co. pp. xx + 251 + 12 pages of "Opinions of the Work."
◇ Errata on p. xx and tipped in facing that page.

Casey, John. See also Euclid 1887.

Casey, Joseph B.
1887. *A Key to the Exercises in the First Six Books of Casey's Elements of Euclid.* Second edition, revised. Dublin: Hodges, Figgis, & Co.; London: Longmans, Green, & Co. pp. vi + 282 + 12 pages of "Opinions of the Work."
◇ The 'Casey' of the title is John Casey (1820–1891).

Cassini de Thury, César François (1717–1784).
1744. *La meridienne de l'Observatoire royal de Paris, vérifiée dans toute l'étendue du royaume par de nouvelles observations; pour en déduire la vraye grandeur des degrés de la terre, tant en longitude qu'en latitude, & pour y assujettir toutes les opérations géométriques faites par ordre du Roy, pour lever une carte générale de la France.* Paris: Hippolyte–Louis Guerin, & Jacques Guerin. pp. [viii] + 292 + ccxxxv + [xvi] + 14 plates.
◇ Pages cvii–ccxxxv are observations on Natural History by LeMonnier.
○ Thayer binding.

Caswell, John (1654 or 1655–1712).
1685. *A Brief (but full) Account of the Doctrine of Trigonometry, both Plain and Spherical.* London: John Playford. pp. [I] + 17.
○ Bound with Wallis 1685.

Caswell, John. See also Wallis 1693.

Catalan, Eugène (1814–1894).
1857; 1861. *Traité élémentaire de géométrie descriptive.* Première partie. Du point, de la droite et du plan. Paris: Victor Dalmont, successeur de Carilian–Goeury et Vor Dalmont. pp. 139 + 18 plates.

   ◦   The "Atlas" or plates are from the second edition, published by Dunod in 1861, but they are bound together with the first part.

1860. *Traité élémentaire des séries*. Paris: Leiber et Faraguet. pp. viii + 132 + 4.
   ◇   The last four pages list some other mathematics books published by Leiber and Faraguet.
   ◦   Signed "c. Dem...[?], USMA, Dec 23, 1881". "Presented to Dept of Math, Dec, 1910".

Cauchy, Augustin–Louis (1789–1857).
1826; 1827; 1828; 1829; 1830. *Exercices de mathématiques.* 5 vols. Paris: de Bure Frères. pp. ii + 357 + [iii]; ii + 376 + [iv]; 368 + [iii]; 310 + [v]; 72.
   ◇   After the first, the volumes are designated 2nd year, 3rd year, 4th year, and 5th year. The 4th and 5th volumes are bound together as one book. For volumes 1–4, the errata come at the end of the volume.

1827. *Cours d'analyse, premiere année, Ecole polytechnique de France.*   Paris: unpaginated, approximately 200 pages.
   ◇   This is a manuscript of handwritten course notes.

1830. *Mémoire sur la dispersion de la lumière*. Paris: De Bure Frères. pp. 24.
   ◇   An entirely mathematical work.

1835; 1836. *Nouveaux exercices de mathématiques.*   Bound together with *Mémoire sur la dispersion de la lumière.*   Publié par la Société royale des sciences de Prague. Prague: J. G. Calve. pp. iv + 204 + iv + 205–236.

Charruit, N.
1897. *Problèmes et épures de géométrie descriptive et de géométrie cotée.*   Tome 1 à l'usage des candidats aux Écoles de Saint–Cyr et Navale, à l'Institut Agronomique et des aspirants au baccalauréat. Deuxième édition. Paris: Nony & Cie. pp. 282 + 2.

Chase, Stephen (1813–1851).
1849. *A Treatise on Algebra. For the Use of Schools and Colleges.* Philadelphia: Appleton.

Chasles, Michel (1793–1880).
1852. *Traité de géométrie supérieure.*   Paris: Bachelier. pp. lxxxiii + 603 + 12 plates.
   ◇   Errata on p. 603.

1860. *Les trois livres de porismes d'Euclide, rétablis pour la première fois, d'après la notice et les lemmes de Pappus, et conformément au sentiment de R. Simson sur la forme des énoncés ces propositions.* Paris: Mallet–Bachelier. pp. ix + 324.
   ◇   Errata on p. 324.

1889. *Aperçu historique sur l'origine et le développement des méthodes en géométrie particulièrement de celles qui se rapportent al la géométrie moderne suivi d'un* *mémoire de géométrie sur deux principes généraux de la science la dualité et l'homographie.* Troisiéme édition, conforme a la première. Paris: Gauthier–Villars et Fils. pp. [ii] + [851].

⋄ The "Historical Outline" is pp. 1–572 and the "Memoir on Duality and Homography" is pp. 573–851.

Chasles, Michel. See also Lagrange 1797, Euclid 1860.

Chaulnes, the Duke of [Michel–Ferdinand d'Albert d'Ailly] (1741-1793).
1768a. *Nouvelle méthode pour diviser les instruments de mathématique et d'astronomie.* [Paris]: L. F. Delatour. pp. 44 + 15 plates.

○ Corrections and additions are on the verso of the title page. Bound with his 1768b.

1768b. *Description d'un microscope, et de différents micrometres destinés à mesurer des parties circulaires ou droites, avec la plus grande précision.* [Paris]: L. F. Delatour. pp. 18 + 6 plates.
○ Bound with his 1768a.

Chauvenet, William (1820–1870).
1862. *A Treatise on Plane and Spherical Trigonometry.* Fifth edition. Philadelphia: J. B. Lippencott & Co. pp. 256.
○ This copy of the book was owned by John Pitman, Jr., Troy, NY.
⋄ Copyright 1850.

1879. *A Treatise on the Method of Least Squares or the Application of the Theory of Probabilities in the Combination of Observations. Being the Appendix to the Author's Manual of Spherical and Practical Astronomy.* Philadelphia: J. B. Lippincott and Co. pp. [iii] + 469–599.
⋄ This book was used as a text at USMA from 1879 until 1890.
○ The library has two copies. One copy signed by "John Biddle."
○ Second copy contains course notes and syllabus.

Chladni, Ernst Florens Friedrich (1756–1827).
1809. *Traité d'acoustique.* Paris: Courcier. pp. [iii] + xxvii + 375 + 8 plates.
⋄ Dedicated to Napoleon. Corrections and Observations on p. xxviii.
○ Thayer binding.

Chomé, F.
1893. *Cours de géométrie descriptive de l'École militaire.* Volume 1 only. Seconde édition, entièrement revue, corrigée et augmentée, contenant les prescriptions à observer pour l'exécution des épures. Bruxelles [Brussels]: Alfred Castaigne; Paris: Gauthier–Villars. pp. [vii] + [156] + [37] descriptive geometry plates.

1896. *Éléments de géométrie descriptive (Point, droite et plan) à l'usage des candidats à l'École militaire et aux écoles spéciales des universités.* Édition conforme au programme officiel et contenant les prescriptions à observer pour l'exécution des épures. Bruxelles [Brussels]: Alfred Castaigne. pp. xii + 158 + [2].

⋄ This book is the text only, no plates are included in this volume. Errata on the unnumbered page following p. 158.

○ Copy was rebound in 1952. Has a War Department stamp and U.S. Legation stamp. There is a War Department letter pasted opposite the title page to Prof. E. W. Bass, USMA.

1898; 1899; 1904. *Cours de géométrie descriptive de l'École militaire.* 3 vols. Troisième édition, entièrement revue, corrigée et augmentée, contenant les prescriptions à observer pour l'exécution des épures. Bruxelles [Brussels]: Alfred Castaigne; Paris: Gauthier–Villars. pp. [vii] + [155] + [37] plates; xx + 161–498 + 86 plates; vi + 171 + 36 plates.

⋄ Volume 2 contains "Additions and modifications to the first volume" (pp. xvii–xx); and the errata pp. 495–496. Volumes 2 and 3 are the First Part and Second Part, respectively, for the use of the students of the higher curriculum.

Chompré, Nicolas Maurice (1750–1825). See Cagnoli 1808.

Choquet, Charles (1798–?).
1856. *Traité d'algèbre.* Paris: Mallet–Bachelier. pp. xvi + 551.

Church, Albert Ensign (1807–1878; USMA, 1828).
See Church (Trigonometry), Church (Analytical Geometry), Church (Descriptive Geometry), and Church (Calculus)
⋄ Church served as Department Head for USMA Mathematics Department from 1837 to 1878.

Church, Albert Ensign (Trigonometry).
1869. *Plane and Spherical Trigonometry.* New York: A. S. Barnes. pp. 80.
⋄ This book was used as a text at USMA from 1879 to 1887.
○ This copy is signed by "Hugh T. Reed, class '73, Cadet USMA March 1878." There are a few annotations in the book.
○ This copy has a cardboard cover and is stored in a bound box.

n.d. *Plane and Spherical Trigonometry.* New York: A. S. Barnes & Co. pp. 13 blank + 3–80 + 13 blank.
○ This book was received in 1880. Notes made by Church (initials on pp. 22–23) presumably for the next edition. Notes throughout, several extra pages glued in at pages 22, 23, and 71.

Church, Albert Ensign (Analytical Geometry).
[1851]a. *Elements of Analytical Geometry.* New York: A. S. Barnes. pp. ix + 297.
⋄ This book was used as a text at USMA from 1852 to 1898.
⋄ Date determined from preface.
○ Library has two copies. One has endpaper signed "B. B. Bassette (U.S.M.A.), West Point, NY, January 11, 1890." Page 1 is annotated "commence Saturday, April 19, 1890. 3rd Section, Lt. Jackson, Inst., E. Bass, Professor." Each day's lesson is marked. Several printed and handwritten slips are typed in.
○ Second copy is clean.

Church, Albert Ensign (Analytical Geometry – Continued).

[1851]b. *Elements of Analytical Geometry.* New York: American Book Co. pp. ix + 297.

◇   Date determined from preface. Identical with Church (Analytical Geometry) 1851a, 1851c, and 1851d, except for different publisher.

1851c. *Elements of Analytical Geometry.* New York: Geo. P. Putnam. pp. ix + 297.

◇   Identical with Church (Analytical Geometry) 1851a, 1851b, and 1851d, except for different publisher and date explicitly written on title page.

○   Rebound in 1994.

[1851]d. *Elements of Analytical Geometry.* Third edition. New York: A. S. Barnes. pp. ix + 297.

◇   Date determined from preface.

○   Library has two copies. One copy signed by J. S. Winn. Copy contains numerous hand-written annotation, additions, and solved problems.

○   Second copy is stamped "Drawing Academy, USMA."

1860. *Elements of Analytical Geometry.* Second edition. New York: A. S. Barnes and Co. pp. ix + 297.

○   Copy contains notes from William Reiley [USMA 1863] and was rebound in 1994.

1862. *Elements of Analytical Geometry.* Second edition. New York: A. S. Barnes and Burr. pp. ix + 297.

○   Signed "Cadet A. Macomb Miller, 3rd class 1862–63." The 14 students in the 1st section are listed on the front end cover. Extensive formulas written on the end papers.

1866. *Elements of Analytical Geometry.* Second edition. New York: A. S. Barnes and Co. pp. ix + 297.

1867. *Elements of Analytical Geometry.* Third edition. New York: A. S. Barnes and Co. pp. ix + 297.

○   Numerous notes, printed paste-ins, and annotations in this copy.

1870. *Elements of Analytical Geometry.* Third edition. New York: A. S. Barnes and Co. pp. ix + 297.

○   This copy contains student notes throughout. Additional sheet added at page 156 by H. T. Reed, October 13, 1870. Section rosters from September 1, 1870 to Spring 1871 are listed near the back cover.

1873. *Elements of Analytical Geometry.* Third edition. New York: A. S. Barnes and Co. pp. ix + 297.

○   Signed "W. Harman Wills, '82 S.S." Otherwise, this is a clean copy.

1875. *Elements of Analytical Geometry.* Third edition. New York: A. S. Barnes and Co. pp. ix + 297.

◇   This edition appears to be identical with all the previous editions of Church's *Element of Analytical Geometry.*

○   Signed "Cadet L. W. V. Kennson, USMA. Aug 31st, 77."

Church, Albert Ensign (Descriptive Geometry).

1864a. *Elements of Descriptive Geometry, with its Applications to Spherical Projections, Shades and Shadows, Perspective and Isometric Projections.* New York: American Book Co. pp. vii + 192.

⋄ This book was used as a text at USMA from 1864 to 1929.

○ Three copies of this book are present in the library.

○ One copy is stamped "Dept of Drawing USMA Feb 15 1893". It has no other annotations.

○ Another copy is signed "M. M. Macomb, 1876". It is very carefully annotated with numerous corrections and taped in printed sheets.

○ The third copy has numerous notes and rewriting throughout. It has several pasted–in pages. It is stamped "United States Military Academy, Department of Drawing".

1864b. *Elements of Descriptive Geometry.* New York: A. S. Barnes and Co. pp. vii + 192.

⋄ Identical with Church (Descriptive Geometry) 1864a, except different publisher.

○ Train schedule written inside front cover for West Point and Fishkill. (Fishkill is across the Hudson River from West Point.)

1865a. *Plates to "Descriptive Geometry".* New York: A. S. Barnes & Burr. pp. 132 figures.

⋄ There are numerous blank pages bound in book.

○ This copy is signed by Donald Brown and David Floyd.

1865b. *Plates to "Descriptive Geometry".* New York: A. S. Barnes & Burr. pp. 83 figures.

⋄ This book has fewer plates than Church (Descriptive Geometry) 1865a, otherwise appears to be the same.

○ This copy has copious notes and drawings throughout the book. It was a gift of D. W. De Grange, Office of Chief of Engineers.

1865c. *Plates to "Descriptive Geometry".* New York: A. S. Barnes & Burr. pp. 102 figures on 21 plates.

○ This copy is signed "Cadet C. W. Whipple, USCC, April 14, 1865". There is a portrait of an unidentified man sketched in the copy. Also contains "1st section Math, 3rd class," with a list of 14 names of cadets (mostly USMA graduates of 1868). Charles William Whipple, USMA 1868, is listed last in the section. Edgar Wales Bass, USMA 1868, Professor of Mathematics and Department Head, USMA, 1878–1898, was also in this same section.

○ Copy donated by Dudley W. Stoddard.

1867a. *Elements of Descriptive Geometry, with its Applications to Spherical Projections, Shades and Shadows, Perspective and Isometric Projections.* New York: A. S. Barnes & Co. pp. vi + 192.

○ This copy has numerous amusing notes and drawings by a student.

Church, Albert Ensign (Descriptive Geometry – Continued).

1867b. *Plates to Descriptive Geometry. Shades, Shadows and Linear Perspective.* New York: A. S. Barnes & Co. pp. 132 figures.

1870a. *Elements of Descriptive Geometry, with its Applications to Spherical Projections, Shades and Shadows, Perspective and Isometric Projections.* New York: American Book Co. pp. viii + 192 + [8].
- Copy was rebound in 1994.

1870b. *Plates to Descriptive Geometry. Shades, Shadows and Linear Perspective.* New York: A. S. Barnes & Co. pp. 132 figures.

1877. *Elements of Descriptive Geometry, with its Applications to Spherical Projections, Shades and Shadows, Perspective and Isometric Projections.* New York: A. S. Barnes. pp. vii + 192.
- Corrections are pasted into this copy. Numerous notes are written in the margins.
- Rules for use of colored chalk for student boardwork at USMA described on back matter.

[1892]a. *Elements of Descriptive Geometry, with its Applications to Spherical Projections, Perspective, and Isometric Projections and to Shades and Shadows.* New York: American Book Co. pp. vii + 192 + [8].
- Four copies are in the library.
- One copy has numerous notes throughout book. Numerous corrections pasted in. Copy signed by Lewis Turtle. Student text with omitted sections marked and lesson written on back pages. Rules for use of colored chalk for student boardwork at USMA described on back matter.
- One copy signed by "G. R. Alley."
- One copy signed by "Charles Brook Clark."
- See Figure 30 for a photograph of marginal notes found in one copy.

1892b. *Plates to Descriptive Geometry. Shades, Shadows and Linear Perspective.* New York: American Book Co. pp. 132 figures on unnumbered pages.
- Nine copies of this book are in the library.
- One copy signed by "Winston." Some notes written in the copy.
- One copy is a gift from Arthur R. Hercz and has handwritten notes and editing throughout.
- One copy has a plate with 2 figures taped inside front cover.
- One copy belonged to Cadet William "Cooper" Foote, issued December 17, 1909. It has notes and pasted–in figures. It also has two ink drawings in Foote's hand and several ditto sheets taped in.
- One copy signed by "Heiberg."
- One copy has numerous blank pages at the end.
- One copy signed by "C. P. Nicholas," Nicholas was Professor and Department Head, Mathematics Department, USMA, 1959–1967.
- One copy signed by "Turtle."

Church, Albert Ensign (Descriptive Geometry – Continued).

[1902]. *Elements of Descriptive Geometry with Applications to Spherical, Perspective, and Isometric Projections, and to Shades, and Shadows.* Second edition. New York: American Book Co. pp. i + 214.
- Nine copies of this book are in the library.
- Several copies have numerous corrections pasted in books.
- One copy annotated in blue, red, and green to indicate coverage by lower, upper, and middle sections. This book also has course lesson assignments and syllabus on the back of the front cover.
- One copy signed "William Cooper Foote ... Issued December 17, 1909". Pages covered in each lesson written on endpapers.
- Another copy has handwitten notes throughout the book. Course syllabus on back inside cover. Numerous printed slips taped in. This copy was a gift from COL Arthur R. Hercz.
- One copy signed by "J. K. Mitchell."
- Several copies have 10 extra pages of advertising at the end.
- One copy is clean with no annotations or marks.

Church, Albert Ensign (Descriptive Geometry). See also Church and Bartlett 1911.

Church, Albert Ensign (Calculus).
1842. *Elements of the Differential and Integral Calculus.* New York: Wiley and Putnam. pp. viii + 318.
- This book was used as a text at USMA from 1843 to 1895 and by instructors only as a reference from 1896 to 1899.
- Two copies are in the library.
- One copy has marginal notes and annotations.
- Second copy is clean.

1850. *Elements of the Differential and Integral Calculus. Improved Edition, Containing the Elements of the Calculus of Variations.* New York: Geo. P. Putnam. pp. vii + 344.
- Copy signed by "L. L. Langdon" in 1852. Copy contains some annotations and marginal notes.

1851. *Elements of the Differential and Integral Calculus. Improved Edition, Containing the Elements of the Calculus of Variations.* New York: Geo. P. Putnam. pp. vii + 344.
- Copy has a 1-page paste-in.

1858. *Elements of the Differential and Integral Calculus. Improved Edition, Containing the Elements of the Calculus of Variations.* New York: A. S. Barnes & Co. pp. xii + 344.
- Copy contains 20 names of Dickinson College Junior Class.

Church, Albert Ensign (Calculus – Continued).

1861. *Elements of the Differential and Integral Calculus. Revised Edition, Containing the Elements of the Calculus of Variations.* New York: A. S. Barnes and Burr. pp. xii + 369.

- Four copies are present in the library.
- One copy is signed by A. Macomb Miller [USMA 1862]. Notes on blank pages. Section roster in book.
- Another copy is signed by "William Reilly." It contains numerous drawings of individuals and poems. It also has numerous annotations and notes placed in margins throughout the book.
- One copy signed by "Buell B. Bassette."
- One copy signed by "Chas. P. Echols," department head at USMA.

1870. *Elements of the Differential and Integral Calculus. Revised Edition, Containing the Elements of the Calculus of Variations.* New York: A. S. Barnes and Co. pp. xi + 369.

- Copy is bound with a special binding of Major E. C. Boynton.

1874. *Elements of the Differential and Integral Calculus. Revised Edition, Containing the Elements of the Calculus of Variations.* New York: A. S. Barnes & Co. pp. xi + 369.

- Library has two copies.
- One copy is clean.
- Second copy signed by "Francis H. French."
- See Figure 29 for a photograph of an advertisement from this volume.

1876. *Elements of the Differential and Integral Calculus. Revised Edition, Containing the Elements of the Calculus of Variations.* New York: A. S. Barnes and Co. pp. xi + 369.

- The library has two copies.
- In one copy, Part I is a clean copy. Part II (pp. 209–348) on the integral calculus is marked up as if for a later edition. Part III (pp. 349–369) on the calculus of variations is to be omitted.
- The second copy has many notes in the margins.

Church, Albert Ensign, and Bartlett, George Miller (1873–1936).

1911. *Elements of Descriptive Geometry with Applications to Spherical and Isometric Projections, Shades, and Shadows, and Perspective.* New York: American Book Co. pp. 286 + 2.

- Three copies are in the library.

Clairaut, Alexis Claude (1713–1765).

1765. *Théorie de la lune, déduite du seul principe de l'attraction réciproquement proportionnelle aux quarrés des distances.* Seconde édition, a laquelle on a joint des tables de la lune, construites sur une nouvelle révision de toutes les espéces de calcul dont leurs équations dépendent. Paris: Dessaint & Saillant. pp. [vi] + 161 + 1.

- Dedicated to the Duke of Choiseul. 1 plate after p. 118 and before p. 119. This work won the prize proposed in 1750 by the St. Petersburg Academy (according to the title page). The first edition was published in St. Petersburg in 1752.

1801. *Élémens d'algebre, avec des notes et des additions très–étendues, par le citoyen Garnier, professeur d'analyse à l'Ecole polytechnique. (Avec figures.) Précédés d'un traité d'arithmétique, par Théveneau, avec une instruction sur les nouveaux poids et mesures.* 2 vols. Sixieme edition. Paris: Courcier. pp. viii + cccvii + 120 + [iii] + 3 plates; ix + 448 + 8 plates.
◊   One of the first textbooks on the metric system.

1808. *Théorie de la figure de la terre, tirée des principes de l'hydrostatique.* Second édition. Paris: Courcier. pp. xl + [ii] + 308.
◊   Dedicated to Count Maurepas.
○   Thayer binding.

Clark, Thomas.
1815. *A New System of Arithmetic; including Specimens of a Method by which most Arithmetical Operations may be Performed without a Knowledge of the Rules of Three; and Followed by Strictures on the Nature of the Elementary Instruction contained in English Treatises on that Science.* Second edition. London: Printed for William Baynes. pp. xxxiv + 430 + [ii].
◊   Errata and five corrected sheets are bound between pp. xxxii and xxxiii.

Clarke, Henry (1743-1818). See Lorgna 1779.

Clarke, Samuel (1675–1729). See Newton 1706.

Clermont, de.
1733a. *L'arithmetique militaire, ou l'arithmetique pratique de l'ingenieur et de l'officier; Divisée en trois parties, ouvrage également necessaire aux officiers, aux ingénieurs & aux commerçans.* Troisiéme edition, corrigée & de beaucoup augmentée. Paris: Pierre Witte; Didot. pp. [vi] + 170.
◊   Clermont 1733b follows in the same binding.
○   Title page signed "De Riverson."

1733b. *La geometrie pratique de l'ingenieur, ou l'art de mesurer, ouvrage également necessaire aux ingenieurs, aux toiseurs & aux arpenteurs. Divisé en huit livres, dont les titres sont à la page suivante.* Paris: Pierre Witte; Didot. pp. xx + 262 + 27 plates.
◊   Dedicated to de Vauban. The dedications are dated 1688. Bound with Clermont 1733a.
◊   Page 163 is incorrectly numbered 613.

Clevenger, Shobal V. (1843–1920).
1883. *A Treatise on the Method of Government Surveying as Prescribed by the United States Congress and Commissioner of the General Land Office. With Complete Mathematical, Astronomical and Practical Instructions, for Use of United States Surveyors in the Field, and Students who Contemplate Engaging in the*

Business of Land Surveying. Third edition. New York: D. Van Nostrand. pp. [i] + 5–200.

Cloquet, J.–B.
1823. *Nouveau traité élémentaire de perspective à l'usage des artistes et des personnes qui s'occupent du dessin, précédé des premières notions de la géométrie élémentaire, de la géométrie descriptive, de l'optique et de la projection des ombres.* Paris: Bachelier. pp. viii + 333.
◇ Errata follow p. 333. Plate 34 is cited on p. 320, but there are no plates in this volume.

Cocker, Edward (1631–1675).
1702. *Cocker's Decimal Arithmetick, Wherein is Shewed the Nature and Use of Decimal Fractions in the Usual Rules of Arithmetick, and the Mensuration of Plains and Solids. Together with Tables of Interest and Rebate for the Valuation of Leases and Annuities, Present, or in Reversion, and Rules for Calculating those Tables. Whereunto is Added his Artificial Arthmetick, Shewing the Genesis or Fabrick of the Logarithms, and Their Use in the Extraction of Roots, the Solving of Questions in Anatocism, and in Other Arithmetical Rules in a Method not usually Practised. Also his Algebraical Arithmetick, Containing the Doctrine of Composing and Resolving an Equation; with all Other Rules Requisite for the Understanding of that Mysterious Art, according to the Method used by Mr. John Kersey in his Incomparable Treatise of Algebra.* Third edition. London: Perused, Corrected and Published by John Hawkins, Printed for George Strawbridge. pp. [xvi] + 436.
◇ Dedicated to Sir Peter Daniel and Peter Rich, and to Thomas Lee and James Reading. Some prefatory material is dated 1684. Arithmetic refers to logarithms, and in addition to some algebra, the book also contains material on interest and mensuration.

Coffin, James Henry (1806–1873).
1849. *Elements of Conic Sections and Analytical Geometry.* Revised and stereotyped edition. New York: Collins and Brother. pp. 158.
○ Copyright 1848.
◇ The author has prepared this work for his students at Lafayette College. It treats the conics by both the "geometrical" and "analytical (often called the French method)."

Colburn, Warren (1793–1833).
1830. *An Introduction to Algebra upon the Inductive Method of Instruction.* Boston: Hilliard, Gray, Little, and Wilkins. pp. 276 + 2 page description of the work.

1845. *Arithmetic Upon the Inductive Method of Instruction being a Sequel to Intellectual Arithmetic.* Boston: Jordan, Swift, & Wiley. pp. [i] + 245 + 5 + 12 advertisements.

Colby, Thomas (1784–1852). See Mudge and Dalby 1799; 1801; 1811.

Colebrook, Henry Thomas (1765–1837) (translator).
1817. *Algebra, with Arithmetic and Mensuration From the Sanscrit of Brah-megupta [sic] and Bháscara.* London: John Murray. pp. lxxxiv + 378.

Collins, John (1625–1683).
1725. *Commercium epistolicum de varia re mathematica, inter celeberrimos prae-sentis seculi mathematicos, viz.* D^{num} *Isaacum Newtonum equitem auratum,* D^{num} *Isaacum Barrow,* D^{num} *Jacobum Gregorium,* D^{num} *Johannem Wallisium,* D^{num} *J. Keillium,* D^{num} *J. Collinium,* D^{num} *Gulielmum Leibnitium,* D^{num} *Henricum Oldenbourgum,* D^{num} *Franciscum Slusium et alios jusse; Societatis Regiae in lucem editum; et jam una cum recensione praemissa insignis controversiae inter Leibni-tium & Keillium de primo inventore methodi fluxionum; & judicio primarii, ut ferebatur, mathematici subjuncto, iterum impressum.* Londini [London]: Impensis J. Tonson & J. Watts. pp. [viii] + 250.
◇ Errata follow p. 250.

Colson, John (1680–1760). See Agnesi 1801; Newton 1736; Taylor 1811.

Commandino, Federico (1509–1575). See Euclid 1802; Pappus 1660.

Comstock, George C. (1855–1934).
1890. *An Elementary Treatise upon the Method of Least Squares with Numerical Examples of its Applications.* Boston: Ginn and Company. pp. vii + 67 + 1.

Comte, Auguste (1798–1857).
March 1843. *Traité élémentaire de géométrie analytique a deux et a trois dimen-sions, contenant toutes les théories générales de géométrie accessibles a l'analyse ordinaire.* Paris: Carilian–Goeury et Vor Dalmont. pp. vii + 598 + 3 plates.
◇ The author is known for his *System of Positive Philosophy*.

Condorcet, Marie Jean Antoine Nicholas de Caritat (1743–1794).
1768. *Essais d'analyse.* Paris: Didot. pp. li + vii + 94 + [iv] + 83 + 93.
◇ Corrections and additions are on pp. xlvi–li. Errata follow p. 94. This work consists of 3 separate parts, each with its own title page: *Du calcul integral*, 1765; *Du probleme de trois corps*, 1767; *Le Marquis de Condorcet a Mr. d'Alembert, sur le systêm du monde et sur calcul intégral*, 1768. The library has only volume 1.
o This copy was purchased for USMA in 1825 by Sylvanus Thayer from Kilian in Paris.

1800. *Moyens d'apprendre a compter surement et avec facilité.* Seconde édition. Paris: Bachelier. pp. 132.

Copernicus, Nicolas (1473–1543).

1543. *Nicolai Copernici Torinensis de revolutionibus orbium coelestium, libri VI. Habes in hoc opera iam recens nato, & aeditio, studiose lector, motus stellarum, tam fixarum, quàm erraticarum, cum ex ueteribus, tum etiem ex recentibus observationibus restitutos: & nouis insuper ac admirabilibus hypothesibus ornatos. Habes etiam tabulas expeditissimas, ex quibus eosdem ad quoduis tempus quàm facilli me calculare poteris. Igiture eme, lege, fruere.* Αʼγεωμέ [?]ητος γʼδεὶς εʼιacuteιτω. *Norimbergae apud Ioh. Petreium.* Norimbergae [Nuremburg]: Petreium. pp. [xii] + 196; only the recto leaves are numbered.

○ There are annotations on pages 1v (v = verso), 50, 128v, 187v, 191, 191v, 192v, 194v, 195; possibly in two hands. There are several signatures on cover two and the front endpaper, including that of Joannis Stadius (1527–1575) who taught at Louvain (1565–1569) and preceded Kepler at Graz.

1617. *Nicolai Copernici Torinensis astronomia instaurata, libris sex comprehensa, qui de revolutionibus orbium coelestium inscribuntur. Nunc demum post 75 ab obita authoris annum integritati suae restituta, notisque illustrata, opera & studio D. Nicolai Mulerii Medicinae ac matheseos professoris ordinarij in nova academia quae est Groningae. Amstelrodami, Excudebat Wilhelmus Iansonius, sub Solari aureo.* Amsterlrodami [Amsterdam]: W. Iansonius. pp. [xxii] + 487. Errata on p. [xxii].

◇ The third edition of the *De revolutionibus* of Copernicus. There are (printed) notes throughout by Mulerius.

○ The top portion of pp. 483–487 have been torn off.

Copernicus, Nicolas. See also Gassendi 1653b; Kepler 1635.

Cor, N. and Riemann, J.

1898. *Traité d'algèbre élémentaire à l'usage des élèves de mathématiques élémentaires des aspirants au baccalauréat de l'enseignement classique (2e série) et au baccalauréat de l'enseignement moderne (2e et 3e séries) et des candidats aux écoles du gouvernement.* Paris: Nony & Cie. pp. iii + 460.

Coriolis, Gaspard-Gustav de (1792–1843).

1835. *Théorie mathématique des effets du jeu de billard.* Paris: Carilian–Goeury. pp. viii + 174 + 12 plates.

◇ A heavy mathematical treatment of the game of billiards.

○ Pages i-ii missing. Title page ([iii]) is present.

Cotes, Roger (1682–1716).

1722a. *Harmonia mensurarum, sive analysis & synthesis per rationum & angulorum mensuras promotae: accedunt alia opuscula mathematica*; bound together with *Theoremata tum logometrica tum trigonometrica datarum fluxionum fluentes exhibentia, per methodum mensurarum ulterius extensam.* Cantabrigiae [Cambridge]: Edidit et auxit Robertus Smith. pp. xviii + 109 + 1 fold out table; 111–249.

◇ Dedicated to Richard Mead, M.D.

○ Bound with Cotes 1722b. Binding is marked "Opera Miscellanae."

1722b. *Aestimatio errorum in mixta mathesi, per variationes partium trianguli plani et sphaerici;* bound together with *De methodo differentiali Newtoniana;* and with *Canonotechnia sive constuctio tabularum per differentias;* and with *De descensu gravium. De motu pendulorum in cycloide. De motu projectilium;* and with *Editoris notae ad harmoniam mensurarum.* pp. 22; 23–33; 35–71; 73–91; 93–125.

◇ Corrigenda follow p. 125. No publisher or place of publication stated.

○ Bound with Cotes 1722a.

1770. *De descensu gravium de moto pendulorum in cycloide et de motu projectilium.* Cantabrigiae [Cambridge]: Typis excudebant T. Fletcher & F. Hodson. Impensis J. Nicholson. pp. iv + [5]–37 + 1 plate.

◇ List of subscribers on pp. i–iii. Errata on p. [iv]. Bound together with Morgan 1770. First published as part of Cotes 1722a.

Cotes, Roger. See Newton 1822.

Courtenay, Edward Henry (1803–1853; USMA 1821).

1833. *An Elementary Treatise on Mechanics. Translated from the French of M. Boucharlat with Additions and Emendations, Designed to Adapt it to the Use of the Cadets of the U. S. Military Academy.* Translated by Courtenay. New York: J. & J. Harper. pp. 432 + 9 plates.

◇ Courtney served as Department Head of Natural and Experimental Philosophy, USMA (1829 – 1834) and Professor of Mathematics, University of Virginia (1842 – 1853).

◇ Used as a textbook at USMA. Based on the works of Poisson, Francoeur, Navier, Persy, Genieys, and Gregory.

○ The library has three copies.

○ Plates are loose in one copy. Boucharlet's name appears on the spine.

1836. *An Elementary Treatise on Mechanics. Translated from the French of M. Boucharlat with Additions and Emendations, Designed to Adapt it to the Use of the Cadets of the U. S. Military Academy.* Translated by Courtenay. New York: Harper & Brothers. pp. 432 + 9 plates.

○ The library has two copies.

○ Student notes and drawings throughout one copy.

1848. *An Elementary Treatise on Mechanics. Translated from the French of M. Boucharlat with Additions and Emendations, Designed to Adapt it to the Use of the Cadets of the U. S. Military Academy.* Translated by Courtenay. New York: Harper & Brothers. pp. 432 + 9 plates.

◇ Used as a textbook at USMA, 1833–1850.

○ Many blank pages are bound into this copy; many student notes on these pages.

1855. *Treatise on the Differential and Integral Calculus, and on the Calculus of Variations.* New York: A. S. Barnes & Co. pp. 4 blank + i–xviii + 13–501.

◇ "Late Professor of Mathematics in the University of Virginia" under the author's name. Publisher's inscribed front cover. Small biography of Courtenay. Written while he was the Chairman of the Department of Mathematics, University of Virginia.

○ Torn front cover.

1873. *A Treatise on the Differential and Integral Calculus and the Calculus of Variations.* New York: A. S. Barnes and Company. pp. xviii + 13–501.
- ◇ Biography of author is contained in the preface.
- ◇ Tables of integrals written on the back pages.

Cousin, Jacques Antoine Joseph, (1739–1800).

1796. *Traité de calcul différentiel et de calcul intégral. Première partie.* and *Traité de calcul différentiel et de calcul intégral. Deuxième partie.* Paris: Régent & Bernard. pp. xxiv + vii + 319 + 284 + 6 plates.
- ◇ Two volumes bound together as one.
- ○ Title page is torn and part of it is missing.

Cousinery, Barthelemy Edward (1790–1851).

1851. *Géométrie élémentaire du compas exposant les divers systèms de tracé que comporte l'emploi exclusif du compas, tant les rigoreux que les approximatifs, pour servir de prolégomènes au rapporteur de précision.* Paris: Carilian–Goeury et Vor Dalmont. pp. 216 + 5 plates.

Craig, Thomas (1855–1900).

1879. *Elements of the Mathematical Theory of Fluid Motion. On the Motion of a Solid in a Fluid, and the Vibrations of Liquid Spheroids.* Van Nostrand's Science Series. New York: D. Van Nostrand. pp. 208.

Crefcoeur, Albert J. M. (1864–?).

1902. *Cours d'analyse. Calcul différentiel et calcul intégral. Méthode simple pour apprendre ces branches des mathématiques supérieures.* Anvers [Antwerp]: Vlijt. pp. 336.

Crelle, August Leopold (1780–1855).

1880. *Tables de calcul òu se trouvent les multiplications et divisions toutes faites de tous les nombres au dessous de mille et qui facilitent et assurent le calcul.* Cinquième édition stéréotype. Berlin: G. Reimer. pp. x + unpaginated tables.
- ◇ There is a separate tile page in German. The front matter is in both French and German.
- ◇ The foreword is by C. Bremiker.

Cremona, Luigi (1830–1903).

1893. *Elements of Projective Geometry.* Second edition. Translated by Charles Leudesdorf. Oxford: Clarendon Press. pp. xxiii + 310 + 72.
- ○ Binding is taped together.

Cresswell, Daniel (1776–1844).

1811. *The Elements of Linear Perspective, Designed for the Use of Students in the*

University. Cambridge: Printed by Francis Hodson for J. Deighton. pp. xi + 64 + 9 plates.

◦ On one endpaper inscribed "Col. Thayer L.L.D.," but writing looks like student work.

1816. *A Treatise on Spherics Comprising the Elements of Spherical Geometry and of Plane and Spherical Trigonometry, together with a Series of Trigonometrical Tables.* Cambridge: J. Mawman. pp. xi + 294.

◇ Errata follow p. xi.

1817. *An Elementary Treatise on the Geometrical and Algebraical Investigation of Maxima and Minima, being the Substance of a Course of Lectures Delivered Conformably to the Will of Lady Sadler: to which is Added, a Selection of Propositions Deducible from Euclid's Elements.* Second edition, corrected and enlarged. Cambridge: Printed by J. Smith for Deighton and Sons. pp. iv + 298 + 138.

◇ Dedicated to Rt. Rev. William Lort Mansel. The last 138 pages contain two appendices: I. on propositions deducible from the first six books of Euclid; and II. on some propositions in natural philosophy. These are Sadlerian Lectures.

Cresswell, Daniel. See also Euclid 1819; Venturoli 1823.

Crockett, Charles Winthrop (1862–1936).
1896a. *Elements of Plane and Spherical Trigonometry.* New York: American Book Company. pp. 192.

◇ This book was used as a text at USMA from 1907 until 1919 and from 1923 until 1936.

◦ There are four copies of this book present in the library. One copy contains numerous instructor notes in red, blue, and green. Description of colors is on p. 7.

◦ The second copy was owned by Robert Tate, USMA. Errata taped into book and footnotes pasted in. Some notes in the margins. Assignments are listed on the inside of the back cover.

◦ The third copy was owned by Arthur R. Hercz, a cadet. Contains pasted in footnotes, notes in margins, and assignments inside back cover.

◦ The fourth copy was signed by LT C. P. Nicholas.

1896b. *Elements of Plane and Spherical Trigonometry.* New York: American Book Company, pp. 192 + 103 pages of tables + xv pages of explanation of the tables.

◇ This is the same book as Crockett 1896a with the tables included.

Crozet, Claudius (1790?–1864).
1821. *A Treatise on Descriptive Geometry for the Use of the Cadets of the United States Military Academy. Part I. Containing the Elementary Principles of Descriptive Geometry, and its Application to Spherics and Conic Sections.* New York: A. T. Goodrich and Co. pp. xii + 150 + 19 plates.

◇ Claudius Crozet was a Professor of Engineering at the Academy from 1817 to 1823. This was the first math text designed for cadets and emanating from West Point. It was used as a text at USMA from 1824 until 1831.

◦ Book presented to C. Canda as a mark of regard and friendship by the author.

1858. *An Arithmetic for Colleges and Schools.* Improved edition. Richmond: A. Morris. pp. viii + 314.

⋄ The multiplication table (through 12 × 12) on p. [ii] is called a "Table of Pythagoras." Some "mistakes are purposely introduced in some of the answers, in order that the pupil may learn to be sure of himself, and not force his results." [p. viii].

Cullum, George W. (1809–1892; USMA 1833).

1832. *Problems of Descriptive Geometry Referred to One Plane of Projection* and *Defilement of Works of Permanent Fortifications.* Lithographed. pp. 8 + plates + 20.

⋄ Cullum served as Superintendent, USMA, 1864–1866. He is the founding author of the *Biographical Register of the Officers and Graduates of the United States Military Academy.*

⋄ Dactitil plates.

○ Book plate, USMA, gift of Mrs. C. A. Dempsey, 1941. Signed by Cullum, 1832.

Cullum, George W. See Anonymous 1833.

Cunha, Joseph–Anastase da (1744–1787).

1811. *Principes mathématiques.* Traduits littéralement du Portugais par J. M. D'Abreu. Bordeaux: D'André Racle. pp. viii + 299 + 6 plates.

⋄ Errata follow p. 299.

D'Abreu, J. M. See Cunha 1811.

Dana, Edward Salisbury (1849–1935).

1908. *A Text Book of Elementary Mechanics for the Use of Colleges and Schools.* Twelfth edition. New York: John Wiley & Sons. pp. xiv + 291 + 19.

⋄ Copyright 1881.

⋄ List of the publications of John Wiley & Sons on the last 19 pages.

○ Gift of Col. Walter Krueger, Jr (USMA 1931).

Danti, Ignazio (1537–1586). See Vignola 1611.

Darboux, Gaston (1842–1917).

1896. *Leçons sur la théorie général des surfaces et les applications géométriques du calcul infinitésimal. Quatrième partie. Déformation infiniment petite et représentation sphérique.* Cours de géométrie de la faculté des sciences. Paris: Gauthier–Villars et Fils. pp. [viii] + [548].

⋄ Errata for all four parts pp. vii–viii.

1917. *Principes de géométrie analytique.* Cours de géométrie de la faculté des sciences. Paris: Gauthier–Villars et Cie. pp. vi + 519.

Davidson, John (fl. 1817–1835).
1824. *A System of Practical Mathematics; Containing Geometrical Problems, Plane Trigonometry, the Mensuration of Heights and Distances,–of Surfaces and Solids, –of the Conic Sections,–of Specific Gravity,–of Ships, Cattle and Hay,–of Land and Wood, and –of Artificers' Works; Gauging, Gunnery, Spherical Trigonometry, Geographical, Geodetical, & Astronomical Problems; and a Collection of Accurate Stereotyped Tables; comprising the Logarithms of Numbers; and of Sines, Tangents, and Secants; Natural Sines; and the Requisite Astronomical Tables. To which are Prefixed, Elements of Algebra. For the Use of Schools.* Second edition, improved, and greatly enlarged. Edinburgh: Printed for Bell and Bradfute. pp. xvi + 453 + 80 +6 plates.
  ◊   Errata on p. xvi.
  ◊   First edition was 1817.

Davies, Charles (1798–1876; USMA 1815).
See Davies (Arithmetic), Davies (Algebra), Davies (Geometry and Trigonometry), Davies (Analytical Geometry), Davies (Linear Perspective), Davies (Desriptive Geometry), Davies (Surveying), Davies (Calculus), and Davies (General Mathematics).
  ◊   Davies served as Department Head for USMA Mathematics Department from 1823 to 1837.

Davies, Charles (Arithmetic).
1833. *The Common School Arithmetic, Prepared for the Use of Academies and Common Schools in the United States, and also for the Use of the Young Gentlemen who may be Preparing to enter the Military Academy at West Point.* Hartford: H. F. Sumner. pp. 270.
  ◊   This is the first edition of this book.
  ○   Signed on Flyleaf: "The Property of Mary C. Saunders, Pilead, 1877." Otherwise, a clean copy.

1836. *The Common School Arithmetic, Prepared for Use of Academies and Common Schools in the United States, and Also for the Use of the Young Gentlemen Who May be Preparing to Enter the Military Academy at West Point.* Buffalo: O. G. Steele. pp. 7 + xxiv + 270.
  ○   This copy has a bookplate: "For Thayer Public Library, presented by GEN Thayer, 1884." Also has a second bookplate: "West Point" and the seal of Thayer Public Library, 1874.

1840. *First Lessons in Arithmetic, Designed for Beginners.* Hartford: A. S. Barnes & Co. pp. iv + 6–132.
  ◊   An advertisement for 11 other books by Davies is listed on the back cover.

1866. *University Arithmetic, Embracing the Science of Numbers, and General Rules for Their Application.* New York: A. S. Barnes & Co. pp. 454.
  ○   This copy was rebound in 1952.

Davies, Charles (Algebra).

1835. *Elements of Algebra: Translated from the French of M. Bourdon. Revised and Adapted to the Course of Mathematical Instruction in the United States.* Revised and adapted edition by Charles Davies. New York: Wiley and Long. pp. viii + 9–353.

◇ This book was used as a textbook at USMA from 1839 to 1899.

○ Stamped "McRee Swift, New Brunswick, N.J." on the title page.

1836. *Elements of Algebra: Translated from the French of M. Bourdon. Revised and Adapted to the Course of Mathematical Instruction in the United States.* New York: Wiley and Long. pp. viii + 9–354.

○ Copy has numerous annotations and its title page is signed "John McSweeney, Cincinnati, Ohio, 1840."

1838. *Elements of Algebra: Translated from the French of M. Bourdon. Revised and Adapted to the Course of Mathematical Instruction in the United States.* Revised edition, by Charles Davies. Hartford: A. S. Barnes. pp. x + 11–356.

1839. *Elements of Algebra: Translated from the French of M. Bourdon. Revised and Adapted to the Course of Mathematical Instruction in the United States.* Revised edition. Hartford: A. S. Barnes. pp. x + 355.

◇ List of other works by Davies at the end of the book (p. [356]).

○ "Textbook U.S.M.A. 1839 to 1899" stamped in the gutter of title page.

1846. *Elements of Algebra: Including Sturms' Theorem. Translated from the French of M. Bourdon. Adapted to the Course of Mathematical Instruction in the United States.* Davies' Course of Mathematics Series. New York: A. S. Barnes & Co. pp. 368.

○ The library has two copies.

○ One copy is signed "Cas S. Lee, Allentown 1847 April 16th." This copy has a listing of exam questions on p. 170.

○ The second copy is clean.

c.1852. *Elementary Algebra: Embracing the First Principle of the Science.* Davies' Course of Mathematics Series. New York: A. S. Barnes & Burr. pp. vii + 303.

○ A few notes in the margins showing what needs to be covered in class indicate this may have been an instructor's text.

1860. *Elements of Algebra: on the Basis of M. Bourdon: Embracing Sturm's and Horner's Theorems, and Practical Examples.* New York: A. S. Barnes & Burr. pp. 400.

○ Signed "Malcolm McArthur [USMA 1865], West Point, Sept 5th 1861, Military Academy." The endpaper is dated Jan 4th, 1862 and lists the 14 students in "3rd Section, Math 4th Class." Spine reads "Davies' Bourdon."

1870. *Key to Davies' Bourdon, with many Additional Examples, Illustrating the Algebraic Analysis: also, a Solution of all the Difficult Examples in Davies' Legendre.* New York: A. S. Barnes & Co. pp. iv + 205.

◇ Copyright 1856.

Davies, Charles (Algebra – Continued).

1875. *Elements of Algebra: on the Basis of M. Bourdon: Embracing Sturm's and Horner's Theorems.* Revised and re–written. New York: A. S. Barnes and Co. pp. xiv + [15]–406.

⋄ Preface dated June 1873.

⋄ This version of Davies' Bourdon was the text used at USMA from 1874 to 1899.

○ Spine reads "Elements of Algebra–Davies." "Abbot" stamped on cover. Copy is heavily annotated and numerous extra printed and handwritten sheets have been pasted in.

[1877]a. *Elements of Algebra: on the Basis of M. Bourdon: Embracing Sturm's and Horner's Theorems.* Revised and re-written. New York: A. S. Barnes & Co. pp. xiv + 15–406.

⋄ Contains prefaces of June 1873 and of April 1877 (pp. [3–4]). Peck is thanked in the former and the later is written by A. E. Church who joined Davies in editing in 1875.

⋄ Pagination in book does not match the table of contents.

⋄ This is part of Davies' Collegiate Course Collection.

○ Four copies of this book are present in the library.

○ One copy has spine marked "Davies' Bourdon." Copy has flyleaf signed "Buell B. Bassett," [USMA 1893].

○ Another copy is signed "J. W. Acton, West Point, USMA, Sep 2nd 1878." On the facing of the title page is a sketch of a bearded 3–star general.

○ Another copy contains marginal notes and inserts throughout.

[1877]b. *Elements of Algebra: on the Basis of M. Bourdon: Embracing Sturm's and Horner's Theorems.* Revised and re-written. New York: American Book Co. pp. xiv + 406.

⋄ Contains prefaces of June 1873 and of April 1877 (pp. [3–4]). Peck is thanked in the former and the later is written by A. E. Church who joined Davies in editing in 1875.

⋄ Pagination in book does not match the table of contents.

○ Four copies of this book are present in the library.

○ The first copy has "Grant, US 1903" stamped on cover. Signed by Ulysses S. Grant. Contains notes throughout. USMA bookplate. Cadet poem on back cover. This book was a gift of Mrs. Paul E. Rueston, daughter of Frederick Dent Grant. A section roster including the name Lewis Turtle is written on the front cover. A problem set is written in the front cover. Corrections are pasted within.

○ The second copy is signed by Lewis Turtle. It contains notes written throughout, course outline and problem list in back pages, and hand written notes pasted in.

○ One copy is signed by A. N. Miller, Jr., 1 September 1892. There are numerous notes written and pasted throughout the book. Revision notes in margins.

○ One copy is clean.

Davies, Charles (Algebra). See also Bourdon 1831a; Ross 1831.

Davies, Charles (Geometry and Trigonometry).

[1834]a. *Elements of Geometry and Trigonometry. Translated from the French of A. M. Legendre, by David Brewster. Revised and Adapted to the Course of Instruction in the United States.* New York: A. S. Barnes. pp. vii + 297 + 62.
- ◇ This book was used as a textbook at USMA from 1841 until 1902.
- ◇ Date established from the preface.
- ○ Book plate of the McSweeny Family pasted inside the fron cover. Copy signed by "John McSweeny."

[1834]b. *Elements of Geometry and Trigonometry. Translated from the French of A. M. Legendre, by David Brewster. Revised and Adapted to the Course of Instruction in the United States.* Fifth edition. New York: Wiley and Long, et.al. pp. vii + [9]–297 + 62 pages of tables.
- ◇ Date established from the preface.

1834c. *Elements of Geometry and Trigonometry. Translated from the French of A. M. Legendre, by David Brewster. Revised and Abridged by Charles Davies.* Fourth edition. New York: Harper and Brothers. pp. vi + 297 + 62 pages of tables + [6] pages of advertisements.
- ◇ Slightly abridged version with a slightly different title than earlier versions.
- ○ This copy is from the library of Col. Duncan.
- ○ Copy rebound in 1994.

1836 *Elements of Geometry and Trigonometry. Translated from the French of A. M. Legendre, by David Brewster. Revised and Abridged by Charles Davies.* Fifth edition. New York: Wiley and Long, et.al. pp. vii + [9]–297 + 62 pages of tables.
- ○ The library has two copies of this book.
- ○ One copy was presented by Gen. S. Thayer in 1874 to the Thayer Public Library.

1839. *Elements of Geometry and Trigonometry. Translated from the French of A. M. Legendre, by David Brewster. Revised and Adapted to the Course of Mathematical Instruction in the United States.* Hartford: A. S. Barnes & Co, et. al. pp. vii + 297 + 62.
- ◇ Preface dated 1834.
- ○ Spine marked "Elements of Geometry and Trigonometry—Legendre."

1848. *Elements of Geometry and Trigonometry. Translated from the French of A. M. Legendre, by David Brewster. Revised and Adapted to the Course of Mathematical Instruction in the United States.* New York: A. S. Barnes & Co. pp. vi + 297 + 62 pages of tables.

1851. *Elements of Geometry and Trigonometry, from the Works of A. M. Legendre. Revised and Adapted to the Course of Mathematical Instruction in the United States.* New York: A. S. Barnes & Co. pp. viii + 370 + 62 pages of logarithm and trigonometric tables.
- ◇ Expanded by a considerable amount from the earlier editions.
- ○ Spine marked "Geometry and Trigonometry–Legendre."

Davies, Charles (Geometry and Trigonometry – Continued).

1854. *Elements of Geometry and Trigonometry, from the Works of A. M. Legendre. Revised and Adapted to the Course of Mathematical Instruction in the United States.* New York: A. S. Barnes & Co. pp. vii + 370 + 62 tables of logs/sines/tangents.
  o   This copy signed by W. F. Steele–2d Infantry.

1868. *Elements of Geometry and Trigonometry, From the Works of A. M. Legendre. Adapted to the Course of Mathematical Instruction in the United States.* New York: A. S. Barnes & Co. pp. viii + 9–259 + 1–34 + 62 tables.
  o   Copy stamped "Haldane".

1872. *Elements of Geometry and Trigonometry, from the Works of A. M. Legendre. Adapted to the Course of Mathematical Instruction in the United States.* New York: A. S. Barnes & Co. pp. vii + 259 + 134 + 62.
  o   Signed "J. F. Reynolds Landis, Jan 1875." There are numerous notes on various pages.

1874. *Elements of Geometry and Trigonometry, from the Works of A. M. Legendre. Adapted to the Course of Mathematical Instruction in the United States.* New York: A. S. Barnes & Co. pp. viii + [9]–259 + 134 + 62 pages of log and trigonometric tables.
  ◊   Preface dated "April, 1862."
  o   Spine marked "Elements of Geometry and Trigonometry–1874–Legendre." From the library of Frederick V. Abbot, [USMA, 1879]. The names of the eleven cadet members of Abbot's classroom section in January 1879 are written on the front endpaper.

[1882]a (date from preface). *Elements of Geometry from Davies' Legendre.* New York: A. S. Barnes & Co. pp. vi + [9]–275 + 105–134 + 62.
  ◊   The second pagination is a section on "Mensuration." The third is logarithm and trigonometric tables.
  ◊   The Preface states that this book was edited by J. H. Van Amringe.
  ◊   The library has two copies.
  o   One copy signed by "Buell B. Bassette," dated Jan 11, 1890, when he was a freshman cadet.
  o   The other copy is signed "John Gurney '95 #139."

1882b. *Selected Propositions in Geometrical Constructions and Applications of Algebra to Geometry. Being a Key to the Appendix of Davies' Legendre. (With a Supplement).* New York: A. S. Barnes & Co. pp. iv + 5–181 + 6–61 supplemental advertising.
  ◊   Contains illustrative examples and solutions to the exercises in Davies' Legendre.
  o   Two copies are present in the library.
  o   One copy is annotated, probably by an instructor, with dates of use of problems in class.
  o   The other copy is signed by Charles R. Noyer inside cover.

Davies, Charles (Geometry and Trigonometry – Continued).

[1890]a (Preface dated June 1885). *Elements of Geometry and Trigonometry, from the Works of A. M. Legendre. Adapted to the Course of Mathematical Instruction in the United States.* Edited by J. Howard Van Amringe. New York: American Book Co. pp. viii + [9]–291 + [1]–150 + 71.

◇ This edition differs from the 1890b edition by the lines "From the press of A.S. Barnes & Co." on the title page. No internal differences are noted.

◇ The second pagination is "Trigonometry & Mensuration," the third is tables.

○ The library has five copies of this book.

○ One copy signed "Lewis Turtle, January, 1890."

○ Another copy stamped "Text–Book USMA 1895."

○ A third copy is signed "T.F. Howard, 12/19/91" and is heavily marked. It was a gift of Joseph V. Columbus.

[1890]b (Preface dated June 1885). *Elements of Geometry and Trigonometry, from the Works of A. M. Legendre. Adapted to the Course of Mathematical Instruction in the United States.* Edited by J. Howard Van Amringe. New York: American Book Co. pp. viii + [9]–291 + [1]–150 + 71.

◇ This is two books in one. Pages 1–291 is geometry and pp. 1–150 is trigonometry and mensuration; 16 pp. log tables; 71 pp. log sines and tangents.

○ The library has four copies.

○ Contains one loose printed sheet, dealing with Dependent Equations and Simultaneous Equations. Appears to be from another volume.

[1890]c (Preface dated June 1885). *Elements of Geometry and Trigonometry from the works of A. M. Legendre Adapted to the Course of Mathematical Instruction in the United States.* Edited by J. Howard Van Anringe. New York: American Book Co., pp. viii + 291 + 150 + 7 pages of advertisements.

◇ Facing the title page is a list of books by Davies and the copyright date of 1890. This appears to be a reprint of 1885 with a new title page.

○ Cover is incised "Grant, U.S. West Point NY '02". A poem is on the endcover. This copy was given in memory of Frederick Dent Grant by his daughter Mrs. Paul E. Ruestow.

○ See Figure 28 for a photograph of the inside leaf of the back cover of this book.

Davies, Charles (Geometry and Trigonometry). See also Legendre 1828, 1830, 1832; Davies (Algebra) 1870.

Davies, Charles (Analytical Geometry).

1836. *Elements of Analytical Geometry: Embracing the Equations of the Point, the Straight Line, the Conic Sections, and Surfaces of the First and Second Order.* New York: Wiley & Long. pp. [ii] + 352.

◇ Preface mentions that the treatises of Biot and Bourdon have been freely consulted and the system of Biot has been followed.

◇ This book contains a dedication to Lieutenant Colonel S. Thayer, Late Superintendent of the Military Academy.

○ A gift of LTC Wallace H. Griffith. Contains USMA bookplate.

Davies, Charles (Anaytical Geometry – Continued).

1839. *Elements of Analytical Geometry: Embracing the Equations of the Point, the Straight Line, the Conic Sections, and Surfaces of the First and Second Order.* Second edition revised and corrected. Hartford: A. S. Barnes. pp. 352.

○ Copy contains numerous annotations.

1841. *Elements of Analytical Geometry: Embracing the Equations of the Point, the Straight Line, the Conic Sections, and Surfaces of the First and Second Order.* Second edition. Revised and corrected. Philadelphia: A. S. Barnes and Co. pp. 352.

◇ Dedicated to Lieutenant–Colonel S. Thayer.
◇ Copyright 1836.
○ The library has two copies.
○ One copy is signed by "G. S. Becker."

[1855]. *Elements of Analytical Geometry: Embracing the Equations of the Point, the Straight Line, the Conic Sections, and Surfaces of the First and Second Order.* New York: American Book Company from the press of A. S. Barnes & Co. pp. [2] + 3–352.

◇ Contains a listing of "Davies' Mathematics", 19 different textbooks used at West Point.
◇ Dedicated to LTC Thayer (1836).
○ This copy has been rebound.

1857. *Elements of Analytical Geometry: Embracing the Equations of the Point, the Straight Line, the Conic Sections, and Surfaces of the First and Second Order.* Revised edition. New York: A. S. Barnes & Co. pp. 352.

○ This book contins the dedication to Thayer and the 1836 preface.
○ This copy belonged to George B. West, October 1857.

Davies, Charles (Linear Perspective).

1832. *A Treatise on Shades and Shadows, and Linear Perspective.* New York: J & J Harper, pp. viii + 9–157 + 19 plates throughout.

◇ This was used as a textbook at USMA from 1832 to 1865.
◇ Two copies of this book are in the library.
○ One copy is signed by "A.E. Church, May 1835." It was rebound in 1952 with a red cover. It contains notes on facing pages.
○ The second copy was given by the library of COL Duncan (USMA 1834).

1835. *A Treatise on Shades and Shadows, and Linear Perspective.* Second edition. New York: Wiley & Long, pp. viii + 9–159 + 21 plates throughout the text.

◇ Preface written by the author in 1832.
◇ The publisher for this book is different than the 1832 edition. Extra two pages of text added at the end of the 1832 edition.
◇ The 1839 edition is also a second edition by a different publisher (A. S. Barnes & Co., Hartford).
○ Two signatures of G.C.[?] Davies on title page.

Davies, Charles (Linear Perspective – Continued).

1839. *A Treatise on Shades and Shadows, and Linear Perspective.* Second edition. Hartford: A. S. Barnes & Co. pp. viii + 9–159 + 21 plates throughout text.
 ◇ Preface by author dated 1832.
 ◇ Text identical to 1835 edition by a different publisher.

1851. *A Treatise on Shades and Shadows, and Linear Perspective.* New York: A. S. Barnes & Co. pp. viii + [9]–159 + 21 plates.
 ○ Copy contains a few annotations by a student.

1857. *A Treatise on Shades and Shadows, and Linear Perspective.* New York: A. S. Barnes & Co., pp. viii + 9–159 + 21 plates.
 ○ This copy was was rebound in 1952.
 ○ This copy is signed by "Daniel Washburn, 1863." It contains lesson plans on facing pages.

Davies, Charles (Linear Perspective). See also Davies (Descriptive Geometry) 1867.

Davies, Charles (Descriptive Geometry).
1826. *Elements of Descriptive Geometry, with Their Application to Spherical Trigonometry, Spherical Projections, and Warped Surfaces.* Philadelphia: H. C. Carey and I. Lea. pp. vii + 228 + 28 plates.
 ◇ This book was used as a text at USMA from 1832 to 1864.
 ○ Five copies are present in the library.
 ○ One copy is inscribed "A present to John Bratt from Prof Henry, Late of the Albany Academy but now Professor of Philosophy at Princetown College, Albany, Oct 29, 1832" inside front cover. There is a sketch of an individual on frontpaper and other art work throughout. It appears to be the same individual drawn in *An Elementary Treatise on Plane & Spherical Trigonometry*, Lacroix 1820c, p. 57.
 ○ One copy is signed by "John C. Sprigg". Numerous annotations throughout the book. This copy is stored in a bound box.
 ○ Another copy has two signatures on the front papers, Geo. W. Cullum and Chas. A. Fuller.
 ○ Another copy was a cadet text and contains numerous notes written in the margin, probably written by a student of Albert Church. Copy was signed by "William C. Mosher."
 ○ The title page is missing.

1835. *Elements of Descriptive Geometry: with Their Application to Spherical Trigonometry, Spherical Projections, and Warped Surfaces.* Second edition. New York: Wiley & Long, pp. vii + [9]–174 + 28 plates.
 ◇ Different typeset and pagination from 1826 edition. Preface dated 1826.
 ○ Signed in 1844 by "Fair." Gift of Mrs. Majorie H. Darr.

Davies, Charles (Descriptive Geometry – Continued).

1836. *Elements of Descriptive Geometry, with Their Application to Spherical Trigonometry, Spherical Projections, and Warped Surfaces.* Third edition. New York: Wiley & Long, pp. vii + [9]–174 + 28 plates.

o   Stamped "W.P. Steele".

1844. *Elements of Descriptive Geometry, with Their Application to Spherical Trigonometry, Spherical Projections, and Warped Surfaces.* Philadelphia: A. S. Barnes & Co. pp. vii + [9]–174 + 28 plates (interspersed).

o   This copy belonged to F.E. Prince, dated March 29th, 1847.

1848. *Elements of Descriptive Geometry, with Their Application to Spherical Trigonometry, Spherical Projections, and Warped Surfaces.* New York: A. S. Barnes & Co. pp. vii + 9–174 + 28 plates.

o   This copy was a gift of Mrs. Howard Guthrie. Contains a USMA bookplate. Cover is in very bad condition.

1852. *Elements of Descriptive Geometry, with Their Application to Spherical Trigonometry, Spherical Projections, and Warped Surfaces.* New York: A. S. Barnes & Co. pp. vii + [9] – 174 + 28 plates.

o   This copy has been rebound. Contains notes at the bottom of page 15. Additional notes and additions are throughout the text.

1860. *Elements of Descriptive Geometry, with Their Application to Spherical Trigonometry, Spherical Projections, and Warped Surfaces.* New York: A. S. Barnes & Burr. pp. viii + [9] – 174 + 28 plates.

◇   Contains advertisement of the books in Davies Course of Mathematics. See Figure 26 for a photograph of this advertisement.

o   This copy is signed "Dept of Mathematics."

1866. *Elements of Descriptive Geometry, with Their Application to Spherical Trigonometry, Spherical Projections, and Warped Surfaces.* New York: A. S. Barnes & Co. pp. vii + [9] – 174 + 28 plates.

◇   Preface dated 1826.

o   The library has two copies.

o   One copy is signed by F. W. Robinson, 1868, and W.K.N., '90.

o   Second copy is signed by William C. Mosher and John W. Nichols.

1867. *Elements of Descriptive Geometry, with Its Applications to Spherical Projections, Shades and Shadows, Perspective and Isometric Projections.* New York: A. S. Barnes & Co. pp. vi + 192.

◇   This book is a combination of parts of Davies (Linear Perspective) 1832 and Davies (Descriptive Geometry) 1826. The four major parts (in order) are: Orthographic Projection, Spherical Projection, Shades and Shadows, and Linear Perspective.

◇   This book has no plates.

o   This copy contains numerous student annotations, art work, and poetry. In particular, there are illustrations of "Hope" and "Despair."

Davies, Charles (Surveying).

1830. *Elements of Surveying. With the Necessary Tables.* New York: J. & J. Harper. pp. [1–8] + 2 + 10–147 + 1–62 tables + 1–91 tables + 5 plates.

◇ This book was used as a textbook at USMA from 1841 to 1899.

○ The library has two copies.

○ One copy contains the signature, "Geo. W. Cullum West Point 1831."

○ The second copy has been rebound. It contains notes and signatures of M.S. Mullins.

1836 [1835]. *Elements of Surveying: Including a Description of the Instruments and the Necessary Tables.* Stereotyped. New York: Wiley & Long, et. al., pp. viii + 158 + 62 + 91 + 6 plates.

○ Contains Special Collections bookplate.

○ The library has two copies.

1839. *Elements of Surveying, with a Description of the Instruments and the Necessary Tables, Including a Table of Natural Sines.* Hartford: A. S. Barnes. pp. 170 + 91 tables + 6 plates.

○ A gift of Charles O. Karch, Civil Engineer, Kansas City, Missouri. Inscribed and signed in 1954.

1841. *Elements of Surveying, and Navigation; With a Description of the Instruments and the Necessary Tables.* Fifth edition. Philadelphia: A. S. Barnes & Co. pp. [1–5] + 6–188 + 1–72 tables + 1–100 tables + 6 plates.

○ Circular glued in June 6, 1873 from General Sherman. Stamped "C.A. Coolidge" on p. 3. Textbook of F.P. Cummins, signed December 18, 1842. Contains quote of Cummins "Scientia est Potentia". Word problem written in pencil on p. 2.

1862. *Elements of Surveying, and Navigation, with Descriptions of the Instruments and the Necessary Tables.* Revised edition. New York: A. S. Barnes and Burr. pp. viii + [9]–222 + 100(tables) + 6 plates.

○ Copy signed by "Robert Fletcher, cadet USMA April 9th, 1866."

1874. *Elements of Surveying and Levelling; with Descriptions of the Instruments and the Necessary Tables.* New York: A. S. Barnes & Co. pp. viii + 270 + 161 + 6 plates.

◇ Preface dated 1870.

○ This copy contains USMA special collections bookplate. This appears to be an old student text. Contains some notes and corrections. Page 83 is of interest.

1876. *Elements of Surveying and Levelling; with Descriptions of the Instruments and the Necessary Tables.* New York: A. S. Barnes & Co. pp. 270 + 161(tables) + 6 plates.

○ This copy was used as a text used in 1879. This copy contains some student notes about what was covered and omitted in class.

[1883]a. *Elements of Surveying and Leveling.* Revised by J. Howard Van Amringe. New York: A. S. Barnes & Co. pp. viii + 9–374 + 29 + 161.

◇ Book contains an illustration of collimation.

○ Corrections attached to backing page. Course notes written on facing pages and notes in margins.

Davies, Charles (Surveying – Continued).

1883b. *Elements of Surveying and Leveling.* Revised by J. Howard Van Amringe. New York: American Book Co., from the press of A. S. Barnes & Co. pp. viii + 374 + 29 + 161.
⋄ Preface dated 1870.
○ This book was a gift of Joseph V. Columbus and contains a USMA bookplate. It is signed "12/19/90, T.F. Howard" and stamped "HOWARD, USMA" on cover. Revision notes are written throughout.

1883c. *Elements of Surveying and Leveling.* Revised by J. Howard Van Amringe. New York: American Book Co. pp. viii + 374 + 29 + 161.
⋄ This edition differs from the one above in the "Printed by A. S. Barnes & Co." does not appear on the title page.
○ This copy has a USMA special collection book plate: Alexander Macomb Miller III, class of 1927. Signed by A.M. Miller, 1893. This is an old student textbook with notes and corrections in margins.

1883d. *Elements of Surveying and Leveling.* Revised by J. Howard Van Amringe. New York: A. S. Barnes & Co. pp. viii + 9–374 + 161 appendices + 1 plate (glued in).
○ This book has been rebound. Pages 9–32 have corners cut. Pages 113–130 and 263–296 are cut. Plate is glued in at last page. Pages iii and iv are completely torn but present. Page 165 has a small figure glued in. This copy was received in the department on Feb 8, 1884. This book differs from those above in page count.

1883e. *Elements of Surveying and Leveling.* Revised by J. Howard Van Amringe. New York: American Book Co. pp. viii + 9–374 + 29 + 161 tables.

Davies, Charles (Calculus).
1836. *Elements of the Differential and Integral Calculus.* New York: Wiley and Long; Collins, Keese, and Co.; et. al., pp. 283.
⋄ This is the first edition of this book.
⋄ This book was used as a text at USMA from 1838 until 1842.
○ Four copies are present in the library.
○ One copy has a USMA bookplate. Notes on separate pages are pasted onto p. 23. Contains student notes and artwork . Signed by G.J. Baker and N. H. Churchill (1838).
○ Second copy is in extremely poor condition. Contains notes inside front and back covers. Sketch of a person inside back cover. Notations throughout.
○ Third copy is signed "M. Knowlton, U.S. Army" inside front cover. This copy is in very good condition.
○ One copy is clean.

1838. *Elements of the Differential and Integral Calculus.* Second edition, revised and corrected. Hartford: A. S. Barnes and Co. pp. 283.

1840. *Elements of the Differential and Integral Calculus.* Second edition, revised and corrected. Philadelphia: A. S. Barnes and Co. pp. 283.
○ Notes after last page, notations throughout, stamped "W.F. Steele" inside front and back covers.

Davies, Charles (Calculus). See also Boucharlat 1826.

Davies, Charles (General).
1850. *The Logic and Utility of Mathematics, with the Best Methods of Instruction Explained and Illustrated.* New York: A. S. Barnes. pp. 375 + 9 advertising.

Davies, Charles (General). See also Davies and Peck 1855.

Davies, Charles (1798–1876; USMA 1815) and Peck, William Guy (1820–1892; USMA 1844)
1855. *Mathematical Dictionary and Cyclopedia of Mathematical Science. Comprising Definitions of all the Terms Employed in Mathematics – An Analysis of Each Branch, and of the Whole, as Forming a Single Science.* New York: A. S. Barnes & Co. pp. 592.
○ The library has two copies.
○ One copy has its title page signed by E. W. Bass.

Davies, Thomas Stephens (1795–1851). See Royal Military Academy, Woolwich 1853; 1852.

Davis, Charles Henry (1807–1877). See Gauss 1857.

Davis, William (1771–1807). See Newton 1819.

Day, Jeremiah (1773–1867).
1814. *An Introduction to Algebra, being the First Part of a Course of Mathematics, Adapted to the Method of Instruction in the American Colleges.* New Haven: Howe and DeForest. pp. 8 + 296 + 2 plates.
◇ Errata follow the title page. This was the first Algebra book prepared by a native American.

1815. *A Treatise of Plane Trigonometry. To which is Prefixed, a Summary View of the Nature and Use of Logarithms. Being the Second Part of a Course of Mathematics, Adapted to the Method of Instruction in the American Colleges.* New Haven: Howe and DeForest. pp. [iv] + 126 + 9 tables + 3 plates.
○ "Presented to the United States Military Academy by the Publishers" is on the title page. May be the first edition.

Dealtry, William (1775–1847).
1816. *The Principles of Fluxions: Designed for the Students in the Universities.* Second edition, with corrections and considerable additions. Cambridge: Printed by J. Smith. pp. v + 466.
◇ Errata follow p. v.

de Heusch, F. See Heusch 1888.

Deidier, Antoine (?–1746).
1761. *Elementi generali delle principali parti delle matematiche, necessarj ancora all'artiglieria, e all'arte militare.* Translated by Francese di Arduino and Matteo Dandolo In Venezia [Venice]: Appresso Modesto Fenzo. Three volumes. pp. [xii] + 352 + 9 plates (errata on p. 352); 240 + 28 plates (errata on p. 240); 276 + 16 plates (errata on p. 276).

De Lagny, Thomas Fantet (1660–1734). See Maseres 1795.

Delambre, Jean Baptiste Joseph (1749–1822).
1799. *Méthodes analytiques pour la détermination d'un arc du méridien.*   Paris: Crapelet. pp. xvi + 176 + 17 unnumbered pages of tables + 6 + 1 plate.
   ◇   Errata on p. [xvi]. The book actually begins with "A memoir on the same subject" by A. M. Legendre, pp. 1–16. The last 6 pages contain "Observations on several passages of the preceding memoir by Delambre" by A. M. Legendre.

1801 (An IX). *Tables trigonométriques décimales, ou table des logarithmes des sinus, sécantes et tangentes, suivant la division du quart de cercle en 100 degrés, du degré en 100 minutes, et de la minute en 100 secondes; précédés de la table des logarithmes des nombres depuis dix mille jusqu'à cent mille, et de plusieurs tables subsidiaires: calculées par Ch. Borda, revues, augmentées et publiées par J. B. J. Delambre.*   Paris: De l'Imprimerie de la République. pp. [iii] + 120 + unnumbered pages of tables.
   ◇   Corrections on p. 120.

1810. *Rapport historique sur les progrès des sciences mathématiques depuis 1789, et sur leur état actuel, présenté à sa majesté l'Empereur et Roi, en son conseil d'état, le 6 Février 1808, par la classe des sciences physiques et mathématiques de l'Institut, conformément à l'arrêté du gouvernement du 13 Ventôse an X.* Rédigé par M. Delambre. Imprimé par ordre de sa majesté. Paris: de l'Imprimerie Impériale. pp. vii + 362.
   ◇   Presented by the Minister of the Interior of France. Additions and corrections follow p. vii. An interesting report on the state of the sciences, with a good deal of information about mathematics. Ventôse was the sixth month of the First French Republic.

1817. *Histoire de l'astronomie ancienne.*   2 vols. Paris: Ve Courcier. pp. lxxii + 556 + 1 plate; viii + 639 + 16 plates.
   ◇   Additions and corrections on pp. xl–lii, lxxii of volume I, and on pp. vii–viii, and following p. 639 of volume II.

Delambre, Jean Baptiste Joseph and Burg, Adam.
1806. *Tables astronomiques publiées par le Bureau des Longitudes de France. Premiere partie.*   Paris: Courcier, unpaginated, but over 200 pages.
   ◇   Dedicated to the Emperor and King. The Tables of the Sun are by Delambre and the Tables of the Moon are by Burg.
   ○   Presented by Delambre to Andrew Ellicott. Ellicott has signed the flyleaf.

Delambre, Jean Baptiste Joseph. See also Archimedes 1807.

Delaplaine, Samuel B.
November 20, 1790. [1791] *Samuel B. Delaplaine's Manuscript of Arithmetic.* unpaginated; approximately 200 pages.
◇ This copybook was finished June 29, 1791. Begins with the rule of three and proceeds, by stating a rule and giving several examples, to the arithmetic of decimals.

DeMoivre, Abraham (1667–1754)
1730. *Miscellanea analytica de seriebus et quadraturis. Accessere variae considerationes de methodis comparationum, combinationum & differentiarum, solutiones difficiliorum aliquot problematum ad sortem spectantium, itemque constructiones faciles orbium planetarum, una cum determinatione maximarum & minimarum mutationum quae in motibus corporum coelestium occurrunt.* Londini [London]: Excudebant J. Tonson & J. Watts. pp. [xii] + 250.
◇ Dedicated to Martin Folkes. A list of subscribers is included on pp. [ix]–[xii]. Errata follow p. 250.

1738. *The Doctrine of Chances: or, a Method of Calculating the Probabilities of Events in Play.* Second edition, fuller, clearer, and more correct than the first. London: Printed for the Author by H. Woodfall. pp. xiv + 259.
◇ Dedicated to Lord Carpenter.

DeMoivre, Abraham. See also Saunderson 1740.

De Morgan, Augustus (1806–1871).
1842. *The Differential and Integral Calculus, Containing Differentiation, Integration, Development, Series, Differential Equations, Differences, Summation, Equations of Differences, Calculus of Variations, Definite Integrals, – with Applications to Algebra, Plane Geometry, Solid Geometry, and Mechanics. Also, Elementary Illustrations of the Differential and Integral Calculus.* Library of Useful Knowledge Series. London: Robert Baldwin (Published under the superintendence of the Society for the Diffusion of Useful Knowledge). pp. xx + 1–64 + 3–785 + 2.
◇ Errata on pp. 778–785. On the page following p. 785, there are lists of the "Committee" and of "Local Committees".

1902. *On the Study and Difficulties of Mathematics.* Second reprint edition. Chicago: Open Court. pp. vii + 288.
◇ Frontispiece is a portrait of De Morgan.

Desaguliers, John Theophilus (1683–1744).
1719. *Lectures of Experimental Philosophy. Wherein the Principles of Mechanicks, Hydrostaticks, and Opticks, are Demonstrated and Explained at large, by a Great Number of Curious Experiments: with a Description of the Air-pump, and the Several Experiments Thereon: of the Condensing-Engine; as also of the Different Species of Barometers, Thermometers, and Hygrometers; with Several Experiments to Prove and Explain Sir Isaac Newton's Theory of Light and Colours, as Performed*

in a Course of Mechanical and Experimental Philosophy. To which is Added, a Description of Mr. Rowley's Machine, called the Orrery, which Represents the Motion of the Moon about the Earth, Venus, and Mercury about the Sun, According to the Copernican System: All Carefully Examined and Corrected by Mr. Desaguliers. London: Printed for W. Mears, et. al. pp. [xviii] + 201 + [5] + 10 plates.

◇ Dedicated to Sir Richard Steele by Paul Dawson; the lectures of Desaguliers were "collected' by Dawson with the former's approval. Errata on pp. [203], [204]. Three pages of advertisement for other books follow the errata.

○ Notes in ink scattered throughout.

Descartes, René du Perron (1596–1650).

1649. *Géometria, a Renato Des Cartes Anno 1637 Gallice Edita; Nunc Autem cum Notis Florimondi de Beaune, in Curia Blaesensi Consiliarii Regii, in Linguam Latinam Versa, & commentariis illustrata, Opera atque Studio Francisci a Schooten.* Translated by Schooten. Lvgdvni Batavorvm [Leiden]: Ioannis Maire. pp. [x] + 336 + [ii].

◇ Dedicated to Elizabeth, eldest daughter of King Frederic of Bohemia. Errata on p. 336. Descartes' *Geometria* is on pp. 1–118; the "Brief Notes" of de Beaune are on pp. 119–161; and the commentary of Schooten is on pp. 163–336. Schooten's etching of Descartes is missing. This is the important first Latin translation of the appendix on Geometrie of Descartes' *Discours de la Methode* (1637).

de Sluse, René François (1622–1685). See Collins 1725; Saint Vincent 1647; Guericke 1672.

Dessenon, Ernest (1843–?).

1894. *Éléments de géométrie analytique a l'usage des candidats aux écoles centrale et navale et des élèves de première année de la classe de mathématiques spéciales.* Deuxième édition. Paris: Hachette et cie. pp. x + 525.

◇ The prerequisites in analytic geometry for admission to l'École centrale and to l'École navale are specified on pp. v–vii.

DeVeley, Emmanuel (1764–1839).

1823. *Arithmétique d'Émile, ouvrage adopte par le conseil academique que du canton de Vaud, pour servir a l'usage des colleges.* Troisième édition, corrigée et augmentée. Lausanne: Lacombe et Cie. pp. xvi + 494.

◇ Errata follow p. xiv and p. 494. This elementary arithmetic is unusual in that it incorporates some information about the history of mathematics and also makes references to advanced treatises.

1830. *Éléments de géométrie, distribués dans un ordre naturel et sur un plan absolument neuf.* Troisième édition, conforme aux précédentes pour le plan, perfectionnée dans les détails, et augmentée. Geneva: J. Barbezat et Compagnie. pp. xlii + 327 + 9 plates.

◇ Errata on p. xlii.

DeWitt, Simeon (1756–1834).

1813. *The Elements of Perspective.* Albany: Printed by H. C. Southwick. pp. xxxi + 33– 69 + 12 plates.

○ This book is contained in its own box, which is marked "Presentation copy in ornamental American binding." "Presented to the Military Academy at West Point By the Author" is on the title page. This is marked "No. 2."

Diez de Prado, Manuel

1852. *Lecciones de trigonometria esferica y de geometria analitica, precedidas de algunos ejercicios ó un resúmen de la trigonometria rectilinea, para la enseñanza de los alumnos de la Academia de Ingenieros del Ejercito.* Madrid: Memorial de Ingenieros. pp. x + xxi + 291 + 3 pages of errata + 4 plates.

◇ Errata follow p. 291.

Digges, Leonard (?–1571).

1590. *An Arithmetical Warlike Treatise Named Stratioticos. Compendiously Teaching the Science of Nombers as well in Fractions as Integers, and so much of the Rules and Aequations Algebraicall, and Art of Numbers Cossicall, as are Requisite for the Profession of a Souldier. Together with the Moderne Militare Discipline, Offices, Lawes and Orders in Euery Well Gouerned Campe and Armie Inuiolably Obserued.* First published by Thomas Digges Esquire Anno Salutis 1579 and dedicated vnto the right Honorable Earle of Leicester: lately reuiewed and corrected by the author him self, and also augmented with sundry additions. London: Richard Field. pp. [xiv] + 380 + 2 plates.

◇ Title page states first edition was 1579. Frontispiece. Dedicated to Robert Dudley, Earl of Leicester. The first 78 pages are devoted to a "Booke Arithmeticall" and a "Booke Algebraicall;" the rest of the book contains four groups of mathematical problems encountered by various persons in the military. Leonard Digges' son, Thomas Digges completed this work and published it.

Digges, Thomas (?–1595). See Leonard Digges 1590.

Diophantus of Alexandria (fl. AD 250).

1670. *Diophanti Alexandrini arithmeticorum libri sex, et de numeris multangulis liber unus. Cum commentariis C. G. Bacheti V. C. & observationibus D. P. de Fermat senatoris Tolosani. Accessit doctrinae analyticae inventum novum, collectum ex varijs eiusdem D. de Fermat epistolis.* Tolosae [Toulouse]: Excudebat Bernardus Bosc. pp. [xii] + 64 + 48 + 341.

◇ Dedicated to Joanni Baptistae Colberto. The *Arithmetic* is printed in both Latin and Greek. It is in this work that Fermat's Last Theorem is printed for the first time.

Diophantus of Alexandria. See also Fermat 1891; Stevin 1634.

Ditton, Humphrey (1675–1715).

1726. *An Institution of Fluxions: Containing the First Principles, the Operations,* *with some of the Uses and Applications of that Admirable Method; According to the Scheme Prefix'd to his Tract of Quadratures, by (its First Inventor) the Incomparable Sir Issac Newton.* Second edition, revised, corrected, and improv'd, by John Clarke. London: Printed by W. Boltam for James and John Knapton. pp. [xvi] + 240.

◇ Dedicated to Benjamin Moreland.

○ Signature of Andrew Ellicott on the title page.

d'Ocagne. See under Ocagne.

Dodgson, Charles Lutwidge (Lewis Carroll) (1832–1898).

1867. *An Elementary Treatise on Determinants with Their Application to Simultaneous Linear Equations and Algebraical Geometry.* London: Macmillan and Co. pp. viii + 143.

◇ Corrigenda on p. vi.

Dodson, James (?–1757).

1775. *The Mathematical Repository, Containing Analytical Solutions of near Five Hundred Questions, mostly Selected from Scarce and Valuable Authors. Designed as Examples to Mac–Laurin's and Other Elementary Books of Algebra; and to conduct Beginners to the More Difficult Properties of Numbers.* 3 vols. Second edition. London: Printed for J. Nourse. pp. xi + 336; xxiv + 348; [xii] + 372.

◇ Volume 1 is dedicated to Abraham DeMoivre; volume 2 is dedicated to David Papillon; and volume 3 is dedicated to George Earl of MacClesfield and rest of the Royal Society. The preface to volume 1 is dated 1747, that of volume 2 is dated 1753, and that of volume 3 is dated 1755.

Dodson, James. See Maseres 1795.

Domcke, George Peter (fl. 1730).

1730. *Philosophiae mathematicae Newtonianae illustratae. Tomi duo. Quorum prior tradit elementa mathesews ad comprehendendam demonstrationem hujus philosophiae scitu necessaria: Posterior continet 1) Definitiones & leges motus generaliores; 2) Leges virium centripetarum & theoriam attractionis seu gravitationis corporum in se mutuo; 3) Mundi Systema.* 2 vols. Londini [London]: Tho. Meighan. pp. xiii + [vii] + 189 + 9 plates; 211 + 6 plates.

◇ Corrigenda follow p. 211 of Book II.

Doster, Georges (1820–?).

1877. *Éléments de la théorie des déterminants avec application a l'algèbre, la trigonométrie et la géométrie analytique dans le plan et dans l'espace, a l'usage des classes de mathématiques spéciales.* Paris: Gauthier–Villars. pp. xxxi + 352.

◇ Purchased by USMA in 1881 for $2.70.

○ The library has two copies.

n.d. *Nouvelle détermination analytique des foyers et directrices dans les sections coniques représentées par leurs équations générales; précédées des expressions* *générales des divers éléments, que l'on distingue dans les courbes du second degré; et suivie de la détermination des coniques à centre par leur centre et les extrémités de deux demi-diamétres conjugues.* Paris: Gauthier–Villars; Lieipzig: C. A. Koch. pp. 92.

○ Two copies are present in the library.

Douglass, George.
1805a. *The Art of Drawing in Perspective, from the Mathematical Principles; Shewing how to give Every Object its True Place in the Figure, and Every Part of a Landscape that Proportion in Size, and Distance from One Another, which the Parts They are Intended to Represent Hold in Nature. The Whole Illustrated by Upwards of Fifty Engravings.* Edinburgh: Printed by and for Mundell and Son, et. al. pp xv + 87.

◇ Dedicated to the Countess of London and Moira. Despite the assertions in the title, there are no illustrations or plates in this book. Plate 9 is cited on p. 82.

1805b. *Plates for Illustrating the Art of Drawing in Perspective.* Edinburgh: Mundell and Son, et al. pp. 10 plates.

◇ The author's name is "Douglas" on the title page.

Douglass, George. See also Euclid 1780.

Douliot, Jean Paul (1788–1834).
1825. *Traité spécial de coupe des pierres.* Paris: l'Auteur; Carilian–Geoury. pp. vii + 472 + 99 plates.

Dowling, Daniel (1735–1765).
1824. *A Key to the Course of Mathematics Composed for the Use of the Royal Military Academy, Woolwich, by Charles Hutton, with an Appendix, Containing a Key to the Late Edition of the Second Volume, by Olinthus Gregory.* Second edition. London: Printed for G. and W. B. Whittaker, pp. x + 415.

◇ Dedicated to Lord Viscount Palmerston. Errata follow p. 415.

Dubois, Jean M. (1765–1848). See Boisayme 1811.

Duchesne, Edward Adolphe (1804–1869).
1828. *Élémens de géométrie descriptive, a l'usage des élèves qui se destinent a l'École polytechnique, a l'École militaire, a l'École de marine.* Paris: Malher et Cie. pp. 189.

◇ Errata follow p. 189. Plate 18 is cited on p. 168, but there are no plates in this volume.

Dufour, Guillaume Henri (1787–1875).
1827. *Géométrie perspective, avec ses applications a la recherche des ombres.* 2 vols. Geneva: Barbezat et Delarue; Paris: Bachelier. pp. viii + 84 + 22 plates; 22 plates.

⋄   Another set of the plates from volume 1 is bound separately in volume 2, but are not folded in volume 2. Volume 2 uses a different font for the title page; volume 1 has no separate title page for the plates.

Du Hamel, Jean–Baptiste (1624–1706).

1701. *Regiae scientiarum academiae historia, in qua praeter ipsius academiae originem & progrssus, variasque dissertationes & observationes per triginta quatuor annos factas, quam plurima experimenta & inventa, cum physica, tum mathematica in certum ordinem digeruntur.* Secunde editio priori longe auctior. Parisiis [Paris]: Joannem–Baptistam Delespine. pp. [xx] + 704 + 1 fold–out map.

⋄   Dedicated to J. Paul Bignon. Frontispiece.

Duhamel, Jean Marie Constant (1797–1872).

1847. *Cours d'analyse de l'École polytechnique.* 2 vols. Seconde édition. Paris: Bachelier; Leipzig: L. Michelsen. pp. xi + 365 + 1 plate; viii + 335 + 1 plate.

⋄   Errata for volume 1 on p. 365, and for volume 2 on p. 335.

1853; 1854. *Cours de mécanique.* 2 vols. Seconde édition. Paris: Mallet–Bachelier. pp. xvi + 422 + 2 plates; viii + 364 + 1 plate.

∘   Two copies are in the library.

1856. *Éléments de calcul infinitésimal.* 2 vols. Paris: Mallet–Bachelier. pp. xx + 585; xi + 375.

⋄   Errata page for Volume 1 at the end of Volume 1 before the plates.

1874; 1876. *Éléments de calcul infinitesimal.* 2 vols. Troisième édition, revue et annotée par M. J. Bertrand. Paris: Gauthier–Villars. pp. xiv + 508 + 6 plates; xv + 536 + 1 plate.

Duhamel, Jean Marie Constant (1797–1872). See also Reynaud 1823.

Dupin, Charles (1784–1873).

1813. *Développements de géométrie, avec des applications à la stabilité des vaisseaux, aux déblais et remblais, au défilement, à l'optique, etc.; ouvrage approuvé par l'Institut de France, pour faire suite a la géométrie descriptive et a la géométrie analytique de M. Monge: Theorie.* Paris: Ve Courcier. pp. xx + 373 + 11 plates.

⋄   Dedicated to Monge. Errata follow p. 373.

∘   The library has two copies.

∘   One copy is refurbished with a Thayer binding and is contained in the Thayer Collection.

1822. *Applications de géométrie et de méchanique, à la marine, aux ponts et chaussées, etc., pour faire suite aux développements de géométrie.* Paris: Bachelier. pp. xxxv + 330 + 4 page catalog of books by Bachelier + 17 plates.

∘   It appears that several pages following page 330 were ripped out. Contents of these pages is unknown.

Dupuis, N. F. (1836–1917)
1889. *Elementary Synthetic Geometry of the Point, Line, and Circle in the Plane.*
London: MacMillan and Company. pp. x + 294.

Dupuy. See Anthemius 1777.

Durège, H. (1821–1893).
1896. *Elements of the Theory of Functions of a Complex Variable with Especial
Reference to the Methods of Riemann.* Translated by George Fisher and Isaac
Schwatt. Philadelphia: Fisher and Schwatt. pp. xiii + 288.

Echols, William Holding (1859–1934).
1908. *An Elementary Text-Book on the Differential and Integral Calculus.* New
York: Henry Holt and Company. pp. xii + 482.

Eddy, Henry Turner (1844–1921).
1874. *A Treatise on the Principles and Applications of Analytic Geometry.* Philadelphia: Cowperwaith & Company. pp. 200.
   o   Signed "E.W. Bass, West Point, 1879".

Edwards, Edward (1738–1806).
1803. *A Practical Treatise of Perspective, on the Principles of Dr. Brook Taylor.*
London: Printed by Luke Hansard. pp. xii + 316 + 40 plates.
   ◇   See Figure 21 for a photograph of the frontispiece of this book.
   ◇   Dedicated to the King.
   ◇   Errata follow p. 316. A list of references is given on pp. 315–316.
   o   "Military Academy No. 101" written on front paper.

Einstein, Albert (1879–1955). See Hardy n.d.

Ellicott, Andrew (1754–1820). See Biot 1802; Delambre 1806; Ditton 1726;
Halley 1752; Jack 1742; Keill 1726; Keill 1730; LaChapelle 1750; Martin 1740;
Mayer 1770; Taylor 1780; West 1763.

Elliot, Edwin Bailey (1851–?).
1895. *An Introduction to the Algebra of Quantics.* Oxford: Clarendon Press. pp.
xiii + 423 + 16.

Emerson, Frederick (1788–1857).
1835. *The North American Arithmetic, Part Third, for Advanced Scholars.* Emerson's third part. Boston: Russell, Odiorne, and Metcalf. pp. 288.

1847 [or 1834]. *The North American Arithmetic. Part Third for Advanced Scholars.* Emerson's third part. Boston: Jenks, Palmer & Co. pp. 272.
   o   This copy has been rebound.

Emerson, William (1701–1782).

1743. *The Doctrine of Fluxions: Not only Explaining the Elements thereof, but also its Application and Use in Several Parts of Mathematics and Natural Philosophy.* London: Printed by J. Bettenham. pp. xvi + 300 + 10 plates + [iv].

◇ Errata follow the table of contents at the end of the book.

1763a. *Cyclomathesis: or an Easy Introduction to the Several Branches of the Mathematics. Being Principally Designed for the Instruction of Young Students, before They Enter upon the more Abstract and Difficult Parts thereof.* London: Printed for J. Nourse. pp. xvi + 8 + 244.

◇ Errata on p. 244.

1763b. *The Method of Increments. Wherein the Principles are Demonstrated; and the Practice thereof Shewn in the Solution of Problems.* London: Printed for J. Nourse. pp. viii + 147.

◇ In this work Emerson presents an expostion of Taylor's "Method of Increments" and Sterling's "Differential Method," which he views as discrete forms of Newton's Method of Fluxions.

1767. *The Arithmetic of Infinites, and the Differential Method; Illustrated by Examples; The Elements of the Conic Sections; Demonstrated in Three Books. Book I. Of the Ellipsis. Book II. Of the Hyperbola. Book III. Of the Parabola.; The Nature and Properties of Curve Lines. Book I. Of the Conchoid, Cissoid, Cycloid, Quadratrix, Logarithmic Curve; the Spiral of Archimedes, the Logarithmic Spiral, the Hyperbolic Spiral. Book II. Of Curve Lines in general, and Their Affections.* London: Printed for J. Nourse. pp. iv + 44 + 1 plate; ii + 225 + 33 plates; iv + 115 + 8 plates.

◇ This book consists of three publications bound together as one. Errata for all three books follow p. 115 of Curve Lines.

1770a. *Calculation, Libration, and Mensuration; or the Arts of Reckoning, Weighing, and Measuring. Being a Mechanical Work, Adapted to the Business and Practice of Tradesmen and Artificers, in the Shortest Method Possible; and Designed purely for Common Use.* London: Printed for J. Nourse. pp. v + 168 + 4 plates.

1770b. *The Doctrine of Combinations, Permutations, and Compositions, of Quantities, clearly and succinctly Demonstrated.* London: Printed for J. Nourse. pp. [ii] + 29.

◇ The author had intended to include this material in his Treatise on Algebra, but that book was already quite large so this material was published separately (from advertisement on p. [ii]).

1780. *A Treatise of Algebra, A New Edition, in Two Books. Book I. Containing the Fundamental Principles of this Art. Together with All the Practical Rules of Operation. Book II. Containing a Great Variety of Problems, in the Most Important Branches of the Mathematics.* Second edition. London: Printed for J. Nourse. pp. viii + 530 + 13 plates.

1794. *The Elements of Geomtry, in which the Principal Propositions of Euclid, Archimedes, and Others, are Demonstrated after the Most Easy Manner. To which is Added, a Collection of Useful Geometrical Problems. Also, the Doctrine of Proportion, Arithmetical and Geometrical. Together with a General Method of Arguing* *by Proportional Quantities.* New edition. London: Printed for F. Wingrave. pp. viii + 216 + 14 plates.

1808. *A Treatise of Algebra, A New Edition, in Two Books. Book I. Containing the Fundamental Principles of this Art. Together with all the Practical Rules of Operation. Book II. Containing a Great Variety of Problems, in the Most Important Branches of the Mathematics.* New edition carefully revised and corrected. London: Printed for F. Wingrave, pp. viii + 531 + 13 plates.

◇ The preface is dated 1764.

1811. *The Principles of Mechanics. Explaining and Demonstrating the General Laws of Motion, the Laws of Gravity, Motion of Descending Bodies, Projectiles, Mechanic Powers, Pendulums, Centres of Gravity, &c., Strength and Stress of Timber, Hydrostatics, and Construction of Machines. A Work Very Necessary to be Known by All Gentlemen and Others that Desire to Have an Insight into the Works of Nature and Art, and Extremely Useful to All Sorts of Artifers; Particularly to Architects, Engineers, Shipwrights, Millwrights, Watchmakers, &c. or any that Work in a Mechanical Way.* Sixth edition, corrected, and illustrated with forty–three copper plates. London: Printed for G. Robinson, et. al. pp. xii + 286 + 43 plates.

◇ There is a combined index and glossary on pp. 273–284.

Emerson, William (1701–1782). See also Newton 1819.

Engel, Ferdinand (1805–1866).

1855. *Axonometrical Projections of the Most Important Geometrical Surfaces. Drawings in Descriptive Geometry. Serving at the Same Time as a Catalogue of Models Executed According to the aforesaid Projections.* New York: H. Goebeler. pp. [v] + 9 plates + 5–9.

◇ Preface by Dr. F. Joachimsthal for this "Edition for America." These models are designed for the study of Geometry and Optics. There are testimonials from Wolcott Gibbs (Free Academy in N.Y.), Druckenmuller, and Dirichlet. The Preface states that the first model made by Engel was in the possession of Professor Plücker.

Eratosthenes (ca. 276–ca. 195 B.C.). See Euclid 1803.

Euclid (ca. 300 B.C.)

1703. ΕΥΚΛΕΙΔΟΥ ΤΑ ΣΩΖΟΜΕΝΑ. *[Eukleidou ta sozemena]. Euclidis quae supersunt omnia. In recensione Davidis Gregorii.* Oxoniae [Oxford]: E. Theatro Sheldoniano. pp. [xvii] + 686.

◇ See Figures 10 and 11 for photographs of the title page and frontispiece of this volume.

◇ Dedicated to Henry Aldrich.

◇ Errata follow p. 686. The text is in Greek and Latin. Contains the 15 books of Euclid's *Elements*, his *Data, Harmonics, Conic Sections, Phenomena, Optics, Catoptics, The Division of Figures,* and a fragment on *Heavy Bodies.*

Euclid (Continued)

1751. *Euclide's Elements; the Whole Fifteen Books, Compendiously Demonstrated: with Archimedes's Theorems of the Sphere and Cylinder, Investigated by the Method of Indivisibles. Also Euclide's Data, and a Brief Treatise of Regular Solids. By Isaac Barrow. The Whole Carefully Corrected, and Illustrated with Copper Plates. To which is now Added an Appendix, Containing the Nature, Construction, and Application of Logarithms.* London: Printed for W. and J. Mount and T. Page; et. al. pp. [viii] + 384 + 9 plates.

○   Copy owned by Wm. Partridge of U.S. Engineers.

1756. *The Elements of Euclid, viz The First Six Books Together with the Eleventh and Twelfth. In this Edition, the Errors, by which Theon, or others, have long ago Vitiated these Books are Corrected, and Some of Euclid's Demonstrations are Restored.* Translated by Robert Simson. Glascow: Printed by Robert and Andrew Foulis. pp. [viii] + 431.

◇   Dedicated to the Prince of Wales. Errata on p. [viii].

◇   The "Notes Critical and Geometrical" has its own title page, p. [353]. This part contains an account of things which differ from the Greek text and the reasons for the alterations.

1763. *Euclid's Elements of Geometry, the First Six, the Eleventh, and Twelfth Books; Translated into English from Dr. Gregory's Edition, with Notes and Additions. For the Use of the British Youth.* Second edition with additions. London: Printed for Tho. Payne. pp. xv + 464.

◇   Errata follow p. 464. There is a short Biography, in French, of Edmund Stone on pp. 461–464.

◇   Translation from David Gregory's Latin Euclid.

1780. *The Elements of Euclid, in which the Propositions are Demonstrated in a New and Shorter Manner than in former Translations, and the Arrangement of many of Them Altered; to which are Annexed Plain and Spherical Trigonometry, Tables of Logarithms from 1 to 10000, and Tables of Sines, Tangents, and Secants, both Natural and Artificial.* Compiled by George Douglas. Second edition. Edinburgh: Printed for C. Elliot. pp. viii + 172 + 17 plates.

◇   Various tables follow p. 172 and the pages are numbered separately, 1–92.

1802. *Euclidis elementorum libri priores XII ex Commandini et Gregorii versionibus Latinis. In usum juventutis academicae.* Edited by Samuel Horsley [Samuel, Episcopus Roffensis]. Oxonii [Oxford]: E typographeo Clarendoniano. pp. xvii + 504.

◇   Addenda on pp. 503–504.

1803. *Euclidis datorum liber cum additamento, necnon tractatus alii ad geometriam pertinentes. In usum juventutis academicae. Curavit et edidit Samuel, Episcopus Asaphensis.* The volume contains the following items: (1) *Euclidis Data.* (2) *Additamentum libri Datorum Euclidis. Sive liber datorum secundus. Concnnavit et adjecit Samuelis, Episcopi Asaphensis.* (3) *Problematum delectus, cum resolutione eorum geometricâ.* (4) *Sphricorum liber singularis. Ex primo fere et secundo sphaericorum Theodosii studio atque opera Samuel Episcopi Asaphensis.*

Euclid (Continued)

(5) *Archimedis circuli dimensio, cum commentariis Eutocii Ascaloniae. Latine ex interpretatione Torellii emendata.* (6) *Cribrum Eratosthenis ad numeros primos omnes ordine inveniendos. Samuel Episcopi Asaphensis opera restitutum.* (7) *De numeris primis et compositis. Auctore Samule Episcopio Asaphensi.* (8) *Johannis Keillii geometriae olim in academia Oxoniensi Professoris Saliviensis De naturâ et arithmeticâ logarithmorum tractatus brevis. Notas suas de verâ constitutione systematis Brigsiani, de arithmeticâ fractionum logarithmicâ, deque anatucismo, appendicem etieam de problemata quodam sphaerico adjecit Samuel Episcopus Asaphensio.* Edited by Samuel E. Horsley. Oxonii [Oxford]: E typographeo Clarendoniano. pp. ix + 145 + 172 + 4 plates.

1810. *The Elements of Euclid, viz. the First Six Books, together with the Eleventh and Twelfth. The Errors, by which Theon, or others, have long ago Vitiated these Books, are Corrected, and Some of Euclids Demonstrations are Restored. Also, the Book of Euclid's Data, in like Manner Corrected. To this Edition are also Annexed, Elements of Plane and Spherical Trigonometry.* Edited by Robert Simson. Philadelphia: Printed for Conrad and Co. pp. viii + 518 + 3 plates.

◇ "The Elements of Euclid" occupies pp. 1–293. This is followed by three items, each with a special title page: (1) Simpson's "Notes, critical and geometrical: Containing an account of those things which this edition differs from the Greek text; and the reasons of the alternations which have been made. As also observations on some of the propositions," pp. 295–362. (2) "Euclid's data. In this edition several errors are corrected, and some propositions added," pp. 363–477, which is dated 1807 on its title page. (3) Simpson's "The elements of plane and spherical trigonometry," pp. 479–513.

○ The library has two copies. A note in one copy, dated 10-23-37, indicates that Constant Mathieu Eakin (1794–1869; USMA 1817) used this book while teaching mathematics at West Point.

1811. *The Elements of Euclid, viz. the First Six Books, together with the Eleventh and Twelfth. The Errors, by which Theon, or others, have long ago Vitiated these Books, are Corrected, and Some of Euclid's Demonstrations are Restored. Also, the Book of Euclid's Data, in like Manner Corrected. To this Edition are also Annexed, Elements of Plane and Spherical Trigonometry.* Edited by Robert Simson. Philadelphia: Johnson and Warner, pp. viii + 518 + 3 plates.

◇ Appears to be an exact copy of the 1810 edition, except for the foot of the title page.

1814a; 1816; 1818. *Les oeuvres d'Euclide, en Grec, en Latin et en Français, d'après un manuscrit très-ancien qui était resté inconnu jusqu'à nos jours.* 3 vols. Translated by F. Peyrard. Paris: Patris. pp. xliv + 518; xliv + 518; xviii + 614.

◇ Dedicated to the King. Each page contains the Greek, the Latin, and the French versions of the same passage. Errata of volume 1 follow p. 518, of volume 2 follow p. 518, and of volume 3 on p. xviii and following p. 614. Volume 3 contains Euclid's *Data* (pp. 301–480) and Hypsicilis on *The Five Solids* (pp. 481–531).

○ Thayer binding.

Euclid (Continued)

1814b. *The Elements of Plane Geometry: Containing the First Six Books of Euclid, from the Text of Dr. Simson, Emeritus Professor of Mathematics in the University of Glasgow; with Notes Critical and Explanatory. To which are Added Book VII, including Several Important Propositions which are not in Euclid; and Book VIII, Consisting of Practical Geometry: also, Book IX, of Planes and Their Intersections; and Book X, of the Geometry of Solids.* London: Printed for Longman, Hurst, Rees, Orme, and Brown. pp. xv + 398 + 2 pages of advertisement for 5 other books by Keith.

◇   Dedicated to Princess Charlotte of Wales.

1819a. *A Treatise of Geometry, Containing the First Six Books of Euclid's Elements, Methodically Arranged and Concisely Demonstrated. Together with the Elements of Solid Geometry.* Cambridge: Printed by J. Smith. pp. [iv] + xxiii + 499 + an Addendum on an unnumbered page.

◇   There is an index "shewing the order in which the Propositions of Euclid are arranged in this Treatise" on three pages following the Addendum. Errata are on p. [iii]. This work is a revision of Euclid "which aims at new modelling almost the whole system of elementary Geometry."

1819b. *Elements of Geometry: Containing the First Six Books of Euclid with a Supplement on the Quadrature of the Circle and the Geometry of Solids; to which is Added Elements of Plane and Spherical Trigonometry.* From the last London Edition, enlarged (the preface is dated 1813). New York: Collins and Co. pp. xvi+19–333.

◇   Curiously, what we call "Playfair's Axiom" is not here. His Axiom XI (p. 21) is "Two straight lines which intersect one another, cannot be both parallel to the same straight line."

○   Bookplate is marked "From the Library of Col Duncan."

1827. *Euclid's Elements of Plane Geometry; with Copious Notes, Explanatory, Corrective, and Supplementary: to which is Subjoined a Brief Introduction to Plane Trigonometry. With a Trigonometrical Table.* Edited by John Walker. London: Printed by Bradbury and Co. pp. viii + 284 + 9 plates.

◇   Numerous "inaccuracies" in the plates are listed on pp. 283–284. A list of six other works by the same author is on p. [285]. Page [287] contains an "Explanation of Symbols and Abbreviations." There is an Appendix (pp. 271–282) which contains pedagogical issues and discusses a controversial review of the author's work (see the Monthly Review, March 1823).

1845. *Euclid's Elements of Geometry, Chiefly from the Text of Dr. Simson, with Explanatory Notes; Together with a Selection of Geometrical Exercises from the Senate–House and College Examination Papers; to which is Prefixed an Introduction, Containing a brief Outline of the History of Geometry. Designed for the Use of the Higher Forms of Public Schools and Students in the Universities. By Robert Potts, M.A. Trinity College.* Cambridge: Printed at the University Press. pp. vi + xl + 383.

◇   A page of errata precedes the second p. i. The "brief outline" of the history of geometry extends over pp. i–xxxix.

Euclid (Continued)

1860. *Les trois livres de porismes d'Euclide, rétablis pour la première fois, d'après la notice et les lemmes de Pappus, et conformément au sentiment de R. Simson sur la formé des Énoncés ces propositions.* Paris: Mallet–Bachelier. pp. ix + 324.
⋄ Errata on p. 324.

1861, [1845]. *Elements of Geometry Containing the First Six Books Of Euclid with a Supplement on the Quadrature of the Circles, and the Geometry of Solids to which are Added Elements of Plane and Sphericale Trigonometry.* Dean's Stereotype Edition, From the last London edition, enlarged. Philadelphia: J.B. Lippincott and Co. pp. [iv] + 318.
○ Red/Blue annotations on several pages. Signed by Alfred M. Mayer, 1 Nov 1866.

1869. *The Elements of Euclid for the Use of Schools and Colleges; Comprising the First Six Books and Portions of the Eleventh and Twelfth Books; with Notes, an Appendix, and Exercises.* Translated by Isaac Todhunter. New edition. London and Cambridge: Macmillan and Co. pp. xvi + 400.

1887. *The First Six Books of the Elements of Euclid, and Propositions I.–XXI. of Book XI, and an Appendix on the Cylinder, Sphere, Cone, etc., with Copious Annotations and Numerous Exercises. by John Casey.* Fifth edition, revised and enlarged. Dublin: Hodges, Figgis, & Co. pp. xi + 315 + 12 pages of "Opinions of the Work."

1908. *The Thirteen Books of Euclid's Elements, tr. from the Text of Heiberg, with Introduction and Commentary by T. L. Heath.* Cambridge: The University Press. 3 vols. pp. xi + 432; 436; 546.

Euclid. See also Bland 1842; Bonnycastle 1818b; Casey 1888; Chasles 1860; Cresswell 1817; Emerson 1794; Kepler 1653; Keith 1814; Ludlam 1809; Maseres 1791; Proclus 1792; Simpson 1821; Watts 1716; Whewell 1838.

Euler, Johann Albrecht (1734–1800). See Leonhard Euler 1772.

Euler, Leonhard (1707–1783).

1744. *Methodus inveniendi lineas curvas maximi minimive proprietate gaudentes, sive solutio problematis isoperimetrici latissimo sensu accepti.* Lausannae [Lausanne] & Genevae [Geneva]: Marcum–Michaelem Bousquet & Socios. pp. 322 + 5 plates.
○ See Figure 17 for photograph of title page of this volume.

1749. *Scientia navalis seu tractatus de construendis ac dirigendis navibvs.* The two volumes have separate subtitles: (1) *Pars prior complectens theoriam universam de situ ac motu corporum aquae innatantium.* (2) *Pars posterior in qua rationes ac praecepta navium construendarum et gubernandarum fusius exponuntur.* Each volume carries the comment *Istar supplementi ad tom. I. novorum commentar. acad. scient. imper.* Petropoli [St. Petersburg]: Typis academiae scientiarum. pp. 444 + 33 plates; 534 + 28 plates.

Euler, Leonhard (Continued)

1768. *Institutionum calculi integralis volumen primum in quo methodus integrandi a primis principiis usque ad integrationem aequationum differentialium primi gradus pertractatur.* Petropoli [St. Petersburg]: Impensis Academae Scientiarum. pp. [iv] + 542.

1769. *Institutionum calculi integralis volumen secundum in quo methodus inveniendi functiones unius variablis ex data relatione differentialium secundi altiorisue gradus pertractatur.* Petropoli [St. Petersburg]: Impensis Academiae Imperialis Scientiarum. pp. [iv] + 526 + 8 pages of errata.

1770. *Institutionum calculi integralis volumen tertium, in quo methodus inveniendi functiones duarum et plurium variabilium, ex data relatione differentialium cuiusvis gradus pertractatur. Una cum appendice de calculo variationum et supplemento, evolutionem casuum prorsus singularium circa integrationem aequationum differentialium continente.* Petropoli [St. Petersburg]: Impensis Academiae Imperialis Scientiarum. pp. [viii] + 639.

1772. *Theoria motuum lunae, nova methodo pertractata una cum tabulis astronomicis, unde ad quoduis tempus loca lunae expedite computari possunt, incredibili studio atque indefesso labore trium academicorum: Johannis Alberti Euler, Wolfgangi Ludovici Krafft, Johannis Andreae Lexell. Opus dirigente Leonhardo Eulero.* Petropoli [St. Petersberg]: Typis Academiae Imperialis Scientiorum. pp. [xvi] + 775 + 1 plate.

1781. *Theoria der Planeten und Cometen. Von Johann Freyherrn von Paccassi übersezt und mit einem Anhange und Tafeln vermehrt.* Wien: Gedruckt bey J. T. Edlen von Trattnern. pp. [8] + 230 + 3 folded leaves of plates.

1788; 1788; 1791. *Einleitung in die Analysis des Unendlichen. Aus dem Lateinischen übersetzt und mit Anmerkungen und Zusätzen begleitet von Johann Andreas Christian Michelsen.* 3 vols. Berlin: Carl Massdorff. pp. xxiv + 3–626; viii + 2–578 + 8 plates; [iv] + 3–530 + 1 plate.

◇   Errata for volume 1 follow p. 626 of that volume. It is often incorrectly stated that this work contains no figures. Yet at the end of volume 2 there are 149 figures on 8 plates.

◇   Translation of Euler's *Introductio in analysin infinitorium* of 1748.

1790; 1790; 1793. *Vollständige Anleitung zur Differenzial-Rechnung. Aus dem Lateinischen übersetzt und mit Anmerkungen und Zusätzen begleitet von Johann Andreas Christian Michelsen.* 3 vols. Berlin und Libau: Lagarde und Friedrich, (volume 3 is published just by Lagarde in Berlin). pp. lxxx + 400; [vi] + 342; [vi] + 322.

◇   There was another edition of this translation in 1798, reprinted by LTR–Verlag, Wiesbaden in 1981. This is a translation of *Institutiones calculi differentialis* of 1755.

Euler, Leonhard (Continued)

1796; 1797. *Introduction à l'analyse infinitésimale, par Léonard Euler; tranduite du Latin en Français, avec des notes & des éclaircissements, par J. B. Labey.* 2 vols. Translated by J. B. Labey. Paris: Barrois. pp. xiv + 364 + 1 fold out table; [x]+ 424 + 16 plates.

◇ Errata from volume 1 follows p. xiv and is repeated in volume 2 on pp. [v]–[vi], and for volume 2 on pp. [ix]–[x]. Volume 2 is dedicated by Labey to Bonaparte. The "notes and explanations" are on pp. 305–364 of volume 1 and pp. 404–424 of volume 2.

◇ See Figure 25 for a photograph of the "Thayer binding" of this copy.

1807. *Élémens d'algèbre, par Léonard Euler, traduits de l'Allemand. Nouvelle édition, revue et augmentée de notes par J. G. Garnier.* 2 vols. Paris: Courcier; Marie. pp. xiv + 563; ii + 485.

◇ Errata for volume 1 follow p. 563. Volume 1 is subtitled "Analyse déterminée," and volume 2 "Analyse indéterminée." The "Notes and Additions" by Garnier occupy pp. 449–563 of volume 1. The "Additions" on pp. 281–495 of volume 2 are anonymous, but are those of Lagrange.

○ Thayer binding.

1812. *Lettres a une princesse d'allemagne, sur divers sujets de physique et de philosophie.* 2 vols. Nouvelle édition, conforme à l'édition originale de l'Académie des sciences de St.–Pétersbourg, revue et augmentée de diverses notes par J.–B. Labey, et précédée de l'éloge d'Euler par de Condorcet. Paris: Ve Courcier; Bacheleir. pp. lviii + 530 + 3 plates; 598 + 6 plates.

◇ Frontispiece portrait of Euler, volume 1. The "Eloge" of Euler by Condorcet is in volume 1, pp. ix–lviii. Errata of volume 1 follow p. 530, and of volume 2 follow p. 598.

○ Thayer binding.

1822. *Elements of Algebra, by Leonard Euler, translated from the French: with the Notes of M. Bernoulli, and the Additions of M. de La Grange. By the Rev. John Hewlett. To which is Prefixed a Memoir of the Life and Character of Euler, by the Late Francis Horner.* Third edition, carefully revised and corrected. London: Printed for Longman, Hurst, Rees, Orme, and Company. pp. xxx + 593.

1835. *Letters of Euler on Different Subjects in Natural Philosophy. Addressed to a German Princess. With Notes, and a Life of Euler, by David Brewster, Containing a Glossary of Scientific Terms, with Additional Notes by John Griscom.* 2 vols. New York: Harper and Brothers. pp. 386 + 1 plate; 436.

◇ Frontispiece is a view of the solar system as it was known in 1835. There is a glossary in volume II, pp. 423–436.

Farrar, John (1779–1853).

1822. *An Elementary Treatise on the Application of Trigonometry: to Orthographic and Stereographic Projection, Dialling, Mensuration of Heights and Distances, Navigation, Nautical Astronomy, Surveying and Levelling; Together with Logarithmic and Other Tables; Designed for the Use of the Students of the University at Cambridge, New England.* Cambridge, New England: Printed at the University Press by Hilliard and Metcalf. pp. vi + 153 + tables + 9 plates.

⋄   Author used Cagnoli 1808a and Bonnycastle 1818a in preparing this work.

○   Stamped on the title page: "textbook U.S.M.A. 1824 to 18__."

1825. *An Elementary Treatise on Mechanics, Comprehending the Doctrine of Equilibrium and Motion, as Applied to Solids and Fluids, Chiefly Compiled, and Designed for the Use of Students of the University at Cambridge, New England.* Boston: Hilliard and Metcalf. pp. vii + 440 + 10 plates. Errata on pp. 439–440.

⋄   Based on the works of Biot, Bezout, Poisson, Francoeur, Gregory, Whewell, and Leslie.

1827. *An Elementary Treatise on Astronomy, Adapted to the Present Improved State of the Science Being the Fourth Part of a Course of Natural Philosophy, Compiled for the Use of the Students of the University at Cambridge, New England.* Cambridge, N.E.: Hilliard, Metcalf, & Co. pp. vii + 420 + 7 plates.

⋄   Errata on p. [viii]. "Selected from Biot's Traité Élémentaire d'Astronomie Physique, 2nd edition" (1811).

1828. *Elementary Treatise on the Applications of Trigonometry to Orthographic and Steriographic Projection, Dialling, Mensuration of Heights and Distances, Navigation, Nautical Astronomy, Surveying and Levelling; Together with Loarithmic and Other Tables; Designed for the Use of the Students of the University at Cambridge, New England.* Second edition. Boston: Hilliard, Gray, Little, and Wilkins. pp. viii + 155 + tables + 9 plates.

⋄   An earlier edition of this book was used as a text at the Academy starting in 1824.

○   From the heirs of GEN J. S. Swift and contains a USMA bookplate.

Farrar, John.   See also Lacroix 1820c; Lacroix 1825b; Lacroix 1826; Legrendre 1825.

Fathers of the Christian Schools (F. J.)

1886. *Compléments de trigonométrie et méthodes pour la résolution des problèmes.* Cours élémentarie de mathématiques. Tours: Alfred Mame & Fils; Paris: Poussielgue Frères. pp. xii and 838.

⋄   The book is in 2 parts – part one is textual material, pp. 1–491, and solutions to problems, pp. 492–838. The introduction is a brief history of trigonometry (pp. III–VIII). Errata follow p. 838.

⋄   These books are members of a series of elementary textbooks, from arithmetic to mechanics, and asserted to be the "Property of the Institute of the Fathers of the Christain Schools."

1893. *Exercises de géométrie descriptive.*   Troisième édition. Cours de mathématiques élémentaries. Tours: Alfred Mame & Fils; Paris: Ch Poussielgue. pp. [viii] + [952] + 5 plates.

1895. *Exercices d'algèbre.*   Quatrième édition. Cours de mathématiques élémentaries. Tours: Alfred Mame & Fils; Paris: Ch. Poussielgue. pp. [iii] + 976.

1896a. *Éléments de géométrie comprenant des notions sur les courbes usuelles un complément sur le déplacement des figures et de nombreux exercises.*   Neuvième édition. Cours de mathématiques élémentaries. Tours: Alfred Mame & Fils; Paris:

Charles Poussielgue. pp. [xi] + [523].

◊ This is a textbook containing the material in geometry required for the baccalaurate in science and the "special" baccalaurate at this time in France.

1896b. *Exercises de géométrie comprenant l'exposé des méthodes géométriques et 2000 questions résolues.* Troisième édition. Cours de mathématiques élémentaries. Tours: Alfred Mame & Fils; Paris: Ch. Poussielgue. pp. [xiv] + [1135].

1910. *Éléments de géométrie descriptive avec de nombreux exercises.* Cours de mathématiques élémentaries. Tours: A. Mame & Fils; Paris: Ch. Poussielgue. pp. 458.

○ Copy has been rebound.
○ Copy signed by C. P. Echols.

n.d. *Éléments d'algébre avec de nombreux exercises.* Neuvième édition. Cours de mathématiques élémentaires. Tours: Alfred Mame & Fils; Paris: Ch. Poussielgue. pp. [viii] + [409].

◊ This is a textbook containing the material in algebra required for the baccalaureate in science and the special diploma for students of secondary education.

Fenn, Joseph (fl. 1768–1769).

n.d. *New and Complete System of Algebra: or Specious Arithmetic. Comprehending all the Fundamental Rules and Operations of that Science, Clearly Explained and Demonstrated; with the Resolution of all Kinds of Equations. Illustrated and Exemplified in the Solution of a Vast Variety of the Most Curious and Interesting Questions. For the Use of Schools.* Dublin: Printed by Alex McCulloh. pp. 306 + [x].

○ This book was Jared Mansfield's copy.

Fenwick, Stephen. See Royal Military Academy, Woolwich 1853; 1852.

Ferguson, James (1710–1776).

1771. *Tables and Tracts Relative to Several Arts and Sciences.* The second edition, with additions. London: Printed for W. Strahan, et. al. pp. xv + 334 + 1 page of advertisements of other books by Ferguson + 3 plates.

○ Two copies in library.
○ One copy has two bookplates: F. S. T. Golden Branch, Phillips Exeter Academy; and Gouverneur Kemble Warren (1830–1882).

1773. *Lectures on Select Subjects in Mechanics, Hydrostatics, Pneumatics, and Optics. With the Use of the Globes, the Art of Dialing, and the Calculation of the Mean Times of the New and Full Moons and Eclipses. With the Supplement.* New edition. London: Printed for W. Strahan, et. al. pp. viii + 252 + 23 plates + [iv] + 40 + 13 plates.

◊ Dedicated to Prince Edward. There is an index on pages [i]–[iv].
○ This copy owned by at least two West Point cadets, who used the book in 1811 and 1817.

Fermat, Pierre de (1601–1665).
1679.   *Varia opera mathematica D. Petri de Fermat, senatoris Tolosani. Accesserunt selectae quaedam ejusdem epistolae, vel ad ipsum à plerisque doctissimis viris Gallicè, Latinè, vel Italicè, de rebus ad mathematicas disciplinas, aut physicam pertinentibus scriptae.*   Edited by Samuel de Fermat. Tolosae [Toulouse]: Apud Joannem Pech. pp. [xii] + 210 + [iii] + 5 plates.
  ◇   Dedicated to Bishop Ferdinand Furstenberg.
  ○   Bookplate of Philip Earl Stanhope inside front cover.

1891.   *Oeuvres de Fermat, publiées par les soins de MM. Paul Tannery et Charles Henry sous les auspices du Ministère de l'Instruction publique.   Tome premier. Oevres mathématiques diverses – observations sur Diophante.*   Volume 1 (only) of 3. Paris: Gauthier–Villars. pp. xxvii + 440.
  ◇   Contains portrait of Fermat inside cover.
  ◇   The title page of {Fermat 1679} is reproduced here, but uses a different printer's device.

Fermat, Pierre de.   See also Diophantus 1670.

Ferroni, Pietro (1744–1825).
1782.   *Magnitudinum exponentialium logarithmorum et trigonometriae sublimis theoria nova methodo pertractata.*   Florentiae [Florence]: Ex typographia Allegriniana. pp. lxvi + 612 + 2 plates.
  ◇   Dedicated to Petro Leopoldo, Archduke of Austria.
  ◇   A history of logarithms is given in pp. xi–lxvi.

Fiedler, Wilhelm (1832–1912).
1883, 1885, 1888.   *Die darstellende Geometrie in organischer verbindung mit der geometrie der Lage.*   3 vols. Third edition. Leipzig: B. G. Teubner. pp. xxvi + 376 + 6 plates; xxxiii + 560 + 16 plates; xxx + 660 + 1 plate.
  ◇   Corrections on p. x of Volume 1. The three volumes have different title pages, Volume 1: *Die methoden der darstellenden und die elemente projectivischen geometerie fur vorlesungen und zum selbststudium;* Volume 2: *Die darstellende geometrie der krummen linien und flachen fur vorlesungen und zum selbststudium;* Volume 3: *Die construierende und analytische geometrie der lage fur vorlesungen und zum selbststudium.*

Finck, Pierre Joseph Étienne.
1834.   *Traité élémentaire d'analyse infinitésimale. Première partie, comprenant le calcul différentiel et ses principales applications géométriques.*   Paris: Bachelier. pp. xi + 271 + [1].
  ◇   Five unnumbered pages of errata are placed between the Table of Contents and page 1.
  ◇   There is an unnumbered page of notes following p. 271.

Fine, Henry Burchard (1858–1928).
1891. *The Number–System of Algebra Treated Theoretically and Historically.* Boston: Leach, Shewell & Sanborn. pp. ix + 131.

Fine, Henry Burchard (1858–1928), and Thompson, Henry Dallas (1864–1927).
1916. *Coordinate Geometry.* New York: MacMillan. pp. [ix] + 300 + 9 figures + 5 pages of advertisements.
◇ The figures are photographs of plaster and string models, probably of Brill. Earlier editions of this book were used as a text at the Academy in 1911 and 1912.
◇ Copyright 1907.
○ The library has two copies.
○ One copy is stamped "M. C. Grenata." It is a clean copy.
○ The second copy differs from the first in that the publisher's name and address are not printed on the page facing the title page.

Fink, Karl (1851–1898).
1903. *A Brief History of Mathematics. An Authorized Translation of Dr. Karl Fink's Geschichte der Elementar–Mathematick.* Translated by Wooster Woodruff Beman and David Eugene Smith. Second, revised edition. Chicago: Open Court. pp. xii + 333.

Fischer, William Lewis Ferdinand. See Vega 1857; Vega 1859; Vega 1883.

Fisher, George (ca. 1675–ca. 1732).
1780. *Arithmetic in the Plainest and most Concise Methods Hitherto Extant. With new Improvements for the Dispatch of Business in all the Several Rules. As also Fractions, Vulgar and Decimal, Wrought Together after a new Method, that Renders Both Easy to be Understood in Their Nature and Use. The Whole Pers'd and Approved of, by the Most Eminent Accomptants in the Several Offices of the Revenue, viz. Customs, Excise, etc. as the only Work of its Kind, for Variety of Rules and Brevity of Work.* With considerable Additions, and curious Improvements, by the Author. Paisley: Printed by Alexander Weir. pp. [x] + 11–334.
◇ Dedicated to Sir Samuel Clarke.
○ Gift of COL C. W. Larned.

Fisher, George E. (1863–1920), and Schwatt, Isaac J. (1867–?).
1906. *Complete Secondary Algebra. Quadratics and Beyond.* New York: The Macmillan Co. pp. 277–564 + xviii.
○ Contains some marginal notes.

Fisher, Irving (1867–1947). See Phillips and Fisher 1896; Phillips and Fisher 1898.

F.J. See Fathers of the Christian Schools.

Flamsteed, John (1646–1719).
1725. *Historiae coelestis Britannicae, tribus voluminibus contenta.* Three different volumes bound together with distinct titles. The first is entitled *Historiae coelestis Britannicae volumen primum. Complectens stellarum fixarum nec non planetarum omnium observationes sextante, micrometro, &c. peractas. Quibus subjuncta sunt planetarum loca ab iisdem observationibus deducta. Observante Joanne Flamsteedio, A.R. in Observatorio Regio Grenovicensi continua serie ab anno 1675, ad annum 1689.* Londini [London]: Typis H. Meere. pp. [12] + 40 + 396 + 2 pages of errata + several plates. The second volume is entitled: *Historiae coelestis Britannicae volumen secundum. Exhibens fixarum stellarum planetarumque omnium transitus per planum arcus meridionalis, et distantias eorum a vertice. Nec non de Solis, Lunae, Jovisque satellitum eclipsibus observationes arcu meridionali, aliisq; idoneis instrumentis, habitas. Quibus adnectuntur planetarum loca ab iisdem observationibus derivata. Observante Joanne Flamsteedio, A.R. in Observatorio Regio Grenovicensi. Ab anno 1689, ad annum 1720.* Londini [London]: Typis H. Meere. pp. 573 + 70. The third volume is entitled: *Historiae coelestis Britannicae volumen tertium. Complectens praefationem spatiosam (sive in stellarum fixarum catalogum prolegomena) quae brevem astronomiae historiam praebet, atque descriptionem observationum peractarum, & organorum adhibitorum tum a pristinis astronomis, tum in Observatorio Regio Grenovicensi. Deinceps fixarum catalogum a Ptolemaeo, Uleg Beig, Tychone Brahaeo, Gulielmo Hessiae Landtgravio, ac Hevelio constructum. Demum novum ac amplissimum catalogum, rectarum adscensionum, distantiarum a polo, longitudinum, latitudinum, & stellarum fixarum magnitudinum: una cum variationibus rectarum adscensionum, & polarium distantiarum, dum suas mutant longitudines uno gradu. Quibus adnexus est fixarum quarundam Australium catalogus, in nostro hemisphaerio non adspectabilium. Denique tabulae amplissimae ad inveniendas fixarum, & planetarum longitudines, & latitudines per inspectionem, datis, duntaxat, earum rectis adscensionibus, ac distantiis a polo. A Joanne Flamsteedio, A.R.* Londini [London]: Typis H. Meere. pp. 164 + [2 plates facing p. 100 and p. 104] + 76 + 83 + 103.
◇   Errata for each volume are printed on the last unnumbered page of the book.
◇   Portrait of Flamsteed on frontispiece.

1776. *Atlas céleste de Flamstéed, approuvé par l'Académie royal des sciences, et publié sous le privilege de cette compagnie.* Paris: Chez F. G. Deschamps [et chez] l'auteur. Edited by J. Fortin. pp. viii + 40 + 30 plates.

Fontenelle, Bernard le Bouyer de (1657–1757). See L'Hospital 1781.

Foster, Samuel (? –1653). See Gunter 1653.

Fourier, Jean Baptiste Joseph (1768–1830).
1822. *Théorie analytique de la chaleur.*   Paris: Didot, Père et Fils. pp. xxii + 639 + 2 plates.

1835. *Théorie analytique de la chaleur.* Paris: Bachelier. pp. xxii + 639 + 2 plates.
- ◇ First published in 1822.
- ◇ Fourier intended to publish a sequel to this book, but was unable to do so.

1888. *Oeuvres de Fourier.* Volume 1 only. Publiées par les soins de Gaston Darboux. Paris: Gauthier–Villars et Fils. pp. xxviii + 563.
- ◇ Frontispiece portrait of Fourier.
- ◇ The 1822 edition of Fourier's *Analytic Theory of Heat* comprises the whole volume.

Fournié, Victor.

1835. *Programme détaillé de toutes les parties des mathématiques exigées pour l'admission à l'École polytechnique.* Deuxième édition. Paris: Bachelier. pp. [vi] + 223 + 6.
- ◇ Dedicated to the author's students.

Français, Jacques Frédéric (1775–1883).

1813. *Mémoir sur le mouvement de rotation d'un corps solide libre, autour de son centre de masse.* Paris: VeCourcier. pp. 56.
- ◇ Errata on p. 50.
- ◇ There is a separate part entitled "Supplément au mémoire précédent" on pp. 51–56.

Francoeur, Louis Benjamin (1773–1849).

1807. *Traité élémentaire de mécanique, adopté dans l'instruction publique.* Quadrième édition. Paris: Bernard. pp. xii + 508 + 9 plates.
- ◇ Dedicated to Laplace.
- ◇ There is an appendix treating calculus of variations (pp. 485–505).
- ◇ This book was used as a textbook at USMA in the French form according to R. Ernest Dupuy.

1825. *Traité de mécanique élémentaire.* Cinquième édition. Paris: Bachelier. pp. xiv + 524 + 4 + 9 plates.
- ◇ Dedicated to Laplace.
- ○ Donated by William C. Trimble.

1829. *Complete Course of Pure Mathematics.* 2 vols. Translated by R. Blakelock. Cambridge: W. P. Grant. pp. iv + 431 + 6 plates + [ix]; v + 480 + 2 plates.
- ◇ At the end of volume 1, there is an alphabetical index for both volumes.

1837. *Cours complet de mathématiques pures, ouvrage destiné aux élèves des Écoles normale et polytechnique et aux candidats qui se préparent a y ètre admis.* 2 vols. Quatrième édition, revue et augmentée. Paris: Bachelier. pp. x + 505 + 8 plates; 609 + 3 plates.
- ◇ Errata follow p. x in volume 1 and p. 609 in volume 2.

Frend, William (1757–1841).
1796. *The Principles of Algebra.* London: Printed by J. Davis. pp. xvi + 518 + 1 plate.

◇ Frontispiece. Errata on p. xvi.

◇ This volume also contains "An Appendix, by Francis Maseres," is on pp. 211–456; "Observations on Mr. Raphson's Method of Resolving affected Equations of all degrees by Approximation," is on pp. 457–501; "An Explication of Simon Stevin's General Rule, to extract one root out of any possible equation in numbers, either exactly, or very nearly true. By John Kersey. Being the 10th chapter of the second book of Mr. Kersey's Elements of Algebra," is on pp. 502–518.

◇ Contains a few historical remarks.

Frezier, Amédée François (1682–1773).
1737; 1738; 1739. *La theorie et la pratique de la coupe des pierres et des bois, pour la construction des voutes et autres parties des bâtimens civils & militaires, ou traité de stereotomie a l'usage de l'architecture.* 3 vols. Strasbourg: Jean Daniel Doulsseker. pp. [xiii] + xvi + 424 + 27 plates; [xx] + 503 + plates; [xvi] + 417 + 65 + plates.

◇ Frontispiece. Dedicated to the Marquis d'Asfeld. Errata for volume 1 are on p. [xiii], for volume 2 are on pp. [vii]–[xi], and for volume 3 are on pp. [xiii]–[xvi] and following p. 65. There is a glossary in volume 1, pp. 389–410.

○ Thayer binding.

1738. *Dissertation sur les ordres d'architecture.* Strasbourg: Jean–Daniel Doulsseker le Fils. pp. 65 + 114 plates + 2 pp. errata.

◇ Bound with Frezier 1737, vol. 3.

Frost, Percival (1817–1898).
1886. *Solid Geometry.* London: MacMillan and Company. pp. xv + 408 + 3.

Fryer, Michael (1774?–1844). See Ludlam 1809.

Furlong, Lawrence (1734–1806).
1806. *The American Coast Pilot.* Fifth edition. Newburyport. pp. xvi + [17]–172 + [8] + [177]–239.

Galbraith, William (1786–1850).
1827. *Mathematical and Astronomical Tables, for the Use of Students of Mathematics, Practical Astronomers, Surveyors, Engineers, and Navigators; with an Introduction, Containing the Explanation and Use of the Tables, Illustrated by Numerous Examples.* Edinburgh: Oliver and Boyd. pp. xii + 168 + 42 + 112 pages of tables.

◇ Dedicated to Sir George Clerk. Errata immediately precede the tables.

Galileo, Galilei (1564–1642).

1653. *Sidereus nuncius, magma longeque admirabilia spectacula pandens, suspiciendaque proponens unicuique, prae sertim vero philosophis atque astronomis, quae à Galileo Galileo, Patritio Florentino, Patavini Gymnasii Publico Mathematico, Perspicilli nuper à se reperti beneficio, sunt observata in lunae facie fixis innumeris, lacteo circulo, stellis nebulosis, apprime vero-in quatuor planetis circa jovis stellam disparibus intervallis atque periodis celeritate mirabili circumvolutis; quos, nemini in hanc usque diem cognitos, novissime autor deprehendid primus, atque medicea sidera nuncupandos decrevit.* Londini [London]: Typis Jacobi Flesher: Prostat apud Cornelium Bee, in vic. vulg. voc. Little-Britain. pp. 50 + 4 plates.

◇ Dedication to Cosmo II Medici is dated 1610. This is the second part of Gassendi 1653.

1718. *Opere di Galileo Galilei.* Firenze, Nella Stamp. di S.A.R. per Gio: Gaetano Tartini, e Santi Franchi. Nuova edizione coll'aggiunta di varj trattati dell'istesso autore non più dati alle stampe. 3 vols.

1880. *The Sidereal Messenger of Galileo Galilei: and a Part of the Preface to Kepler's Dioptrics Containing the Original Account of Galileo's Astronomical Discoveries.* London: Rivingtons. A translation with introduction and notes by Edward Stafford Carlos. pp. xi + 111.

Gardner, William. See Sherwin 1761.

Garnier, Jean–Guillaume (1766–1840).

1811a. *Leçons de calcul différentiel.* Troisième édition. Paris: Ve Courcier. pp. xxvi + 474 + 4 plates.

◇ Errata on pp. xxiv–xxvi.

1811b. *Élémens d'algèbre a l'usage des aspirans a l'École polytechnique.* Première section. Troisième édition. Paris: Courcier. pp. xxviii + 508 + 1 plate.

1812. *Leçons de calcul intégral.* Troisième édition. Paris: Ve Courcier. pp. xi + 643 + 2 plates.

◇ Errata following p. 643.

1813. *Géométrie analytique ou application de l'algèbre a la géométrie.* Deuxième édition. Paris: Ve Courcier. pp. vi + 448 + 14 plates.

◇ Errata follow p. 448.

○ Used as a text at USMA from 1823 to 1824.

○ Thayer binding.

1814. *Analyse algébrique, faisant suite a la première section de l'algèbre.* Deuxième édition, revue et considérablement augmentée. Paris: Ve Courcier. pp. xvi + 688 + 1 plate.

◇ A fold–out faces p. 567.

◇ A list of works by the same author is on p. [ii].

Garnier, Jean–Guillaume. See also Azemar 1809; Clairaut 1801; Leonhard Euler 1807.

Gassendi, Pierre (1592–1655).
1653a. *Petri Gassendi institutio astonomica, juxta hypotheses tam veterum quam recentiorum. Cui acceserunt Galilei Galilei nunctius sidereus, et Johannis Kepleri dioprtrice.* Secunda editio priori correctior. Londini [London]: Typis Jacobi Flesher. Prostant apud Gulielmum Morden, Bibliopolam Cantabrigiensem. pp. [xvi] + 199; 50 + 4 plates; 51-173.
◇ Contains Gassendi 1653b, Galileo 1653, and Kepler 1653, each with a separate title page.
◇ Errata on p. [xvi].

1653b. *Institutio astronomica, juxta hypotheseis tam veterum, quam Copernici et Tychonis, dictata à Petro Gassendo Regio Matheseos Professore.* Ejusdem oratio inauguralis iteratò edita. Londini [London]: Typis Jacobi Flesher. London: Flesher. [xvi] + 199.
◇ This is the first item in Gassendi 1753a. Dedicated to Cardinal Ludovico Alphonso.
◇ The "Oratio inauguralis habita in regio collegio" that Gassendi gave on November 23 is on pp. 179–199.

Gaubil, Antoine (1689–1759). See Souciet 1729.

Gaudin, Antoine Pierre.
1825. *Développement d'une pensée de d'Alembert ou introduction a l'application de l'algèbre a la géométrie.* Paris: Bachelier. pp. 34 + 1 plate.
○ The author's initials appear handwritten on p. 2.

Gauss, Carl Friedrich (1777–1855).
1807. *Recherches arthimétiques.* Translated by A.–C.–M. Poullet–Delisle. Paris: Courcier. pp. xx + 502.
◇ Dedicated to Laplace by the translator.
◇ Dedicated by the author to Charles–William Ferdinand, the Duke of Brunswick.
◇ Errata follow p. xx.
◇ This is the first edition of the first translation, into any other language, of Gauss' famous *Disquisitions Arithmeticae* (1801).

1809. *Theoria motus corporum coelestium in sectionibus conicis solem ambientium.* Hamburgi [Hamburg]: Frid. Perthes et I. H. Besser. pp. xii + 227 + 1 plate + 20 pages of tables.
◇ Errata on one page following p. 227.

1855. *Méthode des moindres carrés. Mémoires sur la combinaison des observations.* Translated by J. Bertrand. Paris: Mallet–Bachelier. pp. 167.

◇ This is a collection of excerpts from previously published works of Gauss on least squares.

○ Signed "Dept. of Mathematics–West Point, NY, Feb. 1881".

1857. *Theory of the Motion of the Heavenly Bodies Moving about the Sun in Conic Sections: A Translation of Gauss's "Theoria Motus." With an Appendix by Charles Henry Davis.* Boston: Little, Brown and Company. pp. xvii + 326 + 39 pages of tables + 1 page of constants + 8 plates.

◇ The appendix includes pp. 279–326. Published under the authority of the Navy Department by the Nautical Almanac and Smithsonian Institution.

Gellibrand, Henry (1597–1636). See Briggs 1633.

Gentry, (Abbe).

1784. *L'influence de Fermat sur son siecle, relativement aux progrès de la haute géométrie & du calcul, & l'avantage que les mathématiques ont retiré depuis & peuvent retirer encore de ses ouvrages; discours qui a remporté le prix double à l'Académie royale des sciences, inscriptions & belles–lettres de Toulouse, en 1783.* Orléans: Couret de Villeneuve. pp. vii + 144.

◇ Dedicated to Monseigneur de Jarente d'Orgeval.

Ghetaldi, Marino (ca. 1566–1626). See Apollonius 1764.

Gibbs, Josiah Willard (1839–1903). See Wilson 1902.

Gibson, Robert (1715–1761).

1808. *A Treatise of Practical Surveying; which is Demonstrated from its First Principles. Wherein Every Thing that is Useful and Curious in that Art, is fully Considered and Explained, Particularly Three New and very Concise Methods for Determining the Areas of Right–lined Figures, Arithmetically or by Calculation, as well as the Geometrical ones heretofore Treated of. The Whole Illustrated with Copper–plates.* Ninth edition, with alterations and amendments, adapted to the use of American surveyors. Philadelphia: Printed by Joseph Crukshank. pp. viii + 288 + 152 + 13 plates.

◇ The book also contains *Tables of Difference of Latitude and Departure: Constructed to every Quarter of a Degree of the Quadrant, and Continued from One, to the Distance of One Hundred Miles or Chains:– Of Logarithms from 1 to 10,000:– and of Artificial Sines, Tangents, and Secants,* Carefully revised and corrected, Philadelphia, 1803.

◇ This is an American edition of an English work; popular in the United States until the early 1900's.

○ The title page signed by "Tho. Ragland, July 28th 1816."

Girard, Albert (1595–1632). See Stevin 1634.

Girard, Pierre Simon (1765–1836).
1798 (An vi). *Traité analytique de la résistance des solides, et des solides d'égale résistance. Auquel on a joint une suite de nouvelles expériences sur la force, et l'élasticité spécifiques des bois de chêne et de sapin.* Paris: Didot; DuPont. pp. lv + 238 + 48 pages of tables + 9 plates.
⋄   Errata on p. [239].

Gompertz, Benjamin (1779–1865).
1817; 1818. *The Principles and Application of Imaginary Quantities, Book I; to which are Added Some Observations on Porisms; being the First of a Series of Original Tracts on Various Parts of the Mathematics* and *The Principles and Application of Imaginary Quantities, Book II; Derived from a Particular Case of Functional Projections: being the Second of a Series of Original Tracts on Various Parts of the Mathematics.* London: Scientific Press. Printed and Published for the Author. pp. viii + 9–35; x + 11–44.
⋄   The two books are bound together in one volume. Errata for Book I are on p. 35 and on p. iv of Book II. Errata for Book II are on p. 44.

Goodman, John. See Barker 1902.

Gordon, George. See Ward 1730.

Gordon, William Brandon (1853–1938; USMA 1877).
1914a. *A Text-book on Mechanics.* West Point, NY: United States Military Academy Printing Office. pp. vii + 276.
⋄   Author was Professor of Natural and Experiemental Philosophy at USMA from 1901 to 1917.
○   Gift of COL G. Gordon Bartlett, Jr.

1914b. *A Text-book on Mechanics.* West Point, NY: United States Military Academy Printing Office. pp. 247 columns.
⋄   No title page. Different book than Gordon 1914a. Each page contains 2 columns (recto) and 2 columns (verso).
○   Two copies. One copy was the property of William Cooper Foote. A few marginal notes.
○   The second copy has typewritten correction, with page references, on loose sheets stapled onto the front page of the book.

Gordon, William Brandon. See also Young, Gordon 1912.

Gouré, Edouard.
1842. *Théories générales de géométrie analytique. Appliquées à la discussion des*

courbes algébriques de degrés supérieurs, a l'usage des candidats a l'École polytechnique. Paris: Bachelier. pp. 88 + 2 plates.
- ○ Copy presented by the author to USMA.

Gournerie, Jules de la (1814–1883). See La Gournerie, Jules de 1859.

Goursat, Edouard (1858–1936).
1902; 1905. *Cours d'analyse mathématique.* 2 vols.   Cours de la faculté des sciences de Paris. Tome 1. *Dérivées et différentielles, intégrales définies, – développments en séries, applications géométriques.* Tome 2. *Théorie des fonctions analytiques, – équations différentielles, équations aux dérivées partielles, éléments du calcul des variations.* Paris: Gauthier–Villars. pp. vi + 620; vi + 640.

Gow, James (1854–1923).
1884. *A Short History of Greek Mathematics.* Cambridge: University Press. pp. xvi + 323.

Grandi, Guido (1671–1742). See Huygens 1728.

Granville, William Anthony (1863–1943) and Smith, Percey Franklyn (1867–1956).
1904. *Elements of the Differential and Integral Calculus.* New York: Ginn. pp. xiv + 463.
- ◇ This book was used as a text at USMA from 1907 until 1913.
- ○ The library has two copies.
- ○ One copy is a clean copy.
- ○ Second copy has marginal notes.

1911. *Elements of the Differential and Integral Calculus.* Revised edition [Second]. Boston: Ginn and Co. pp. xv + 463.
- ◇ Portraits of Newton (frontispiece) and Leibniz (facing p. 1). Many changes from the first edition.
- ○ The library has two copies.
- ○ One copy has book plates and was a gift from Mr. Arthur R. Hercz. Homework is inserted and corrections are pasted in (throughout). Correction booklet (in back). There are problem sets written inside the front cover, notes in pencil throughout, and problems marked for the 1928 and 1929 "turnout" exams. The course outline is on the back cover.
- ○ Second copy is clean.

Gratry, J.–B.
1855a. *Cours de perspective linéaire, a l'usage des artists, des peintres, des architectes, des graveurs, des décorateurs, et de toutes les personnes qui s'occupent du dessin.* Bruxelles [Brussels]: C. Muquardt. pp. vi + 69.
- ◇ Dedicated to Charles van den Berghen. Reference is made to 88 figures, but none are included in this book. They are bound in Gratry 1855b.
- ○ Book received at USMA Jan 11, 1886.

1855b. *Cours de perspective linéaire, planches.* Bruxelles [Brussels]: C. Muquardt. pp. 8 plates.

Graves, Robert Perceval (1810–1893).
1882. *Life of Sir William Rowan Hamilton including Selections from his Poems, Correspondence, and Miscellaneous Writings.* 3 vols. Dublin University Press Series. Dublin: Hodges, Figgis, & Co. pp. xix + 698; xv + 1 errata + 719; xxxv + 2 errata + 673.
  ◊ Each volume has a portrait of Hamilton on the frontispiece.
  ○ The library has two copies of this book.

'sGravesande, Willem Jacob van (1688–1742).
1784. *Physices elementa mathematica, experimentis confirmata. Sive intruductio ad philosophiam Newtonianam.* Editio Quarta 2 vols. Leide [Leiden]: Fratres Murray.   pp. [4] + lxxxviii + 572 + 62 interspersed plates; pp. 573–1073 + 41 pages of "Index rerum" + plates 63–127 interspersed.
  ◊ Volume one contains the prefaces to the first edition (1719), pp. i–xi; the second edition, pp. xii–xiv; and the third edition xv–xxxviii. There is also the "Oratio de evidentia," pp. xxxix–lxi; and the "Monitum de demonstrationibus quae quantitates infinitè exiguas pro fundamento habent," pp. lxii–lxiv. Three pages of addenda, corrigenda and "in margine" follow the "Index rerum" at the end of volume 2.

'sGravesande, Willem Jacob van. See also Huygens 1728.

Gray, George John (1863–?).
1907. *A Bibliography of the Works of Sir Isaac Newton Together with a List of Books Illustrating His Works.* Second edition, revised and enlarged. Cambridge: Bowes and Bowes. pp. [viii] + 80.
  ◊ Frontispiece has a steel engraving (1838) of a statue of Newton by Roubiliac which is in the Chapel of Trinity College.

Green, George (1793–1841).
1828. *An Essay on the Application of Mathematical Analysis to the Theories of Electricity and Magnetism.* Nottingham: Printed for the Author. pp. ix + 64.
  ◊ Dedicated to the Duke of Newcastle. List of subscribers is on p. ix.
  ○ There are pages missing at the end of the work.

Gregorius a St. Vincentio (1584–1667). See Saint Vincent.

Gregory, David (1659–1708).
1726. *Astronomiae physicae & geometricae elementa.*   2 vols. Secunda editio revisa & correcta. Genevae [Geneva]: Marcum–Michaelem Bousquet. pp. [xx] + xcvi + 427 + 26 plates; [ii] + 429–751 + 75 appendices and index + [i] + xli tabs (plates).
  ◊ Contains a five page dedication.
  ◊ See Figure 13 for a photograph of the frontipiece of this volume.

○ Both volumes have been rebound.

○ The first volume has a large sketch in front. The name G. M. Lowiz is printed and stamped on the title page.

1769. *A Treatise of Practical Geometry: in Three Parts.* Translated from the Latin, with additions. Seventh Edition. Edinburgh: Printed by Balfour, Auld, and Smellie for John Balfour. pp. iv + 144 + 5 plates.

◇ The Latin original was about 60 years earlier.

○ Preface signed "Col. M'Laurin" and dated May 1, 1745.

○ The title page bears the signature of Alexander Robinson and the date 1774. The end papers indicate that the book was given to him by the Earl of Kinnoull in 1773. They also contain the inscriptions "James Anderson School Master in Carlisle" and "Military Academy '03". This book is listed in the 1810 catalog of the library and so is one of the earliest books in the library. There are some signs of fire damage on the edges of the cover (but not on the edges of the pages); therefore, this is probably one of the volumes saved from the library fire of 1838.

○ The volume has recently been rebacked and is in a slipcase.

Gregory, David. See also Euclid 1703; Euclid 1763; Euclid 1802.

Gregory, John. See Collins 1725.

Gregory, Olinthus Gilbert (1774–1841).
1815. *A Treatise of Mechanics, Theoretical, Practical, and Descriptive. Vol. I. Containing the Theory of Statics, Dynamics, Hydrostatics, Hydrodynamics, and Pneumatics. Vol. II. Containing remarks on the Nature, Construction, and Simplification of Machinery, on Friction, Rigidity of Cords, First Mover &c. and Descriptions of many Curious and Useful Machines.* 2 vols. Third edition, corrected and improved. London: Printed for F. C. and J. Rivington; et al. pp. xx + 569; vii + 565; 40 plates are bound separately.

◇ Errata for volume I are on p. [571]; errata for volume II follow p. vii.

○ The library has two copies of volume II, and only one copy of volume I.

Gregory, Olinthus. See Dowling 1824; Hutton 1827.

Grévy, Auguste (1865–?).
1905. *Traité d'algèbre a l'usage des élèves de mathématiques A et B (Programme du 31 mai 1902).* Paris: Vuibert et Nony. pp. [iv] + 498.

Griess, J. See Weber 1898b.

Griscom, John (1774–1852). See Leonhard Euler 1835.

Grund, Francis Joseph (1805–1863).
1830. *First Lessons in Plane Geometry. Together with an Application of Them to the Solution of Problems. Simplified for Boys not Versed in Algebra.* Boston: Carter and Hendee. pp. [vi] + 255.
⋄ Errata on p. 255.

Grund, Francis Joseph. See also Hirsch 1831.

Guericke, Otto von (1602–1686).
1672. *Ottonis de Guericke experimenta nova (ut vocantur) Magdeburgica de vacuo spatio primùm à R. P. Gaspare Schotto, è Societate Jesu, & Herbipolitanae Academiae Matheseos Professore: nunc verò ab ipso auctore perfectiùs edita, variisque aliis experimentis aucta. Quibus accesserunt simul certa quaedam de aeris pondere circa terram; de virtutibus mundanis, & systemate mundi planetario; sicut & de stellis fixis, ac spatio illo immenso, quod tàm intra quam extra eas funditur.* Amstelodami [Amsterdam]: Apud Joannem Janssonium à Waesberge. pp. [xiv] + 244.
⋄ Frontispiece depicts one of the experiments. Dedicated to Friderico Wilhelmo. Protrait of Otto von Guericke facing p. 7. The index follows p. 244. Errata follow the index.
○ Copy signed by François de Sluse.

Guillemin, Auguste (?–1914).
1906. *Tableaux logarithmiques A et B equivalent a des tables de logarithmes a 6 et a 9 décimales et notice explicative donnant la théorie et le mode d'emploi des tableaux.* Paris: Félix Alcan. pp. 32 + 2 fold–out tables.

Guilmin, Adrien (1812–1884).
n.d. *Algèbre élémentaire (No. 2) a l'usage des classes de lettres et des classes d'enseignement primaire supérieur renfermant un très grand nombre d'exercices de calcul algébrique et de questions usuelles de tout genre.* Dix-neuvième édition corrigée et augmentée. Paris: Belin Frères. pp. xii + 200.
⋄ The 15th edition was 1882.
○ Rebound in 1952.

Gunter, Edmund (1581–1626).
1653. *The Works of Edmund Gunter: Conteining the Description and Use of the Sector, Cross–staff, and Other Instruments, with a Canon of Artificiall Signes and Tangents of Radius of 100000000 Parts, and the Use Thereof in Astronomie, Navigation, Dialling, &c. Much Enlarged by the Author through the Whole Work in his Life Time, Together with a New Treatment of Fortification. Whereunto is now Added the further Use of the Quadrant Fitted for Daily Practise, with New Lines Serving the Former Uses and many Other Purposes more Accurately, by Samuel Foster.* Third edition diligently perused, corrected and amended by Henry Bond practitioner in the Mathematicks, in the Bulwark neer the Tower. London: Printed by F. N. for Francis Eglesfield. pp. [ix] + 164 + 306 + 75 + [4] + unnumbered pages of tables + 2 plates.
⋄ Dedicated to John Comit. Gunter was a professor of astronomy at Gresham

College, London. He constructed the first logarithmic line and scale. This was not a slide rule because it consisted of just one scale and had no sliding parts, however, the Gunter scale was a forerunner of the slide rule. Edited by Samuel Foster.

◦ Contains USMA bookplate. Signed by Moore Smith.

Gwilt, Joseph (1784–1863).
1811. *A Treatise on the Equilibrium of Arches, in which the Theory is Demonstrated upon Familiar Mathematical Principles. Also the Method of Finding the Drift or Shoot of an Arch. Interspersed with Practical Observations and Deductions.* London: Printed for the author. pp. xvi + 80 + 3 plates.

◇ Errata on p. xvi.

Gwilt, Joseph. See also Simpson 1821.

Gwynne, Lawrence. See Robertson 1805.

Haag, Paul (1843–1911).
1893. *Cours de calcul différentiel et intégral.* Paris: Ch. Dunod. pp. v + 612.
◇ Errata follow p. v.
◦ Copy signed by E. W. Bass.

Hachette, Jean Nicolas Pierre (1769–1834).
1817. *Éléments de géométrie a trois dimensions. Partie synthétique. Théorie des lignes et des surfaces courbes; Partie algébrique. Traité des surfaces du second degré.* Paris: The author and Courcier. pp. xii + 146 + 2 plates + x + 274 + xx + 3 plates.

◇ Dedicated to the Duke of Angouleme.

1818. *Second supplément de la géométrie descriptive, suivi de l'analyse géométrique de M. John Leslie.* Paris: Firmin Didot. pp. xvi + 164 + 8 plates + 3 plates.
◇ Pages 33–164 deal with "Analyse géométrique."

Hackley, Charles William (1809–1861; USMA 1829).
1838. *Elements of Trigonometry, Plane and Spherical. Adapted to the Present State of Analysis. To which is Added, Their Application to the Principles of Navigation and Nautical Astronomy. With Logarithmic, Trigonometrical, and Nautical Tables. For the Use of Colleges and Academies.* New York: Wiley and Putnam, Collins, Kease & Co; et al. pp. xii + 207 + 100 pages of Tables.

1847. *Elementary Course of Geometry.* New York: Harper and Brothers. pp. xi + 103 + 4 + 7 + 15 + 7 + 20 + 18 + 2 + 1 + 10 + 15 + 13.
◇ Following p. 103, there are several appendices and independent chapters on geometric topics, such as symmetry, mensuration, etc. Two pages of "critical notices" for his Treatise on Algebra and this volume.

1851. *A Treatise on Trigonometry, Plane and Spherical, with its Application to Navigation and Surveying, Nautical and Practical Astronomy and Geodesy, with Logarithmic, Trigonometrical, and Nautical Tables. For the Use of Schools and*

Colleges. New edition with extensive additions and improvements. New York: George P. Putnam. pp. xix + 372 + 238 pages of tables.

1856. *A Treatise on Algebra Containing the Latest Improvements Adapted to the Use of Schools and Colleges.* New York: Harper & Brothers. pp. xv + 504.

Hadamard, Jacques (1865–1919).
1898. *Leçons de géométrie élémentaire (Géométrie plane).* Cours complet de mathématiques élémentairies publié sous la direction de M. Darboux. Paris: Armand Colin & Cie. pp. xvi + 308.

1901. *Leçons de géométrie élémentaire II (Géométrie dans l'espace).* Cours complet de mathématiques élémentairies publié sous la direction de M. Darboux. Paris: Libraire Armand Colin. pp. xxii + 582.

1910. *Leçons sur le calcul des variations. Tome premier. La variation première et les conditions du premier ordre, les conditions de l'extremum libre.* Cours du College de France. Recueillies par M. Frechet. Paris: A. Hermann et Fils. pp. viii + 520.
  ◇ Volume 1 only. Errata on page vi which is the verso of the title page.

Haddon, James.
1851. *Examples and Solutions in the Differential Calculus.* London: John Weale. pp. vi + 162.
  ◇ There is no Table of Contents. At the end of the book, there are 24 pages of advertisements for other books published by John Weale, including some mathematics books.

Hagadorn, Charles Baldwin (1866–1918; USMA 1889).
1897. *Descriptive Geometry Problems.* West Point: United States Military Academy Printing Office. Pages unnumbered. 13 pages of text. Problems 1–13 face plates I–XIII. Plates XIV and XV have no corresponding problems.

Hagen, John [Johann] George (1847–1930).
1891; 1894. *Synopsis der hoeheren Mathematik. Erster Band. Arithmetische und algebraische Analyse. Zweiter Band. Geometrie der algebaischen Gebilde.* 2 volumes of 4. Berlin: Felix L. Dames. pp. viii + 398; [vii] + 416.
  ◇ Errata in Volume 1 follow p. 398, and in Volume 2 are on pp. [vi]–[vii].
  ◇ The level of mathematics in this work is very high. The author was the director of the observatory at Georgetown College, Washington, D.C.

Hall, William Shaffer (1861–1948).
1903. *Descriptive Geometry with Numerous Problems and Practical Applications.* New York: D. Van Nostrand Company. pp. 12 + 18 plates.
  ○ Not bound.

Halley, Edmond (1656–1742).
1752. *Astronomical Tables with Precepts both in English and Latin for Computing*

the Places of the Sun, Moon, Planets, and Comets. London: William Innys. 57 unnumbered pages of English text + 29 unnumbered pages of Latin text + more than 200 unnumbered pages of tables.

◇ Handsome frontispiece portrait of Halley.

○ Signature of Andrew Ellicott Junior.

Halley, Edmond (1656–1742). See Newton 1822; Sherwin 1761.

Halliwell–Phillipps, James Orchard (Editor) (1820–1889).
1839. *Rara mathematica: or, a Collection of Treatises on the Mathematics and Subjects Connected with Them, from Ancient Inedited Manuscripts.* London: John William Parker. pp. viii + 120.

◇ Dedicated to Thomas Stevens Davies. One page of corrections and additions follows p. 120. There are two appendices: "A few observations on the numerical contractions found in some manuscripts of the treatise on geometry by Boetius," pp. 106–111; "Notes and early almanacs," pp. 112–120.

Hamel, Jean–Baptiste du (1624–1706). See Du Hamel 1701.

Hamilton, Hugh (1729–1805).
1773. *A Geometrical Tretise of the Conic Sections, in which the Properties of the Sections are Derived from the Nature of the Cone, In an Easy Manner, and by a New Method, Translated from the Latin Original into English.* London: Printed for J. Nourse. pp. xv + 211 + 17 plates.

◇ The Latin edition was published in 1758. The translator is not specified.

Hamilton, Henry Parr (1794–1880).
1826. *The Principles of Analytical Geometry designed for the Use of Students in the University.* Cambridge: Printed by J. Smith. pp. viii + 326 + 9 plates.

◇ Errata follow p. 326.

Hamilton, William Rowan (1805–1865).
1853. *Lectures on Quaternions: Containing a Systematic Statement of a New Mathematical Method; of which the Principles were Communicated in 1843 to the Royal Irish Academy; which has since formed the Subject of Successive Courses of Lectures, Delivered in 1848 and Subsequent Years, in the Halls of Trinity College, Dublin: Numerous Illustrative Diagrams, and with some Geometrical and Physical Applications.* Dublin: Hodges and Smith. pp. [iii] + 64 + lxxii + 736 + 1 page of errata.

Hamilton, William Rowan. See also Graves 1882.

Hammond, Nathaniel (1725–1776).
1764. *The Elements of Algebra in a New and Easy Method; with Their Use and Application, in the Solution of a Great Variety of Arithmetical and Geometrical Questions; by General and Universal Rules. To which is Prefixed an Introduction,*

Containing a Succint History of this Science. Third edition, corrected. London: Printed for the Author. pp. xxiv + 328.

◇   The prefixed nationalistic history of algebra includes the repetition of Wallis's claim that Descartes plagarized Harriot.

Hanus, Paul Henry (1855–1941).

1886. *An Elementary Treatise on the Theory of Determinants, a Textbook for Colleges.* Boston: Ginn and Co. pp. viii + 217.

Hardy, Arthur Sherburne (1847–1930; USMA 1869).

1881. *Elements of Quaternions.* Boston: Ginn, Heath and Co. pp. viii + 230.

○   "Received Oct 12, 1881" is on the inside of the front cover.

1897. *Elements of Analytic Geometry.* Boston: Ginn and Co. pp. xii + 229.

n.d. *Einstein and the Unskilled Mathematician* (typed manuscript). pp. 13.

○   Book plate: Given by Mrs. Tillman Martin.

Hardy, Arthur Sherburne. See also Argand 1881.

Hartwig, Theodor.

n.d. *Schule der Mathematik zum Selbstunterrichte. II. Band. Analytische Geometrie der Ebene und des Raumes.*   Volume 2 only. Vienna: Moritz Perles. pp. 199.

Hassler, Ferdinand Rudolph (1770–1843).

1826. *Elements of Analytic Trigonometry, Plane and Spherical.* New York: Published by the Author. Printed by James Bloomfield. pp. iv + 5– 190 + 2 errata + 1 plate.

◇   The first American text for teaching Analytic Trigonometry originally published in 1809. Hassler was head of USMA Mathematics Department from 1807 to 1810.

○   The library has three copies.

○   One copy donated in memory of CAPT Dan Allen Brookhart [USMA 1958]. The spine is stamped "New York 1865", which is a rebinding date. A new sheet at the front is signed "Charles D. Elliot." Loose clippings from the Boston Sunday Globe of Feb 23, 1903 about the handwriting of the lawyer Rufus Choate. The following items are bound into this volume: 1. Original copyright for the book, dated October 10, 1826. Signed by H. Clay. 2. Letter dated March 2, 1807 from H. Dearborn of the War Department acknowledging Hassler's acceptance of the "temporary appointment of Mathematical Teacher in the Military Academy at West Point." The envelope with seal is included. 3. Letter of 1842 to Hassler recommending an unnamed individual. 4. Original copyright for Hassler's *Tablas Logarithmicas y Trigonometricas para las siete Decimales.* Corregidas pars F. R. Hassler, M. A. P. S., dated June 2, 1830. Signed by M. VanBuren.

1828a. *Elements of the Geometry of Planes and Solids With Four Plates.* Richmond: Published by the author, Shepherd and Company. pp. 160 + 4 plates.

◇   The table of contents is on p. 160. Errata on p. 159.

1828b. *Plates and Tables to the Popular Exposition of the System of the Universe.* New York: G. & C. Carvill. pp. Title page + 5 plates + 5 tables.

◇   This book consists entirely of plates and tables. There is no text.

1830. *Logarithmic and Trigonometric Tables; to Seven Places of Decimals, in a Pocket Form. In which the Errors of Former Tables are Corrected.* New York: C., G., and H. Carvill. pp. 10 + 2 pages of formulas + unnumbered pages of tables.

◇   There were editions of this book made in French, German, Spanish, and Latin. The work was reviewed by Gauss in the *Gottingische Gelehrte Anzeigen* (1831).

Hauxley, Edward (1715–ca 1751).
1743. *Navigation Unvail'd; or, a New and Complete System of Navigation in All its Branches.* 2 vols. London: Printed for the Author. pp. viii + 464 + 2 plates; [iv] + 236 + 3 plates + 199.

◇   Volume 2 contains a Table of Brigg's Logarithms from 1 to 10000. The first volume deals with arithmetic, geometry, and trigonometry.

○   Copy was owned by William Greenough in 1772.

Hayward, James (1780–1866).
1829. *Elements of Geometry Upon the Inductive Method. To which is Added an Introduction to Descriptive Geometry.* Cambridge: Hillard and Brown. pp. xvi + 172 + 11 plates.

◇   Errata follow p. xvi. The portion on descriptive geometry. pp. 121–172, is based on writings of Monge, Hachette, Lacroix, "and also the treatises upon the subject of Crozet and Professor Davies of West Point."

Heath, Thomas Little (1861–1940).
1913. *Aristarchus of Samos, the Ancient Copernicus; a History of Greek Astronomy to Aristarchus, Together with Aristarchus's Treatise on the Sizes and Distances of the Sun and Moon, a new Greek Text with Translation and Notes.* Oxford: Clarendon Press. pp. vi + [2] + 425 + 1 plate.

Heath, Thomas Little. See also Archimedes 1897; Euclid 1908.

Heather, John Fry (?–1886).
1851. *An Elementary Treatise on Descriptive Geometry, with a Theory of Shadows and of Perspective: Extracted from the French of G. Monge. To Which is Added, a Description of the Principles and Practice of Isometrical Projection. The Whole being Intended as an Introduction to the Application of Descriptive Geometry to Various Branches of the Arts.* 2 vols. London: John Weale. pp. vi + 137; 14 plates.

◇   Volume 1 is the text, and volume 2 is the plates. Volume 1 contains a 24 page list of works printed by Weale, many of which are mathematical.

Heiberg, Johan Ludwig (1854–1928) and Zeuthen, Hieronymous Georg (1839–1920).
1912. *Paul Tannery – Mémoires scientifiques. I. Sciences exactes dans l'antiquité 1876–1884.* Toulouse: Édourd Privat; Paris: Gauthier–Villars. pp. [xix] + [465].
◇ Contains portrait of Paul Tannery.

Heiberg, Johan Ludwig. See also Euclid 1908.

Hellins, John (? –1827). See Agnesi 1801.

Henderson, Robert (1871–1942).
1915. *Mortality Laws and Statistics.* First edition, first thousand. New York: J. Wiley. pp. v + 111.

Henry.
1810. *Mémoire sur la projection des cartes géographiques, adoptée au Dépôt général de la guerre; Publié par ordre de S. E. M. le Duc de Feltre, Ministre de la Guerre, pour faire suite au mémorial topographique et militaire. Topographie.* Paris: l'Imprimerie Impériale. pp. iv + 220 + 4 plates.
◇ Errata follow the plates.
○ This copy of the book was presented to Colonel Tranchot by the author. Thayer binding.

Henry, Charles (1859–1926). See Fermat 1891.

Herman, Jacob (1678–1733) and Del'isle, Joseph Nicolas (1688–1768).
1728. *Abregé des mathematiques pour l'usage de sa Majesté Imperiale de toutes les Russies.* Volumes 1 and 2 bound together as one book. St. Petersbourg: L'Academie imperial des Sciences. pp. [ix] + 127 + 8 plates; [xi] + 83.
◇ Volume 1 is by Herman, and volume 2 is by Del'isle. Volume 1 contains arithmetic, geometry and trigonometry; and volume 2 contains astronomy and geography. Written in catechism form. Volume 1 is dedicated to Baron D'Osterman and volume 2 is dedicated to Le Compt de Golowkin.
○ Herman's signature is on the title page of volume 1.

Heron (ca. 75 A.D.). See Thévenot 1693.

Herschel, John Frederick William (1792–1871).
1820. *A Collection of Examples of the Applications of the Calculus of Finite Differences.* Cambridge: Printed by J. Smith. pp. v + 171.
◇ One copy is bound together with Babbage's *Solutions of Functional Equations*, and a second copy is bound by itself. See also Peacock 1820. This is referred to as "Part III" on p. 1. Parts I and II, problems in differential and integral calculus, respectively, are contained in Peacock 1820. Thus, Peacock's book and this book form a two volume set.

Herschel, John Frederick William. See also Lacroix 1816b; Spence 1820; Peacock 1820.

Herttenstein, Johann Heinrich (1676–1741).
1737. *Cahiers de mathematique a l'usage de messieurs les officiers de l'Ecole royale d'artillerie de Strasbourg.* Strasbourg: Jean–Renauld Doulssecker, le Pere. pp. [iv] + 702 + 23 plates.
◇ Errata follow p. 702.

Heusch, F., de
1888. *Cours d'analyse, calcul différentiel.* pp. vi+275.
◇ Lithographed. Ecole militaire [de Belgique], section de l'artillerie et du génie.

1889. *Cours d'analyse, calcul intégral.* pp. viii + 366.
◇ Lithographed. Ecole militaire [de Belgique], section de l'artillerie et du génie.

Hewlett, John (1762–1844). See Leonhard Euler 1822.

Hill, George William (1838–1914)
1905; 1906; 1906; 1907. *The Collected Mathematical Works of George William Hill.* 4 vols. Washington: Carnegie Institution. pp. xviii+363; [vi]+339; [ii]+577; [vi]+460.
◇ Errata page for volume 2 not bound. One page Addenda for volume 3 on p. 577. Index for all four volumes is contained in volume 4, pp. 453–457. Errata for all four volumes are contained in volume 4, pp. 458–460. Portrait of Hill is the frontispiece of volume 1. Introduction to volume 1 by H. Poincaré, pp. vii–xviii.

Hirsch, Meier (1765–1851).
1810. *Integraltafeln, oder Sammlung von Integralformeln.* Berlin: Duncker und Humblot. pp. [x] + [302].
◇ Consists of integral tables.
○ Signed by E. W. Bass. Volume contains his bookplates.

1823. *Integral Tables, or a Collection of Integral Formulae. By Meyer Hirsch.* Translated from German. London: William Baynes & Son. pp. xxx + 279. Corrections on p.[iv].

1831. *A Collection of Arithematical and Algebraic Problems and Formulae; by Meyer Hirsch. Translated from the Original German, and Adapted to the Use of the American Student.* Translated by Francis J. Grund. Boston: Carter, Hendee and Babcock. pp. xii + 340 + 2 errata.
◇ The translation is dedicated to Nathaniel Bowditch. Errata follow p. 340.
○ The library has three copies. One copy is signed "J. M. Roberts book, West Point, New York, Sept 24, 1832" inside front cover and signed "M. Roberts, Buffalo, NY, November 22, 1832" inside back cover.
○ Another copy signed "From Prof. H. S. Kentrick, November 1881 to E. W. Bass" and "To Dept of Math from E. W. Bass" inside the front cover.

Hobert, Johann Philipp (1759–1826) and Ideler, Ludewig (1766–1846).
1799. *Neue trigonometrische Tafeln für die Decimaleintheilung des Quadranten.*
Berlin: Verlage der Realschulbuchhandlung. pp. lxxii + 351.
 ◇ Errata follow p. 351. This work was published simultaneously in French.

Hooper, William (fl. 1770).
1783; 1782; 1774; 1774. *Rational Recreations, In which the Principles of Numbers and Natural Philosophy are Clearly and Copiously Elucidated, by a Series of Easy, Entertaining, Interesting Experiments. Among which are All Those Commonly Performed with the Cards.* 4 vols. Second edition, Corrected. London: Printed for L. Davis, et. al. pp. xvi + 267 + 12 plates; xi + 256 + 19 plates; xii + 296 + 20 plates; xi + 280 + 14 plates.
 ○ It appears that several pages at the end of vol. II have been lost.

Horner, Francis (1778–1817). See Leonhard Euler 1822.

Horner, W. G. (1786–1837).
1819. "New Methods of Solving Numerical Equations of All Orders by Continuous Approximation."
 ◇ A paper, not a book, bound together with other papers in Anonymous n.d.
 ○ Title page inscribed "With the Authors Compliments."

Horrebow, Peder (1679–1764).
1740. *Operum mathematico–physicorum tomus primus. Continens elementa matheseos, in progressionem harmonicam mathemata, clavem astronomiae altero tanto auctiorem.* Havniae [Copenhagen]: Sumptibus Jacobi Preussii. pp. [viii] + 398 + [x].
 ◇ Frontispiece of an astronomical observatory. Dedicated to King Christian VI.

1741a. *Operum mathematico–physicorum tomus secundus. Continens historiam reformationis calendarii inter evangelicos. Annotationes in monumenta paschalia antiquorum patrum. M.C.H. vindicias aerae Dionysianae. Methodum paschalem, et secundum hanc computum ecclesiasticum.* Havniae [Copenhagen]: Jacobi Preussii. pp. [xvi] + 499.
 ◇ Dedicated to Domino Friderico, Prince of Denmark and Norway. Errata follow p. 499. Bound with Horrebow 1740.

1741b. *Operum mathematico–physicorum tomus tertius. Continens basin astronomiae cum Triduo Roemeriano. Copernicum triumphantem cum vindiciis. Atrium astronomiae. Artem interpolandi.* Havniae [Copenhagen]: Jacobi Preussii. pp. [xvi] + 422 + 12 plates.
 ◇ Dedicated to Vicentio deLerche. Errata follow p. 422. Bound with Horrebow 1740.

[Horsley, Samuel (1733–1806)].
1801. *Elementary Treatises on the Fundamental Principles of Practical Mathematics. For the Use of Students.* Oxford: Clarendon Press. pp. xvii + 398.
 ◇ Errata follow p. xvii.

Horsley, Samuel. See also Euclid 1802; Euclid 1803; Newton 1779.

Hoüel, J. (1823–1886).
1878; 1879; 1880; 1881. *Cours de calcul infinitésimal.* 4 vols. Paris: Gauthier–Villars. pp. xv + 511; 479; 306; vi + 306.
◇ Errata in volume 1 on pp. 509–511 and in volume 2 on pp. 477–479.

Hoüel, J. See also Argand 1881.

Howison, George Holmes (1834–1917).
1869. *A Treatise on Analytic Geometry, especially as applied to the Properties of Conics: Including the Modern Methods of Abridged Notation. Written for the Mathematical Course of Joseph Ray.* Eclectic Educational Series. Cincinnati: Wilson, Hinkle, & Co. pp. xxxiv + 574.
○ "Received May 28, 1880" on front paper.

Huilier. See under L'Huilier.

Hulburt, Lorrain Sherman (1858–?).
1912. *Differential and Integral Calculus. An Introductory Course for Colleges and Engineering Schools.* New York: Longmans, Green, and Co. pp. xviii + 481.
◇ Contains an index.

Humbert, Georges (1859–1921).
1903; 1904. *Cours d'analyse professé à l'École polytechnique. Tome 1 Calcul différential, principes du calcul intégral, applications géométriques. Tome 2 Compléments du calcul intégral, fonctions analytiques et elliptiques, équations différentialles.* 2 vols. Paris: Gauthier–Villars. pp. xvi + 483; xi + 493.

Hutton, Charles (1737–1823).
1775a. *The Diarian Miscellany: Consisting of all the Useful and Entertaining Parts, both Mathematical and Poetical, Extracted from the Ladies' Diary, from the Beginning of that Work in the Year 1704, Down to the End of the Year 1773. With many Additional Solutions and Improvements.* 5 vols. London: Printed for G. Robinson and R. Baldwin. pp. vi + 364; 392; 424; 396; 364.
◇ Volume 2. pp. 411–424, contains an index of "persons who have proposed and answered questions." An Appendix to Volume 3. pp. 391–409, contains errata for the first 3 volumes and additional solutions. Volume 5. pp. 359–364, contains a "list of all the Enigmas," by year.

1775b. *Miscellanea mathematica: Consisting of a Large Collection of Curious Mathematical Problems, and Their Solutions. Together with many other Important Disquisitions in Various Branches of the Mathematics. Being the Literary Correspondence of Several Eminent Mathematicians.* London: Printed for G. Robinson and R. Baldwin. pp. iv + 342.
◇ Errata follow p. 342.

Hutton, Charles (Continued).

1781. *Tables of the Products and Powers of Numbers. Namely, 1st, the Products of all Numbers to 1000 by 100, 2nd, the Squares of all Numbers to 25400, 3d, the Cubes of all Numbers to 10000, 4th, the First Ten Powers of all Numbers to 100, 5th, Tables for Reducing Money, Weights, and Measures from one Denomination to Another. With an Introduction, Explaining and Illustrating the Use of the Tables.* London: Printed by William Richardson. pp. 7–103; all pages of tables.

◇ Errata follow p. 7.

1796; 1795. *A Mathematical and Philosophical Dictionary: Containing an Explanation of the Terms, and an Account of the Several Subjects, Comprized under the Heads Mathematics, Astronomy, and Philosophy both Natural and Experimental: with an Historical Account of the Rise, Progress, and Present State of these Sciences: also Memoirs of the Lives and Writings of the Most Eminent Authors, both Ancient and Modern, who by Their Discoveries or Improvements have Contributed to the Advancement of Them. In Two Volumes. With Many Cuts and Copper-Plates.* 2 vols. London: J. Davis. pp. viii + 650 + 12 plates; 756 + plates 13–37.

◇ An extremely useful work with many biographical entries.

1798. *A Course of Mathematics in Two Volumes: Composed, and more Especially Designed, for the Use of the Gentlemen Cadets in the Royal Military Academy at Woolwich.* Volume 2 of 2 Volumes. London: Printed for G. G. and J. Robinson. pp. iv + 364 + 2 plates.

◇ Errata follow p. 364. This book was used as a text at USMA from 1802 until 1823.

○ Book belonged to Milo Mason (USMA 1808), when he was a West Point cadet in 1807, and to General James Gibson (USMA 1808).

1800; 1801. *A Course of Mathematics, in two Volumes: Composed, and More Especially Designed, for the Use of the Gentlemen Cadets in the Royal Military Academy at Woolwich.* Third edition, enlarged and corrected. 2 vols. London: Printed for the author. pp. iv + 372; iv + 364.

○ There may be pages missing at the end of volume 1. Numerous signatures in volume 1.

1803. *Recreations in Mathematics and Natural Philosophy Containing Amusing Dissertations and Enquiries Concerning a Variety of Subjects the Most Remarkable and Proper to Excite Curiosity and Attention to the Whole Range of the Mathematical and Philosophical Sciences.* 4 vols. Hutton's Mathematics series. London: T. Davison. pp. xv + xxii + xvi + 447 + 17 plates between pp. 417 and 418; xiii + 464; xi + 501; xii + 516.

◇ Used as a text at USMA for a short period.

◇ Contains plates for arithmetic, geometry; mechanics, optics accoustics; astronomy, dialling; and physics.

○ This copy is rebound.

Hutton, Charles (Continued)

1804a. *Mathematical Tables; Containing Common, Hyperbolic and Logistic Logarithms. Also Sines, Tangents, Secants and Versed Sines, Both Natural and Logarithmic. Together with Several Other Tables Useful in Mathematical Calculations. To which is Prefixed a Large and Original History of the Discoveries and Writings Relating to those Subjects; with the Complete Description and Use of the Tables.* Fourth edition. London: Printed for G. and J. Robinson and R. Baldwin. pp. xi + 179 + 344.

◇ This work was dedicated to Nevil Maskelyne, the Astronomer Royal, and its motivation was the correction of the best tables that had been previously published (Briggs, Vlacq, Callet, Gardiner, and Sherwin).

◇ The author lists the errors that he has detected in several other books of tables on pp. 342–344.

◇ As an introduction to the book, Hutton wrote a history of trigonometry and logarithm tables. pp. 1–41.

1804b. *A Course of Mathematics in Two Volumes: Composed, and more Especially Designed, for the Use of the Gentlemen Cadets in the Royal Military Academy at Woolwich.* Volume 2 of 2 Volumes. Fourth edition, enlarged and corrected. London: Printed for the author and for G. and J. Robinson. pp. iv + 415.

○ This book belonged to Henry Smith (USMA 1815), when he was a cadet in 1814.

1807. *A Course of Mathematics in Two Volumes, for the Use of Academies as well as Private Tuition.* Volume 2. Fifth edition, enlarged and corrected. London: Printed for J. Johnson; et. al. pp. iv + 414.

1811. *Mathematical Tables: Containing the Common, Hyperbolic and Logistic Logarithms. Also Sines, Tangents, Secants and Versed Sines, Both Natural and Logarithmic. Together with Several Other Tables Useful in Mathematical Calculations. To which is prefixed, a Large and Original History of the Discoveries and Writing Relating to those Subjects; with the Complete Description and Use of the Tables.* Fifth edition. London: F. C. and J. Rivington, et. al. pp. viii + 179 + 344 pages of tables.

◇ Errata on p. vii. This work was dedicated to Nevil Maskelyne, the Astronomer Royal. Errors perceived and corrected from other books of tables are listed on pp. 342–344.

1812a. *Tracts on Mathematical and Philosophical Subjects; Comprising among Numerous Important Articles, the Theory of Bridges, with Several Plans of Recent Improvement. Also the Results of Numerous Experiments on the Force of Gunpowder, with Applications to the Modern Practice of Artillery.* 3 vols. London: Printed for F. C. and J. Rivington; et. al. pp. vii + 485; [iii] + 384 + 6 plates; [iii] + 383 + 4 plates.

◇ Charles Hutton's portrait is the frontispiece of volume I.

◇ Corrections for all three volumes are on p. xii of volume I.

◇ This work consists of 38 "tracts" on a variety of subjects.

○ The library has two copies.

○ One copy appears to have John Bonnycastle's signature on the title page.

Hutton, Charles (Continued)

1812b. *A Course of Mathematics in Two Volumes, for the Use of Academies as Well as Private Tuition by Charles Hutton from the Fifth and Sixth London Editions, Revised and Corrected by Robert Adrain.* 2 vols. New York: Samuel Campbell, et. al. pp. xvi + 583; vii + 558 + unnumbered pages of logarithm and trigonometric tables.

⬦ Errata on p. [xvi] and [vii].

○ One copy is stamped "Textbook USMA 1818 to 18?" in gutter of title page.

○ A second copy of volume 2 has several signatures: Charles Marshall, James Cooke, Edwin B. Newton.

1818; 1816. *A Course of Mathematics in Two Volumes, for the Use of Academies as Well as Private Tuition by Charles Hutton from the Fifth and Sixth London Editions, Revised and Corrected by Robert Adrain.* 2 vols. New York: Samuel Campbell, et. al. pp. xv + 583 + 2 plates; vii + 558 + unnumbered pages of tables.

○ Both volumes belonged to West Point cadets.

1822. *A Course of Mathematics in Two Volumes, for the Use of Academies as well as Private Tuition, in Two Volumes. To which is Added an Elementary Essay on Descriptive Geometry by Robert Adrain.* Third American edition revised, corrected, and improved. New York: Samuel Campbell & Son, et. al. pp. iv + 622 + unnumbered pages of tables.

1825; 1826. *A Course of Mathematics for the Use of Academies, as Well as Private Tuition.* Fourth American edition from the seventh London edition, revised, corrected, and improved. To which is Added An Elementary Essay on Descriptive Geometry by Robert Adrain. New York: S. Campbell & Sons, et. al. pp. xvi + 583; iv + 622 + 21 pages of logarithm tables.

○ Contains handwritten papers inside. Part of the special collection donated by Wm C. Trimble.

1827; 1828; 1827. *A Course of Mathematics in Three Volumes: Composed for the Use of the Royal Military Academy.* 3 vols. Volumes 1 and 2 are the Ninth edition, with many corrections and improvements by Olinthus Gregory; Volume 3 is the Fourth edition, with many corrections and improvements by Olinthus Gregory. London: Printed for C. and J. Rivington, et. al. pp. viii + 432; vii + 416; xi + 413.

⬦ Errata for volume 2 follow p. vii, and for volume 2 follow p. xi.

1843. *A Course of Mathematics in Three Volumes: Composed for the Use of the Royal Military Academy.* A new carefully corrected edition, entirely re–modelled, and adapted to the course of instruction now pursued in the Royal Military Academy, by William Rutherford. London: Printed for Thomas Tegg. pp. vi + 895.

Hutton, Charles. See also Maseres 1791; Ozanam 1803; Lorgna 1779; Leybourn 1817; Dowling 1824.

Huygens, Christiaan (1629–1695).

1724. *Opera varia.* 2 vols. Lugduni Batavorum [Leiden]: Apud Janssonios Vander Aa. pp. [v] + 248 + 28 plates; 249–776 + 28 plates + [xiv].

◇ Frontispiece portrait of Huygens in Volume 1. There is an index to both volumes at the end of Volume 2.

1728. *Opera reliqua.* 2 vols. Amstelodami [Amsterdam]: Apud Janssonio–Waesbergios. pp. 10 + [xxiv] + 315 + 15 plates; [xvi] + 184 + 21 plates.

◇ Each volume has a preface by William 'sGravesande. The first volume contains a geometrical treatise by Guido Grandi. There is a brief biography of Huygens in volume 2, pp. [ix]–[xvi].

Huygens, Christiaan. See also Apollonius 1749; Jakob Bernoulli 1801; Keill 1726; Keill 1745; Leslie 1821.

Ingram, Alexander.

1822. *A Concise System of Mensuration, Adapted to the Use of Schools; Containing Algebra, with Fluxions; Practical Geometry, Trigonometry, Mensuration of Superficies and Solids, Land–Surveying, Gauging, &c. Together with a Large Appendix Containing the Demonstrations of the Rules in the Work.* Edinburgh: Oliver and Boyd. pp. x + 323.

◇ Errata follow p. 323.

Jablonski, E.

1891. *Théorie des équations suite aux compléments d'algébre avec de nombreux exercises.* Paris: Jules Delalain et Fils. pp. xii + 404.

Jack, Richard (1720–1759).

1742. *Elements of Conic Sections in Three Books: in which are Demonstrated the Principal Properties of the Parabola, Ellipse, and Hyperbola.* Edinburgh: Printed by Tho. Wal. and Tho. Ruddimans. pp. xi + 331 + 9 plates.

◇ Dedicated to John, Lord Hope. Errata follow p. xi.

○ The signature of Andrew Ellicott, 1779, is on a front endpaper.

Jacquier, François (1711–1788). See Newton 1822.

Jecker, François-Antoine (1765–1834).

ca 1812. *Raport concernant les travaux de M. Jecker, fabriquant d'instrumens d'astronomie, de géodésie et d'optique.* pp. 21.

◇ Title page is missing.

Jephson, Thomas.

1826. *The Fluxional Calculus, An Elementary Treatise, Designed for the Students of the Universities and for Those who Desire to be Acquainted with the Principles of Analysis.* London: Baldwin, Cradock, and Joy. pp. vii + 448.

◇ Errata follow p. vii. Despite the title, modern calculus notation is used.

Johnson, J. B. (1850–1902).

1898. *The Theory and Practice of Surveying. Designed for the Use of Surveyors and Engineers Generally. But Especially for the Use of Students in Engineering.* Fourteenth edition revised and enlarged. New York: John Wiley and Sons. pp. xxvii + 770 + 4 plates.

⋄ Used as text at USMA 1900–1902.

○ Notes in margin. Page 87 contains a collination error.

Johnson, William Woolsey (1841–1927).

1889. *A Treatise on Ordinary and Partial Differential Equations.* New York: John Wiley and Sons. pp. xii + 368.

⋄ Johnson graduated from Yale (A.B. 1862 and A.M. 1868). He was a Mathematics Professor at the United States Naval Academy (USNA) at Newport, Rhode Island initially and later Annapolis, Maryland (1864–1870 and 1881–1921). He was also an early member of the American Mathematical Society (AMS). He wrote numerous popular college-level mathematics testbooks, including several used at West Point.

○ Copy has several annotations.

1890. *The Theory of Errors and Method of Least Squares.* Baltimore: Issac Friedenwald. pp. vi + 146.

⋄ This book was used as a text at USMA from 1891 until 1932.

○ Lesson plan and syllabus for USMA course is pasted on page vi. Notes in margins throughout the book.

[1892]. *The Theory of Errors and Method of Least Squares.* New York: John Wiley & Sons. pp. x + 152 + unnumbered pages of tables.

⋄ Several printed corrections typed in on the appropriate pages. Text used at USMA until 1932.

⋄ This volume has a different date and publishing information from same work above (Johnson 1890).

1896. *The Theory of Errors and Method of Least Squares.* First edition. New York: John Wiley and Sons. pp. x + 152 + pages of tables + 16 page table of publications of John Wiley & Sons. pp. [iii] + iv–x + 152 + 471–476 + 153 – 179 + 4 blank pages.

⋄ Contains applications of Least Squares to military service along with other classical applications.

○ Copy was used as a cadet textbook. Homework assignments for lessons written on front cover. Numerous glued add–ins: pp. 1, 5, 13, 29, 33, 42, 49, 50, 97, and a 3 page insertion beginning on p. 114.

○ Copy was a gift from COL Michael C. Gerenata [USMA 1918].

1907. *A Treatise on the Integral Calculus Founded on the Method of Rates.* First edition. New York: John Wiley & Sons. pp. xiv + 440.

Jones, William (1675–1749).
1706. *Synopsis palmariorum matheseos: or, a New Introduction to the Mathematics: Containing the Principles of Arithmetic & Geometry Demonstrated in a Short and Easie Method; with Their Application to the most Useful Parts thereof: As, Resolving of Equations, Infinite Series, Making the Logarithms; Interest, Simple and Compound; The Chief Properties of the Conic Sections; Mensuration of Surfaces and Solids; The Fundamental Precepts of Perspective; Trigonometry; The Laws of Motion Apply'd to Mechanic Powers, Gunnery, &c. Design'd for the Benefit, and Adapted to the Capacities of Beginners.* London: F. Matthews. pp. [xii] + 301.
   ◇ Dedicated to William Lowndes. Errata listed on p. [xii].
   ○ Copy has been rebound.

Jones, William. See also Martin 1797.

Jordan, Camille (1838–1922).
1882. *Cours d'analyse de l'École polytechnique.*   Tome premier Calcul différentiel.
Paris: Gauthier–Villars. pp. xvi + 378 + 4.
   ◇ Errata on pp. xv–xvi.

Karsten, Wenceslaus Johann Gustav (1732–1787).
1782, 1786, 1769, 1769, 1770, 1771, 1775, 1777. *Lehrbegriff der gesamten Mathematik.* After information about the author, each of the eight volumes caries a subtitle as follows: (1) *Der erste Theil. Die Rechenkunst und Geometrie. Zweyte Auflage.*   (2) *Des zwenten Theils erste Abtheilung.   Weitere Ausführung der Rechenkunst. Die Algebra mit den vorrehmsten Anwendungen auf Zahlenrechnungen. Die ebene und sphärische Trigonometrie, nebst weiterer Ausführ. der Geometrie. Zweyte Auflage.   (3) Der dritte Theil. Die statischen Wissenshcaften, nebst der ersten Gründen der Mechanik.   (4) Der vierte Theil. Die Mechanik der sesten Körper.   (5) Der fünste Theil. Die Hydraulik.   (6) Der sechste Theil. Veschluss der hydraulik und die Peumatik.   (7) Der siebende Theil. Die Optik und Perspectiv.* (8) *Der achte Theil, Die Photometrie.* Greifswald : Anton Ferdin. Röse. pp. [xx] + 490 + 8 plates; 632 + 4 plates; [xx] + 489 + 8 plates; [xiv] + 527 + 5 plates; [xiv] + 792 + 8 plates; [xxx] + 800 + 8 plates; [xxii] + 937 + 50 plates; [xvi] + 774 + 28 plates.

Kartchser, Maurice. See Lubbe 1832.

Kästner, Abraham Gotthelf (1719–1800).
1770. *Anfangsgründe der Analysis des Unendlichen.* Dritter Theil. Zweyte Abtheilung. Zweyte Auflage.   Series: Der mathematischen Anfangsgrunde. Göttingen: Verlag der Wittme Vandenhoek. pp. [xxxvi] + 612 + [32] + 4 plates.
   ◇ There are two unpaginated indices at the end of the book. Dedicated to Levin Adolph von Hase. The first edition was published in 1761.

1790, 1791. *Geometrische Abhandlungen: Erste [Zweyte] Sammlung Anwendungen der ebenen Geometrie und Trigonometrie.* 2 vols. The first volume is "I. Theils III Abtheil." of *Der mathem. Anfangsgr.* and the second is "I. Theils IV Abtheil."

Göttingen: Verlag bey Vandenhoek und Ruprecht. pp. 580 + 9 plates; [xx] + 620 + 4 plates.

⋄ Errata on p. 580 of Volume 1 and on p. 620 of Volume 2.

1792. *Anfangsgründe der angewandten Mathematik. Der mathematischen Anfangsgründe II. Theil. I. Abtheilung. Mechanische und Optische Wissenschaften.* Fourth edition. Göttingen: Verlag bey Vandenhoek und Ruprecht. pp. [14] + 414 + 9 figures.

1793. *Anfangsgründe der höhern Mechanik welche von der Bewegung fester Körper besonders die praktischen Lehren enthalten. Der Mathem. Anfangsgr. IVter Theil; Iste Abtheil.* Second edition. Göttingen: Verlag bey Vandenhoek und Ruprecht. pp. xxx + 626 + 2 plates.

1794. *Anfangsgründe der Analysis endlicher Grössen. Der mathem. Anfängsgr. IIIter Theil; Iste Abtheil.* Göttingen: Verlag bey Vandenhoek und Ruprecht. pp. xvi + [xii] + 579 + 4 plates.

Keill, John (1671–1721).

1726. *An Introduction to Natural Philosophy: or, Philosophical Lectures Read in the University of Oxford, anno dom. 1700. To which are Added, Demonstrations of the Monsieur Huygens's Theorems, Concerning Centrifigal Force and Circular Motion.* Second edition, translated from the Latin. London: J. Senex, W. and J. Innys, J. Osborn, and T. Longman. pp. xii + 306 + 2 page list of technical books by these printers.

∘ This is Andrew Ellicott's copy and has his autograph on the title page. There are no annotations.

1730. *An Introduction to the True Astronomy: A or T Astronomical Lectures, Read in the Astronomical School of the University of Oxford.* Second edition. London: Printed for Bernard Lintot. pp. [vi] + xiv + [iv] + 396 + 27 plates.

⋄ Dedicated to James, Duke of Chandos. There is a 12 page index following p. 396.

∘ Copy owned by Andrew Ellicott (of Baltimore).

1745. *An Introduction to Natural Philosophy: or Philosophical Lectures Read in the University of Oxford, anno dom. 1700. To which are Added, Demonstrations of the Monsieur Huygens's Theorems, concerning Centrifigal Force and Circular Motion.* Fourth edition, translated from the Latin. London: M. Senex, W. Innys, T. Longman, and T. Shewell. pp. xii + 306 + 2 page list of technical books by these printers.

1790. *Elements of Plain and Spherical Trigonometry; also a Short Treatise of the Nature and Arithmetick of Logarithms.* Translated by Samuel Cunn. And carefully Corrected by S. Fuller. Fourth edition. Dublin: Printed by R. Jackson. pp. 142 + 5 plates.

⋄ The plate facing p. 60 contains figures that fold up to make 3 dimensional diagrams.

Keill, John. See also Collins 1725; Euclid 1803.

Keith, Thomas (1759–1824).
1814. *The Elements of Plane Geometry: Containing the First Six Books of Euclid, from the Text of Dr. Simson, With Notes Critical and Explanatory: to Which are Added Book VII, Including Several Important Propositions Which are not in Euclid; and Book VIII, Consisting of Practical Geometry; also Book IX, of Planes and Their Intersections; and Book X, of the Geometry of Solids.* London: Printed for Longman, Hurst, Rees, Orme, and Brown. pp. xv + 398.

1816. *An Introduction to the Theory and Practice of Plane and Spherical Trigonometry and the Stereographic Projection of the Sphere; including the Theory of Navigation: Comprehending a Variety of Rules, Formulae, & c. with Their Practical Applications to the Mensuration of Heights and Distances; to Determining the Latitude by two Altitudes of the Sun, the Longitude by the Lunar Observations, and to Other Important Problems on the Sphere, and in Nautical Astronomy.* Third edition, corrected and improved. London: Printed for Longman, Hurst, Rees, Orme, and Brown, and for the author. pp. xxvii + 436 + 5 plates.
 ◊ Contains tables of logarithms, sines and tangents.
 ○ Signed by Wm. Graham.

1820. *An Introduction to the Theory and Practice of Plane and Spherical Trigonometry, and the Stereographic Projection of the Sphere; Including the Theory of Navigation: Comprehending a Variety of Rules, Formulae, &c. with Their Practical Applications to the Mensuration of Heights and Distances; to Determining the Latitude by two Altitudes of the Sun, the Longitude by the Lunar Observations, and to Other Important Problems on the Sphere, and on Nautical Astronomy.* Fourth edition, corrected and improved. London: Printed by Longman, Hurst, Rees, Orme, and Brown, and for the author. pp. xxviii + 442 + 5 plates.
 ◊ On pp. xii–xv, Keith discusses his claim that 70 pages of Bonnycastle's *Plane and Spherical Trigonometry* (1806) are copied from an earlier edition of this work.

Keith, Thomas. See also Euclid 1814b.

Kelvin, William Thomson. See Thomson, William.

Kepler, Johannes (1571–1630).
1606. *Joannis Keppleri Sac. Caes. Majest. Mathematici de stella nova in pede serpentarii, et qui sub ejus exortum de novo iniit, trigono igneo. Libellus astronomicis, physicis, metaphysicis, meteorologicis & astrologicis disputationibus, ὀνδόξοις & παραδόξοις plenus. Accesserunt: I. De stella incognita cygni: narratio astronomica. II. De Jesu Christi servatoris vero anno natalitio, consideratio novissimae sententiae Lavrentii Svslygae Poloni, quatuor annos in usitata epoca desiderantis. Cum Privlegio S. C. Majest. ad annos XV. Pragae. Typis Pauli Sessii, impensis Authoris.* Pragae [Prague]: Paul Sess. pp. [xii] + 212 + 1 plate + 35 + [3].
 ◊ Dedicated to Rudolpho II.
 ○ Signed "Louis de Machault. 1618.28.9" in gutter of title page.

Kepler, Johannes (Continued)

1609. *Joannis Keppleri S. C. M^{tis} Mathematici phaenomenon singulare seu Mercurius in Sole. Cum digressione de causis, cur Dionysius Abbas Christianos minus iusto à nativitate Christi Domini numerare docuerit: De capita & anni ecclesiastici 1609. Lipsiae. Impensis Thomae Schureri Bibliopolae.* Lipsiae [Leipzig]: T. Schurer. pp. [xxxvi] + 1 plate.

1611. *Ioannis Kepleris. C. Maiest. Mathematici strena seu de niue sexangula. Cum priuilegio S. Caes. Maiest. ad annos XV. Francofvrti ad moenvm, apud Godefridum Tampach.* Francofvrti [Frankfurt]: G. Tampach. pp. 24.

1624. *Joannis Kepleri Imp. Caes. Ferdinandi II. Mathematici chilias logarithmorum ad totidem numeros rotundos, pramissâ demonstratione legitima ortus logarithmorum eorumą[ue] usus quibus nova traditur arithmetica, seu compendium, quo post numerorum notitiam nullum nec admirabilius, nec utilius solvendi pleraq[ue]; Problemata calculatoria, praesertim in doctrina triangulorum, citra multiplicationis, divisionis, radicumá[ue]; extractionis, in numeris prolixis, labores molestissimos. Ad illustriss. Principem & Dominum, Dn. Philippvm Landgravium Hassiae &c. Cum privilegio Autoris Caesareo Marpurgi Excusa typis Casparis Chemlini.* Marburg: Caspar Chemlin. pp. 55 + 51 pages of unpaginated tables.
  ○ Title page contains "Ex Lib Lud. Roussand" and "Collegu Pari s. Soc. Jesu."

1627. *Tabulae Rudolphinae quibus astronomicae scientiae, temporum longinquitate collapsae restauratio continetur; A phoenice illo astronomorum Tychone ex illustri & generosa Braheorum in Regno Daniae familiâ oriundo equite, primum animo concepta et destinata anno Christi MDLXIV: exinde observationibus siderum accuratissimis, post annum praecipue MDLXXII, quo sidus in cassiopeiae constellatione novum effulsit, serió affectata; variisque operibus, cum mechanicis, túm librariis, impenso patrimonio amplissimo, accedentibus etiam subsidiis Frederici II Daniae Regis, regali magnificentia dignis, tracta per annos XXV, potissimum in insula freti Sundici Huenna, & arce Uraniburgo, in hos usus à fundamentis extructâ: Tandem traducta in Germaniam, inque aulam et Nomen Rudolphi Imp. anno M D IIC. Tabulas ipsas, jam et nuncupatas, et affectas, sed morte authoris sui anno MDCI desertas, Jussu et Stipendiis fretus Trium Imppp. Rudolphi, Matthiae, Ferdinandi, annitentibus haeredibus Braheanis; ex fundamentis observationum relictarum; ad exemplum feré partium iam exstructarum; continuis multorum annorum speculationibus & computationibus, primum Pragae Bohemorum continuauit; deindé Lincii, superioris Austriae metropoli, subsidijs etiam Ill. Provincialium adjutus, perfecit, absolvit, adą[ue] causarum & calculi perennis formulam traduxit Ioannes Keplerus, Tychoni primùm à Rudolpho II. Imp. adjunctus calculi minister; indeá[ue] trium ordine Imppp. Mathematicus: Qui idem de speciali mandato Ferdinandi II Imp. petentibus instantibúsq[ue]; Haeredibus, opus hoc ad usus praesentium & posteritatis, typis, numericis proprijs, caeteris & praelo Jonae Saurii, Reip. Ulinanae Typograhpi, in publicum extulit, & Typographicis operis Ulmae curator assuit. Cum Privilegiis Imp. & Regum Rerúmq[ue]; publ. vivo Tychoni ejúsq[ue]; Haeredibus, & speciali Imperatorio, ipsi Keplero concesso, ad anno XXX.* Nuremburg: [no publisher]. pp. [xvi] + 119 + 125 + [1].
  ◇ The frontispiece is a stunning classic. Dedicated to Ferdinand II.
  ○ Some annotations in the margins.

Kepler, Johannes (Continued).

1635. *Epitome astronomiae Copernicanae usitatâ formâ quaestionnum & responsionum conscripta, inqne VII. libros digesta, quorum tres hi priores sunt de doctrina sphaerica habes, amice lector, hac prima parte, praeter physicam accuratam explicationem motus terrae diurrai, ortusq[ue]; ex eo circulorum sphaerae, totam doctrinam sphaericam nova & concinniori methodo, auctiorem, additis exemplis omnis generis computationum astronomicarum & geographicarum, quae integrarum praeceptionum vim sunt complexa. Authore Ioanne Keplero Imp. Caes. Matthlae, Ordd:q:Illustra. Archiducatus Austriae supra Onasum, mathematico. Cum privilegio Caesareo ad Annos XV. Francofurti, Impensis Ioannis Godefridi Schönwetteri excudebar Iohan. Fridericus Werssius.* Francofurti [Frankfurt]: J. G. Schönwetter. pp. [xvii] + 932 + [14 page index] + 1 plate.

1653. *Joannis Kepleri S^æ. C^æ. Mis Mathematici Dioptrice: seu Demonstratio eorum quae visui & visilibus propter Conspicilla non ita pridem inventa accidunt. Prae missae Epistolae Galilae i de iis quae post editionem Nuncii Siderei, ope Perspicilli, nova & admiranda in œlo derehensa sunt. Item Examen Prae fationis Jaonnis Penae Galli in Optica Euclidis, de usu Optices in Philosophia. Londini, Typris Jacobi Flesher.* Londini [London]: Jacobi Flesher. pp. 51–173 (in Gassendi 1653).

◇ The preface (pp. 59–87) contains two letters from Galileo to Kepler. Kepler's "Dioptrice" begins on p. 89.

◇ This is the third item in Gassendi 1653.

1858-1871. *Joannis Kepleri astronomi opera omnia. Edidit Ch. Frisch.* Frankofurti a.M. [Frankfurt]: Erlangae, Heyder & Zimmer. 8 vols. pp. xiv + 672; [v] + 838 + [3]; [vi] + 746 + [3]; [vi] + 664 + [1]; [vi] + 648; [viii] + 774 + [1]; [vi] + 837 + [1]; [viii] + 1028 + cxv + [1].

◇ Volume 1 contains a frontispiece of a devise showing orbits of the planets. Volume 6 has an elaborate frontispiece engraving astronomy. Volume 8 has a frontispiece portrait of Kepler.

Kepler, Johannes. See also Galileo 1880; Gassendi 1653; Lebon 1899; Maseres 1791.

Kiaes, J.

1866. *Traité élémentaire de géométrie descriptive. Première partie a l'usage des classes de mathématiques élémentaires et des candidats au baccalauréat ès sciences. Planches.* Paris: L. Hachette et Cie. 21 plates.

◇ The plates are marked "Géométrie Descriptive Première partie" at the top of each. The author's name is given as "Kioes" on this volume, but he held precisely the same positions as the "Kiaes" of the next volumes.

1870. *Traité élémenataire de géométrie descriptive. Deuxième partie a l'usage des classes de mathématique spéciales et des candidats a l'École normale supérieure at a l'École polytechnique.* Deuxième édition. Paris: Hachette et Cie. pp. 300.

○ The first part, presumably containing plates, is missing.

1882. *Traité élémenataire de géométrie descriptive. Deuxième partie a l'usage des classes de mathématique spéciales et des candidats a l'École normale supérieure at a l'École polytechnique.*   Cinquième édition. Paris: Hachette et Cie. pp. 395 + 42 plates.

⬦   Two volumes bound as one: textual material followed by 42 plates. There are two title pages, the first is "Texte", the second "Planches". The plates are marked "Géométrie Descriptive Deuxième partie" at the top of each.

1888. *Traité élémentaire de géométrie descriptive. Première partie a l'usage des classes de mathématiques élémentaires et des candidats au baccalauréat ès sciences.* Huitième Édition. Paris: Hachette et Cie. pp. iii + 298; + 39 plates.

⬦   This volume contains two parts; first the text portion and then 39 plates labeled "Géométrie Descriptive. Première partie".

n.d. *Traité élémentaire de géométrie descriptive. D ... a l'usage des ... et des candidats a ...*   [Paris: L. Hachette et Cie.] 32 plates.

⬦   The plates are marked "Géométrie Descriptive Deuxième partie" at the top.

○   Front cover torn badly and repaired. The ellispses above indicate what has been torn off. The repair incorrectly labels this volume as "Atlas". There is no title page.

Kirkby, John (1705–1748). See Barrow 1734.

Klein, Felix (1849–1925).
1893. *On Riemann's Theory of Algebraic Functions and Their Integrals. A Supplement to the Usual Treatises.* Translated from the German with the author's permission by Frances Hardcastle. Cambridge: Macmillan and Bowes. pp. xii + 76.

1897. *Famous Problems of Elementary Geometry: the Duplication of the Cube, the Trisection of an Angle, the Quadrature of the Circle, an Authorized Translation of F. Klein's Vortrage Uber Ausgewahlte Fragen der Elementargemetrie Ausgearbeitet von F. Tagert.* Translated by Wooster Woodruff Berman and David Eugene Smith. Boston: Ginn and Co. pp. ix + 80.

Klein, Felix. See Smith 1897.

Klügel, Georg Simon (1739–1812).
1770. *Analytische Trigonometrie.* Braunschweig: Fürstl. Waisenhausbuchhandlung. pp. [xvi] + 248 + 3 plates.
⬦   Chapter 8 deals with calculus, including differentiation of sine and cosine.

Knight, John George David (1846–1919; USMA 1868).
1880. *Notes on Determinants for the Use of Cadets.* West Point: United States Military Academy Printing Office. pp. [11].
⬦   This book was used as a text at USMA from 1880 until 1887.
○   The library has two copies. One copy has the title page and several initial (unnumbered) pages missing. A handwritten title page bound in.

Krafft, Wolfgang Ludwig. See Leonhard Euler 1772.

Kretz, Xavier [François-Xavier]. See Poncelet 1874.

Labey, J. B. (1750?–1825). See Leonhard Euler 1796, 1797; Leonhard Euler 1812.

La Caille, Nicolas Louis de (1713–1762).

1757. *Astronomiae fundamenta novissimis solis et stellarum observationibus stabilita lutetiae in Collegio Mazarinaeo et in Africa ad caput bonae spei peractis. A Nicolao–Ludovico de la Caille, in almâ studiorum Universitate Parisiensi matheseon professore, Regiae Scientiarum Academiae Astronomo, & earum quae Petropoli, Berolini, Holmiae & Bonnoniae florent, Academiarum Socio. Parisiis, E. Typographiâ J. J. Stephani Collombat, typographi ordinarii regis.* Parisiis [Paris]: Collombat. pp. [viii] + 243 + 1 plate + 27 pages of solar tables (dated 1758).

1765. *Leçons élémentaires de méchanique, ou traité abrégé du mouvement et de l'équilibre.* Nouvelle édition. Paris: H. L. Guerin & L. F. Delatour. pp. viii + 193 + 5 plates.

◇ Pages 177–192 bound between 160 and 161; p. 192 is after the plates.

1794. *Cours élémentaire et complet de mathématiques–pures.* Rédigé par Lacaille, augmenté par Marie, & éclairci par Theveneau. Paris: Courcier. pp. vi + 396 + [4] + 11 plates.

LaChapelle, de (1710–1792).

1750. *Traité des sections coniques, et autres courbes anciennes, appliquées ou appliquables à la practique de différens arts, tels que l'artillerie, l'architecture, la construction de miroirs ardens, des télescopes, des lunettes, des portevoix, des échos, des cornets acoustiques, ou des instrumens qui servent à corriger les défauts de l'Ouïe, etc.. Avec un petit traité de la cycloïde, où l'on fait voir comment cette courbe a contribué à perfectionner les horloges à pendule. Le tout enrichi de notes ou de dissertations historiques & critiques sur l'origine & le progrès des sciences & des arts, qui sont entrés dans le plan de cet ouvrage.* Paris: J. F. Quillau, Fils. pp. iv + xxv + xiv + 330 + 10 plates.

◇ Dedicated to the rector of the University of Paris. Errata are on two unnumbered pages immediately preceeding p. 1.

○ The signature of Andrew Ellicott, 1780, is on the title page. One sheet of notes is tucked in the back.

Lacroix, Silvestre François (1765–1843).

1800. *Traité des différences et des series, faisant suite au traité du calcul différentiel et du calcul intégral.* Paris: J. B. M. Duprat. pp. viii + 582.

◇ There is a table of references on pp. iii–viii. Pages 545–578 is an index of this volume and the 2 volumes on differential and integral calculus by paragraph number. Pages 579–582 are errata.

Lacroix, Silvestre François (Continued)

1804a. *Élémens d'algèbre, a l'usage de l'École centrale des Quatre–Nations.* Cinquième édition, revue et corrigée. Paris: Courcier. pp. xxxii + 358 + 1 fold out page facing 263.
o   Napoleonic Armorial Binding in green slipcase.

1804b. *Complément des élémens d'algèbre, a l'usage de l'École centrale des Quatre–Nations.* Troisième édition, revue et augmentée. Paris: Courcier. pp. viii + 315.
o   Napoleonic Armorial Binding in green slipcase. Follows Lacroix 1804a and precedes Lacroix 1805 in this binding.

1805. *Traité élémentaire d'arithmétique, a l'usage de l'École centrale des Quatre–Nations.* Sixième édition, revue et corrigée. Paris: Courcier. pp. viii + 156 + 1 fold out table facing p 81.
o   Napoleonic Armorial Binding in green slipcase. Follows Lacroix 1804b.

1810; 1814. *Traité du calcul différentiel et du calcul intégral.* 2 vols. Seconde édition, revue et augmentée. Paris: Courcier. pp. lvi + 652; xxi + 816.
◇   In Volume 2, there are two corrections which appear to have been added in proof on an unnumbered page between the Table of Contents and page 1. Following p. 154 of Volume 2, there is a large fold–out "synopsis" table of integrals. At the beginning of each volume there is a table of the principal works consulted for each chapter.

1812. *Essais de géométrie sur les plans et les surfaces courbes, (élémens de géométrie descriptive.)* Quatrième édition, revue et corrigée. Paris: Ve Courcier. pp. viii + 119 + 10 plates.

1813a. *Traité élémentaire de trigonométrie rectiligne et sphérique; et d'application de l'algèbre.* Sixième édition, revue et corrigée. Paris: Ve Courcier. pp. xii + 296 + 5 plates.
o   Owned by Samuel S. Smith, LT., Corps of Artillery.

1813b. *Traité élémentaire d'arithmétique a l'usage de l'École centrale des Quatre–Nations.* Treizième édition, revue et corrigée. Paris: Courcier. pp. viii + 156 + 1 fold out table facing p. 81.

1814. *Élémens de géométrie a l'usage de l'École centrale des Quatre–Nations.* Dixième édition, revue et corrigée. Paris: Courcier. pp. xlviii + 215 + 9 plates.

1815. *Élémens d'algèbre, a l'usage de l'École centrale des Quatre–Nations.* Onzième édition, revue et corrigée. Paris: Ve Courcier. pp. [xii] + 360.

1816a. *Essais sur l'enseignement en général, et sur celui des mathématiques en particulier.* Seconde édition, revue et corrigée. Paris: Ve Courcier. pp. vii + 358 + 6 plates.
o   The library has two copies.
o   One copy has a Thayer binding. Copy contains USMA Special Collections book plate. Copy donated by W. Trimble and is signed by FR Trimble.
o   The second copy has a regular binding.

Lacroix, Silvestre François (Continued)

1816b. *An Elementary Treatise on the Differential and Integral Calculus* [with an Appendix and Notes]. Cambridge: J. Smith for J. Deighton and Sons. pp. viii + 720 + 2 plates.
  ◇ Errata pp. 713–720.
  ○ Thayer binding.

1817. *Complément des éléments d'algèbre a l'usage de l'École centrale des Quatre-Nations.* Quatrième édition, revue et augmentée. Paris: Ve Courcier. pp. xiii + 328.

1819. *Élémens de géométrie, a l'usage de l'École centrale des Quatre-Nations.* Onzième édition, revue et corrigée. Paris: Ve Courcier. pp. xlviii + 216 + 9 plates.
  ◇ This book was "part of the elementary course of mathematics" by Lacroix; the course extended from arithmetic to calculus. The book contains a "Supplement" (pp. xxxii–xlvii) which reviews the arithmetic required for the study of geometry.

1820a. *Élémens d'algèbre, a l'usage de l'École centrale des Quatre-Nations.* Treizième édition, revue et corrigée. Paris: Ve Courcier. pp. xii + 360.
  ◇ Stamped "textbook USMA, 1825 to 18_" in gutter of title page.

1820b. *Traité élémentaire de calcul différentiel et de calcul integral.* Troisième édition, revue, corrigée, et augmentée. Paris: Ve Courcier. pp. xii + 634 + 5 Plates.
  ◇ Errata follow p. 634.
  ○ Thayer binding. Stamped "Text–Book USMA 1824 to 18_".

1820c. *An Elementary Treatise on Plane and Spherical Trigonometry, and on the Application of Algebra to Geometry; from the Mathematics of Lacroix and Bezout. Translated from the French for the Use of the Students of the University at Cambridge, New England.* First edition. Cambridge, N[ew] E[ngland]: Printed by Hilliard and Metcalf. pp. v + 162 + 5 plates.
  ◇ Translated by John Farrar.
  ○ A few annotations in book. A sketch of someone is on p. 57. The same individual is sketched in Davies 1826 *Descriptive Geometry.*

1822a. *Élémens de géométrie, a l'usage de l'École centrale des Quatre-Nations.* Douzieme édition, revue et corrigée. Paris: Bachelier (gendre et successeur de Mme Ve Courcier). pp. xlviii + 216 + 9 plates.

1822b. *Traité élémentaire de trigonométrie rectiligne et sphérique et d'application de l'algèbre a la géométrie.* Septième édition, revue, corrigée, et augmentée. Paris: Bachelier et Huzard, Successeurs de Mme Ve Courcier. pp. [xii] + 308 + 5 plates.

1825a. *Élémens de géométrie, a l'usage de l'École centrale des Quatre-Nations.* Treizième édition, revue et corrigée. Paris: Bachelier. pp. xlviii + 216 + 9 plates.
  ◇ No differences in the texts of Lacroix 1819, 1822a, 1825a were noted.

1825b. *Elements of Algebra, Translated from the French for the Use of the Students of the University of Cambridge, New England.* Second edition. Translated by John Farrar. Cambridge, New England: Hilliard and Metcalf. pp. xii + 276.
  ○ Stamped "Textbook USMA 1823 to ?" and annotated "Copy of William E. Basinger, Sept 1823."

Lacroix, Silvestre François (Continued)

1826. *An Elementary Treatise on Plane and Spherical Trigonometry: and the Applications of Algebra to Geometry; from the Mathematics of Lacroix and Bezout; Translated from the French for the Use of the Students of the University of Cambridge, New England.* Translated by John Farrar. Second edition. Cambridge, New England: Printed by Hilliard and Metcalf. pp. 165.
   o   The Library has two copies.
   o   One copy has "Cadet Jacob Broom Sept 1st 1827" written on the front endpaper. The spine is marked "Lacroix."

1828. *Traité élémentaire de calcul différentiel et de calcul intégral.* Quatrième édition, revue, corrigée et augmentée. Paris: Bachelier (Successeur de Mme Ve Courcier); Brussels: Parisienne. pp. xvi + 685 + 2 plates.
   ◇   Errata follow p. 685. This book was used as a text at USMA from 1820 until 1833.
   o   Two copies present in library.
   o   One copy has binding by A. J. Kilian, Paris.

1837. *Traité élémentaire de calcul différentiel et de calcul intégral.* Cinquième édition, revue, corrigée et augmentée. Paris: Bachelier. pp. xvi + 734 + 5 plates.

1854. *Éléments d'algèbre, a l'usage des candidats aux écoles du gouvernement.* Vingt et unième édition, revue, corrigée et annotée par M. Prouhet. Paris: Mallet–Bachelier. pp. xvi + 502.

1861; 1862. *Traité élémentaire de calcul différentiel et de calcul intégral.* 2 vols. Sixième édition, revue et augmentée de notes par Hermite et J.-A. Serret. Paris: Mallet–Bachelier. pp. xv + 460 + 5 plates; vii + 491.

La Gournerie, Jules de la (1814–1883).

1859. *Traité de perspective linéaire contenant les tracés pour les tableaux plans et courbes, les bas–reliefs et les décorations théatrales, avec une théorie des effets de perspective; ouvrage conforme au cours de perspective qui fait partie de l'enseignement de la géométrie descriptive au Conservatoire impérial des arts et métiers.* Paris: Dalmont et Dunod. pp. xxviii + 280 and 44 plates.
   ◇   Dedicated to General Arther Morin and to the author's collegues at the conservatoire. Reference is made to 45 plates, but none are included in this book.
   ◇   There are two different printings of this work.

Lagrange, Joseph Louis (1736–1813).

1797. *Théorie des fonctions analytiques, contenant les principes du calcul différentiel, dégagés de toute considération d'infiniment petits ou d'énvanouissans, de limites ou de fluxions, et réduits à l'analyse algébrique des quantités finies.* Paris: L'Imprimerie de la République. pp. viii + 277.
   ◇   Errata on p. 277.
   o   This book was owned by Michael Chasles.

1808. *Traité de la résolution des équations numériques de tous les degrés, avec des notes sur plusieurs points de la théorie des équations algébriques.* Nouvelle édition, revue et augmentée par l'auteur. Paris: Courcier. pp. xii + 311.
  ⋄ Errata follow p. 311. There are notes on pp. 101–311.
  ○ Thayer binding.

1811; 1815. *Méchanique analytique.* 2 vols. Nouvelle édition, revue et augmentée par l'auteur. Paris: Ve Courcier. pp. x + 422; viii + 378 (page numbers 367–368 have been used twice, the material is different.)
  ⋄ Errata for volume 1 follows p. 422. In volume 2, pp. 372–378, there is a list of Lagrange's works.
  ○ Thayer binding.

1853; 1855. *Mécanique analytique.* 2 vols. Troisième édition, revue, corrigée et annotée par M. J. Bertrand. Paris: Mallet–Bachelier. pp. xiv + 428; iv + 390 + 1.
  ⋄ Includes in Volume 1, the advertisements to each of the first three editions. At the end of Volume 2, there are "Fragments" (pp. 367–381), a list of Lagrange's works (pp. 383–389), the Report (November 3, 1817) of Lacroix on the manuscripts left by Lagrange (pp. 389–390), and errata for both volumes on p. [391].

Lagrange, Joseph Louis. See also Leonhard Euler 1807; Leonhard Euler 1822; Wronski 1812.

La Hire, Philippe de (1640–1718).
1694. *Memoires de mathématique et de physique.* Paris: L'Imprimerie Royale. pp. [viii] + 302 (except no numbers in the 100s are used).
  ⋄ Following this is bound *"Observations physiques et mathématiques, pour servir a l'histoire naturelle & à la perfection de l'astronomie & de la geographie: Envoyées des Indes et de la Chine à l'Academie royale des sciences à Paris, par les peres jesuites. Avec les reflexions de Mrs de l'Academie, & les notes du P. Goüye, de la compagnie de Jesus* [iv] + 114 + 20, but this not of mathematical interest.

La Hire, Philippe de. See also Mariotte 1700; Thévenot 1693.

Lalande, Joseph Jérôme Lefrançais de (1732–1807).
1803. *Bibliographie astronomique: avec l'histoire de l'astronomie depuis 1781 jusqu'à 1802.* Paris: l'Imprimerie de la République. pp. viii + 966.
  ⋄ The bibliography covers pp. 1–660. The "Histoire abrégée de l'astronomie, depuis 1781 to 1802," chronologically ordered is contained on pp. 661–880. An index of names and dates is presented on pp. 881–915. Finally, there is an index of technical terms correlated to the authors who used them, pp. 917–966.

Lalande, Joseph Jérôme Lefrançais de. See also Montucla 1792a.

Lambert, Gustave (?–1871).
1855. *Lettres sur les mathématiques et l'enseignement.* Paris: Victor Dalmont. pp. 362.

Lambert, Johann Heinrich (1728–1777).
1760. *I. H. Lambert Academiae Scientiarum Electoralis Boicae, et Societatis Physico-Medicae Basiliensis membri, Regiae Societati Scientiarum Goetingensi commercio literario adiuncti Photometria sive de mensura et gradibus luminis, colorum et umbrae.* Augustae Vindelicorum [Augsburg]: Sumptibus Viduae Eberhardi Klett. Typis Christophori Petri Detleffsen. pp. [xvi] + 547 + [13] + 8 plates.
⋄ Contains an index.

Lamé, Gabriel (1795–1870).
1818. *Examen des différents méthodes employées pour résoude les problèmes de géométrie.* Paris: Ve Courcier. pp. xii + 124 + 2 plates.
⋄ The plates follow p. 118, but are bound in the wrong order.

Landen, John (1719–1790).
1755. *Mathematical Lucubrations Containing New Improvements in Various Branches of the Mathematics.* London: Printed for J. Nourse. pp. 156 + 4 plates.

1764. *The Residual Analysis, a New Branch of the Algebraic Art, of very Extensive Use, Both in Pure Mathematics, and Natural Philosophy, Book I.* London: Printed for the Author. pp. viii + 128 + 5 plates.
⋄ The library has only Book I.

Landmann, Isaac (1741–1826).
1805. *Practical Geometry, For the Use of the Royal Military Academy at Woolwich.* Second edition, with considerable additons and improvements. London: Printed by I. Gold. pp. xix + 238 + 27 plates.

Lansberge, Philip van (1561–1632).
1591. *Philippi Lansbergii triangvlorvm geometriae liber qvatvor; in quibus nouâ & perspicuâ methodo, & λ'ποδει'ξει [apodeixei], tota ipsorum trian gulorum doctrina explicatur. Ad Senatum Populumá[ue] Middelburgensem. Lvgdvni Batavorvm, Ex officina Plantiniana, Lvgdvni Batavorvm, Ex officina Plantiniana, Apud Franciscum Raphelengium.* Middelburgi [Middleburg]: Apud Franciscum Raphelengium. pp. [xii] + 207.
⋄ Contains a few marginal notes in Latin.

1616. *Philippi Lansbergii cyclometraiae novae libri duo. Ad illustrissimum principem Mauricium Nassouim et illustres ac potentes Zeelandiae ordd.* Middelburgi [Middleburg]: Ex officina Richardi Schilders. pp. [viii] + 62.
⋄ Dedicated to Mauricio. Errata follow p. 62.
⋄ Introduction by Willebrord Snel.
○ Badly damaged on its outside edge.

Laplace, Pierre Simon (1749–1827).

1799; 1799; 1802; 1805; 1823. *Traité de mécanique céleste.* 5 vols. Paris: J. B. M. Duprat for volumes 1, 2, 3; Courcier for volume 4; Bachelier for volume 5. pp. xxx + 368; [iv] + 382; xxiv + 303 + 24; xl + 347 + [iii] + 65 + 1 plate + 79; [v] + 419 + 80 + 1 plate.

◇ Errata for volumes 1 & 2 follow p. xxx of volume 1; for volume 3 follow p. 303; for volume 4 follow p. 347 and follow p. 79 of the second supplelment; for volume 5 follow p. 419. Volume 3 is dedicated Bonaparte. Volume 3 contains the *Supplément au traité de mécanique céleste; Présenté au Bureau des Longitudes, le 17 Août 1808.* Volume 4 contains *Supplément au dixième livre du traité de mécanique celeste sur l'action capillaire* and the *Supplément a la théorie de l'action capillaire.* Volume 5 contains the *Nouvelles méthodes pour la détermination de l'orbite des comètes.*

○ Thayer binding. Volumes 1–4 were purchased by Thayer and McRee in Paris in 1817; the 5th volume was purchased from Kilian in Paris on February 10, 1826.

1809. *The System of the World.* 2 vols. Translated from the French by J. Pond. London: Printed for Richard Phillips. pp. viii + 379; iii + 380.

◇ Errata for vol. 1 are on p. [ii]. There is a list of books published by Richard Phillips on pp. 377–380.

1813. *Exposition du système du monde.* Quatrième édition, revue et augmentée par l'auteur. Paris: Ve Courcier. pp. viii + 457.

◇ Frontispiece portrait of Laplace. Errata follow p. 457.

○ Thayer binding.

1814a. *Théorie analytique des probablitiés.* Seconde édition, revue et augmentée par l'auteur. Paris: Ve Courcier. pp. cvi + 506.

◇ Errata follow p. 506.

○ Thayer binding.

1814b. *A Treatise upon Analytical Mechanics; being the First Book of the Mechanique Celeste of P. S. Laplace.* Translated and elucidated with Explanatory Notes by John Toplis. Nottingham: Printed by H. Barnett. pp. vii + 286.

◇ One page of errata follow p. 286.

1816. *Essai philosophique sur les probabilitiés.* Troisième édition, revue et augmentée par l'auteur. Paris: Ve Courcier. pp. 223.

◇ Errata on p. 223.

○ Thayer binding.

Laplace, Pierre Simon. See also Todhunter 1873.

Lardner, Dionysius (1793–1859).

1820. *The Elements of the Theory of Central Forces. Designed for the Use of the Students in the University.* Dublin: University Press by D. Graisberry. pp. xviii + 122 + 2 pages of errata.

1823. *A System of Algebraic Geometry.* 2 vols. London: Printed for G. and W. B. Whittaker. pp. lvi + 512.

◇ The library has volume 1 only. Corrigenda on pp. lv. and lvi. This is the first book on analytic geometry in English according to Young 1833, p. 5.

◦ Two sheets of uninteresting notes stuck in at pp. 16–17. The introduction and notes contain numerous historical comments.

1825. *An Elementary Treatise on the Differential and Integral Calculus.* London: Printed for John Taylor. pp. xxxiii + 520.
◇ Errata follow p. xxxiii.

1826. *An Analytical Treatise on Plane and Spherical Trigonometry and the Analysis of Angular Sections.* London: Printed for John Taylor. pp. xxviii + 313 + 9 tables.
◇ Dedicated to Charles Babbage.
◇ Includes a plate (facing p. 216) illustrating the triangularization for measuring the meridian between Dunkirk and Paris.

Launay, L. (1816–1872).
1883. *Premiers élements d'algèbre contenant plus de 700 questions et problems.* Seconde an. Paris: Hachette et Cie. pp. [iii] + 527.

Laurent, Hermann (1841–1908).
1853. *Traité de calcul différentiel a l'usage des aspirants au grade de licencié ès sciences mathématiques.* Paris: Mallet–Bachelier. pp. xvi + 480.
◇ Dedicated to Monseigneur L'Eveque de la Rochelle. Errata on pp. xv–xvi.

1885; 1887; 1888; 1889; 1890; 1890; 1891. *Traité d'analyse.* 7 vols. Paris: Gauthier–Villars. pp .xi + 397; 478; [v] + 513; 456; 417; 339; 342.
◇ Errata: volume 1, pp. 391–397; volume 2, pp. 476–478; volume 3, pp. 509–513; volume 4, pp. 454–456; volume 5, pp. 416–417; volume 6, p. 339; volume 7, pp. 341–342. Dedicated to A. M. Moutard. Volume 1 is subtitled "differential calculus, analytic and geometric applications"; volume 2 is subtitled "differential calculus, geometric applications"; volume 3 is subtitled "integral calculus, definite and indefinite integrals"; volume 4 is subtitled "integral calculus, theory of algebraic functions and their integrals"; volume 5 is subtitled "integral calculus, ordinary differential equations"; volume 6 is subtitled "integral calculus, partial differential equations"; volume 7 is subtitled "integral calculus, geometric applications of the theory of differential equations."

1897. *Traité d'algèbre, a l'usage des candidats aux écoles du gouvernement.* Cinquième édition, en harmonie avec les nouveaux programmes, revue par J.-H. Marchand. Première partie, a l'usage des classes de mathématiques élémentaires. Paris: Gauthier–Villars. pp. xxi + [227] + [270] + [211] + [53].
◇ This is four books in one. Book 1 (1897), book 2 (1894), book 3 (1894), book 4 (1894). Book 4 has a different title: *Traité d'algèbre, compléments.*

Lavit, J. -B. -O.
1804. *Traité de perspective.* 2 vols. Paris: P. Didot l'aine. pp. viii + 274; 50 plates + 59 plates.
◇ The title page for volume 2 is missing. Errata follow p. 274 in volume 1.

Lawson, John. See Apollonius 1764; Bonnycastle 1818b.

Lea, W.

1811. *A Treatise on the Resolution of the Higher Equations in Algebra.* London: Printed for J. Johnson. pp. vii + 40.

◇ Errata follow p. vii.

LeBlond, Guillaume (1704–1781).

1767a. *Abrégé de l'arithmetique et de la géométrie de l'officier; contenant les quatre premieres opérations de l'arithmétique; les regles de trois & de compagnie; les principes de la géométrie les plus utiles pour l'intelligence & la pratique des fortifications, & pour lever des cartes & des plans; le toisé des surfaces, & un abrégé de celui des solides.* Troisieme édition, revue & corrigée. Paris: Ch. Ant. Jombert. pp. vii + 421 + 19 plates.

◇ Presented by the Minister of Public Instruction of France in 1847.

1767b. *L'arithmétique et la géométrie de l'officier, contenant la théorie et la pratique de ces deux sciences, appliquées aux différens emplois de l'homme de guerre.* Seconde édition, corrigée et augmentée. Tome premier. Paris: Charles–Antoine Jombert. pp. xvi + 497 + 18 plates.

◇ Dedicated to the Dauphin. Errata on p. 497.

○ The library has volume 1 only.

Lebon, Ernest (1846–1922).

1899. *Histoire abrégée de l'astronomie.* Paris: Gauthier–Villars. pp. iii + 288.

◇ Frontispiece: H. Faye. Also contains portraits of Copernicus, Galileo, Kepler, Newton, W. Herschel, Laplace, Arago, LeVerrier, J. Janssen, M. Loewy, F. Perrier, S. Newcomb, F. Tisserand, Sophie Kowalevski, and H. Poincaré.

1910. *Émile Picard biographie, bibliographie analytique des écrits.* Savants du jour. Paris: Gauthier–Villars. pp. viii + 80.

◇ Frontispiece portrait of Picard, half–tone. Pages 1–13 are the biography, and pp. 14–80 contain the bibliographies. The bibliographies are arranged by subject and each entry is annotated.

Lee, Chauncey (1763–1842).

1797. *The American Accomptant; Being a Plain, Practical and Systematic Compendium of Federal Arithmetic; in Three Parts: Designed for the Use of Schools, and especially Calculated for the Commercial Meridan of the United States of America.* Lansingburgh, NY: William W. Wands. pp. xlii + 43–298 + 12 pages of subscribers.

◇ The first use of a dollar sign in an American work appears on p. 56.

◇ Frontispiece contains illustrations of both sides of various coins, including the American eagle, the Spanish pistole, coins of French and British Guinea, and others.

Lee, Thomas Jeffferson (1808–1891; USMA 1830).

1853. *A Collection of Tables and Formulae Useful in Surveying, Geodesy, and Practical Astronomy, Including Elements for the Projection of Maps Prepared for the Use of the Corps of Topographical Engineers.* Second edition, with additions.

Washington: Taylor and Maury. pp. xviii + 239.
  ○ Insert contains errata for Lee's Tables by Professor Bartlett with notes in margins p. 3. Autographed by E.W. Bass, Corps of Engineers; notes inside cover.

Lefébure de Fourry, Louis Étienne (1785–1869).
1827. *Leçons de géométrie analytique.* pp. 544 + 11 plates.
  ◇ Title page is missing; title is taken from Larousse. Errata on pp. 543–544.

Lefèvre, A.
1819. *Manuel du trigonomètre, servant de guide aux jeunes ingénieurs qui se destinent aux opérations géodésiques; suivi de diverses solutions de géométrie pratique, de quelques notes et de plusieurs tables.* Paris: Ve Courcier. pp. [iv] + 284 + 2 plates.
  ◇ Errata follow p. 284. Foldout sheet facing p. 207.

Lefèvre, Arthur (1863–1929).
1896. *Number and its Algebra. Syllabus of Lectures on the Theory of Number and its Algebra Introductory to a Collegiate Course in Algebra.* Boston: D. C. Heath and Co. pp. i + 230.
  ◇ Errata on page 231.

Lefèvre, Arthur (not the Arthur Lefèvre listed above). See also L'Hospital 1781.

Lefrançois, F.L. (1732–1807).
1801. *Essais sur la ligne droite et les courbes du second degré.* Paris: Duprat. pp. [vi] + 142 + 2 plates.
  ◇ Errata on p. [v].
  ○ Signature of Andrew Ellicott on a front paper.

Legendre, Adrien Marie (1752–1833).
1806. *Nouvelles méthodes pour la détermination des orbites des comètes; avec un supplément contenant divers perfectionnemens de ces méthodes et leur application aux deux comètes de 1805.* Paris: Courcier. pp. viii + 80 + 1 plate + 55.
  ◇ Errata on p. 55 of the supplement.
  ○ Thayer binding.

1808. *Essai sur la théorie des nombres.* Seconde édition. Paris: Courcier. pp. xxiii + 1 page of errata + 480 + 34 unnumbered pages of tables.
  ◇ Errata follow p. xxiii.
  ○ Library has two copies.
  ○ One copy has a Thayer binding. Bound at the end of this copy is the *Supplément a l'essai sur la théorie des nombres, seconde édition.* Février 1816. Separately paginated: pp. 62 + 1 plate.
  ○ Second copy has 2 pages of advertisements. There is an unreadable signature inside the front cover.

Legendre, Adrien Marie (Continued)

1811; 1817; 1816. *Exercises de calcul intégral sur divers orders de transcendantes et sur les quadratures.* 3 vols. Paris: Ve Courcier. pp. 386 + 51 + plate; [xix] + 544 + 1 plate; 462.

◇ Errata for volume 1 follow p. 386, then there are 51 pages of supplementary exercises. Errata for volume 2 follow p. xix. Also p. 85–95 of volume 2 contain a table of the logarithms of the gamma function, together with first, second and third differences. Errata for the third volume follow p. 462.

◇ Volume 1 begins with elliptic functions. Volume 3 is devoted to the construction of elliptic tables.

1813. *Éléments de géométrie avec des notes.* Dixième édition. Paris: Firmin Didot. pp. [iv] + 431 + 14 Plates.

◇ Pages 337–431 deals with trigonometry.

○ Thayer binding.

1816. *Exercices de calcul intégral. Méthodes diverses pour la construction des tables elliptiques, suivies de la table générale des fonctions complètes, de la table particulière pour le module sin 45°, etc.* Bound together with Legendre 1811. pp. 124 pages + unnumbered pages of tables + 50.

◇ Errata follow p. 50.

1825. *Elements of Geometry, Translated by John Farrar, For the Use of the Students of the University at Cambridge, New England.* Second edition, corrected and enlarged. Cambridge, New England: Hilliard and Metcalf, at the University Press. pp. xv + 224 + 13 plates.

○ Stamped "Textbook USMA 1823 to 18?"

1828. *Elements of Geometry and Trigonometry with Notes. Revised and Altered for Use of the United States Military Academy.* Translated by David Brewster. New York: James Ryan. pp. viii + 316.

◇ Early version of Davies' *Legendre*. Davies refers to himself as "Editor" in the Preface.

○ Three copies are in the library.

○ One copy has book plate "Thayer Pubic Library 1874."

○ Another copy has book plate from USMA.

1829. *Elements of Geometry, by A. M. Legendre, Translated from the French, for the Use of the Students of the University at Cambridge, New England.* Translated by John Farrar. Boston: Hilliard, Gray, Little, and Wilkins. pp. xv + 224 + 13 plates.

○ Copy lent by heirs of Gen. J. G. Swift.

1830. *Elements of Geometry and Trigonometry with Notes. Revised and Altered for Use of the United States Military Academy.* Translated by David Brewster. Second edition. New York: White, Gallaher and White. pp. viii + 2–316.

◇ Early version of Davies *Legendre*. Davies name does not appear.

○ Property of Frederick Picking Drew, Sep 1843, on front matter.

Legendre, Adrien Marie (Continued)

1832. *Elements of Geometry and Trigonometry; with Notes. Translated From the French of A. M. Legendre by David Brewster, Revised and Altered for the Use of The Military Academy at West Point.* Third edition. New York: N. & J. White; Collins & Hannay; Collins & Co.; and James Ryan. pp. xvi + 316.

◇   Early version of Davies *Legendre.* Davies' name does not appear.

◇   There are four "notes" appended to the end of the geometry, pp. 239–256. The geometry is pp. 1–256, and the trigonometry comprises pp. 257–316.

◇   The Preface is dated "West Point 1828." James Ryan claims the copyright.

○   Pages 147–150 torn out and missing.

1833. *Elements of Geometry, by A. M. Legendre, Translated from the French, for the Use of the students of the University at Cambridge, New England.* New edition improved and enlarged, translated by John Farrar. Boston: Hilliard, Gray, and Co. pp. xv + 235 + 16 plates.

○   Signed "Ino Bratt, 1833".

Legendre, Adrien Marie. See also Delambre 1799; Venable 1877, Davies (Geometry and Trigonometry).

Le Gendre, François.

1707. *L'arithmetique en sa perfection, mise en pratique selon l'usage des financiers, banquiers et marchands. Contenant une ample & familier explication de ses principes tant en nombres entiers qu'en fractions. Traité de geometrie pratique appliquée à l'arpentage au toisé tant des superficies que des corps solides. Un abregé d'algebre, suivi de quantité de questions non moins curieuses que necessaires.* Douxième edition. Lyon: Benoist Vignieu. pp. [viii] + 578.

◇   Presented by the Minister of Public Instruction of France.

Leibniz, Gottfried Wilhelm, Freiherr von (1646–1716).

1789. *Opera omnia, nunc primum collecta, in classes distributa, praefationibus et indicibus exornata studio Ludovici Dutens.*   6 vols. Edited by Ludovicus Dutens (1730–1812). Coloniae Allobrog [Geneva]: apud Piestre & Delamolière. Berolini [Berlin]: apud Christianum Frid. Himburg. pp. iv + ccxliv + 790; viii + 400 + 291 + xiv plates; viii + lv + 215 + 285 + 663 + xxv plates; viii + 285 + 647; viii + 632; vi + 344.

◇   The contents of the individual volumes are: (1) theology; (2) logic, metaphysics; general physics, etc; (3) mathematics; (4) philosophy, history, law; (5) philology; (6) philology and etymology. Each volume has its own table of contents. There are indices for volumes 1 and 2, and a combined index for 5 and 6.

◇   Frontispiece in volume 1 is a portrait of Leibniz.

◇   Volume 1 contains the eulogy of Fontenelle (1716) and a vita by Brucker. First edition was 1768.

Leibniz, Gottfried Wilhelm, Freiherr von (1646–1716) and Bernoulli, Johann [Jean] (1667–1748).

1745. *Virorum celeberr. Got. Gul. Leibnitii et Johan. Bernoullii commercium* *philosophicum et mathematicum.* 2 vols. Lausannae et Genevae [Lausanne and Geneva]: Marci–Michaelis Bousquet & Socior. pp. [iv] + xxviii + 484 + 15 plates; 492 + 23 plates.

◇ Dedicated to Ludovic XIV and Ludovic XV. The index for both volumes is in volume 1, pp. iv–xxviii. Volume 1 covers the period 1694–1699, and volume 2 covers 1700–1716. At the end of volume 2, pp. 397–492, there is an alphabetized "Index Rerum Natabiliorum."

◇ See Figure 18 for a photograph of the frontispiece of Johannes Bernoulli.

Leibniz, Gottfried Wilhelm, Freiherr von. See also Collins 1725.

Le Monnier, Pierre Charles (1715–1799).
1774. *Description et usage des principaux instruments d'astrononomie.* Paris?: Publisher unknown. pp. iv + 60 + 14 plates.

◇ Errata on p. iv.
◇ No publisher or place of publication stated.

Lenoir, B. A.
1828. *Calculs faits, a l'usage des industriels en général, et spécialement des mécaniciens, charpentiers, pompiers, serruriers, chaudronniers, toiseurs, etc., etc.* Paris: Librairie Scientifique–Industrielle de Malher et Cie. pp. viii + 262.

Leroy, Charles François Antoine (1780–1854).
1829. *Analyse appliquée a la géométrie des trois dimensions, comprenant les surfaces du second ordre, avec la théorie général des surfaces courbes et des lignes à double courbure.* Paris: Bachelier (Successeur de Mme Ve Courcier); Brussells: Librairie Parisienne. pp. xvi + 308 + 5 plates.

1834. *Traité de géométrie descriptive, avec une collection d'épures, composée de 60 planches.* 2 vols. Paris: Carilian–Goeury. pp. xx + 390; 60 plates.
◇ Volume 1 is the text, and volume 2 is the plates.

1845. *Traité de stéréotomie comprenant les applications de la géométrie descriptive, a la théorie des ombres, la perspective linéaire, la gnomonique, la coupe des pierres et la charpente; avec un atlas composé de 74 planches in–folio.* 2 vols. Liege: Dominique Avanzo et Cie. pp. xvi + 485.
◇ Volume 1 is the text, and volume 2 is the plates.

1896. *Traité de géométrie descriptive, suivi de la méthode des plans cotés et de la théorie des engrenages cylindriques et coniques, avec une collection d'épures, composée de 71 planches.* 2 vols. Quatorzième édition, revue et annotée par E. Martelet. Paris: Gauthier–Villars et Fils. pp. xx + 369.
◇ Volume 1 is the text, and volume 2 is the plates. Rebound in 1952.

Leseur, Thomas (1703–1770). See Newton 1822.

Leslie, John (1766–1832).
1811. *Elements of Geometry, Geometrical Analysis, and Plane Trigonometry with*

an Appendix, and Copious Notes and Illustrations. Second edition, improved and enlarged. Edinburgh: Printed for John Ballantyne and Company. pp. ix + 500.

1820a. *Elements of Geometry and Plane Trigonometry. With an Appendix, and Very Copious Notes and Illustrations.* Fourth edition improved and expanded. Edinburgh: Printed for W. and C. Tait. pp. xii + 465.

⋄   Advertisements for six books by Leslie are followed by four pages of other advertisements. This book is "the first of a projected Course of Mathematical Science" (from the preface, p. v).

1820b. *The Philosophy of Arithmetic Exhibiting A Progressive View of the Theory and Practice of Calculation With Tables for the Multiplication of Numbers as for as One Thousand.* Second edition, improved and enlarged. Edinburgh: Printed by Abernethy and Walker for William and Charles Tait. pp. [iv] + 1 fold–out table + 258.

⋄   This work is a historical introduction to arithmetic.

1821. *Geometrical Analysis and Geometry of Curve Lines, Being Volume Second of a Course of Mathematics, and Designed as an Introduction to the Study of Natural Philosophy.* Edinburgh: Printed for W. and C. Tait. pp. ix + 448 + 24 plates.

⋄   "Chiefly drawn from the writings of Huygens and two Bernoullis" [p. ix].

1823. *Elements of Natural Philosophy, Volume First [No more published], Including Mechanics and Hydrostatics.* Edinburgh and London: W. & C. Tait. pp. lvi + 406 + 2 page list of books + 1 page of errata + 10 plates.

⋄   The introduction contains a history of the field.

Leslie, John. See also Hachette 1818.

Lespinasse, Louis Nicolas de (1734–1808).
1801. *Traité de perspective linéaire à l'usage des artistes: contenant la pratique de cette science, d'apres les meilleurs auteurs; les méthodes les plus simples, pour mettre toutes sortes d'objets en perspective, leurs réflexions dans l'eau, et leurs ombres, tant au soleil qu'au flambeau.* Paris: Magimel. pp. xiv + 231 + 26 plates.

○   Bookplate War Department Library. Rebound by the War Department.

Lexell, Anders Johan (1740–1784). See Leonhard Euler 1772.

Leybourn, Thomas (1770–1840).
1817. *The Mathematical Questions, Proposed in the Ladies Diary, and Their Original Answers, Together with some New Solutions, from its Commencement in the Year 1704 to 1816.* 4 vols. London: J. Mawman, et al. pp. xi + 415; 416; 400; 440.

⋄   Volume 1 includes questions posed 1707–1747; volume 2 questions posed 1748–1755; volume 3 questions posed 1776–1801; and volume 4 questions posed 1802–1815. There is an Appendix (volume 4, pp. 232–414) of additional comments, solutions to some of the problems. Volume 4 features an Index, alphabetized, of the names of persons who have proposed and/or solved problems, pp. 415–436, and an Index in which the problems are classified, pp. 437–440.

L'Hospital, Guillaume François Antoine de (1661–1704).

1720. *Traité analytique des sections coniques et de leur usage pour la resolution des equations dans les problêmes tant déterminez qu'indéterminez.* Ouvrage Posthume. Paris: Montalant. pp. [viii] + 459 + 33 plates.

1723. *An Analytick Treatise of Conick Sections And Their Use for Resolving of Equations in Determinate and Indeterminate Problems. Being The Posthumous Work of the Marquis De L'Hospital.* Translated by E. Stone. London: J. Senex et. al. pp. viii + 351 + 3 plates.

◇   Errata follow p. 351.

◇   See Figure 12 for a photograph of the title page of this volume.

○   This volume has been rudely trimmed and some plates and marginal notes are damaged.

1781. *Analyse des infiniment petits, pour l'intelligence des lignes courbes.* Nouvelle édition, revue & augmentée par M. Le Fevre.   Paris: Alex. Jombert, jeune. pp. xix + 234.

◇   This is a relatively late edition of the first book solely devoted to calculus, L'Hospital's (1696).   The eight notes by LeFevre are set in smaller type.   The preface, which we now know to be by Fontenelle, contains a brief history of calculus up to the time of Newton and Leibniz. L'Hospital's rule occurs on pp. 186–187.

L'Hospital, Guillaume François Antoine de. See also Stone 1730.

L'Huilier, Antoine Jean Simon (1750–1840).

1804. *Élémens raisonnés d'algèbre publiés à l'usage de étudians en philosophie.*   2 vols. Geneva: J. J. Paschoud. pp. iv + 408; 451.

◇   Errata on the verso of the title page of both volumes.

Libes, Antoine (1752–1832).

1810; 1810; 1812; 1813. *Histoire philosophique des progrès de la physique.* 4 vols. Paris: The Author; Courcier; Michand. pp. vii + 291; 303; 330; 272.

◇   Volume 1 covers the period up to Descartes; volume 2 from Descartes to Newton; volume 3 from Newton to "the birth of pneumatic chemistry;" and volume 4 to "our day." Each volume concludes with an extensive (15 page average) index for that volume.

Libri, Guillaume (1803–1869).

1838; 1838; 1840; 1841. *Histoire des sciences mathématiques en Italie, depuis la renaissance des lettres jusqu'a la fin du dix–septieme siécle.* 4 vols. Paris: Jules Renouard et Cei. pp. xxxii + 464; [viii] + 530; [viii] + 461; [viii] + 492.

◇   Dedicated to the author's friends who have left Italy. Errata for volumes 1 and 2 follow p. 530 of volume 2; errata for all four volumes follow p. 492 of volume 4.

Lochée, Lewis (? –1791).

1776. *A System of Military Mathematics.* 2 vols. London: Printed for the author. pp. [iii] + 262 + ix; [iii] + 169 + [xv] + 26 plates.

    ○ Purchased with donations to the Library Gift Fund. Torn page (9), table of contents in the back of both volumes, book worm in second volume.

Lockhart, James (1763–1852).

1813. *A Method of Approximating Towards the Roots of Cubic Equations Belonging to the Irreducible Case.* London: Printed for the Author. pp. vi + 7–87.

    ◇ Dedicated to Isaac Milner. Errata on p. 87.

Loomis, Elias (1811–1889).

1854. *Elements of Analytical Geometry and of the Differential and Integral Calculus.* Sixth edition. New York: Harper and Brothers. pp. xii + 278.

1872, 1881. *The Elements of Analytical Geometry.* New York: Harper and Brothers. pp. viii + 261.

    ○ Rebound in 1938 with *Elements of the Differential and Integral Calculus* by Loomis.

Lorgna, Antonio Maria (1735–1796).

1779. *A Dissertation on the Summation of Infinite Converging Series with Algebraic Divisors. Exhibiting a Method not only Intirely New, but much more General than any Other which has Hitherto Appeared on the Subject. Translated from the Latin of A. M. Lorgna, with Illustrative Notes and Observations. To which is Added, an Appendix; Containing all the most Elegant and Useful Formulae which have been Investigated for the Summing of the Different Orders of Series; with Various Examples to Each.* Translated by H. Clarke. London: Printed for the Author. pp. xx + 221 + 2 plates.

    ◇ Dedicated to Charles Hutton. Lorgna's *Dissertation* occupies pp. 1–127 and Clark's "Appendix" pp. 129–221. Errata follow p. 227.

Love, Augustus Edward Hough (1863–1940).

1909. *Elements of the Differential and Integral Calculus.* Cambridge: University Press. pp. xiii + 207.

    ◇ Corrigenda follow p. xiii.

Lubbe, Samuel–Ferdinand (1786–1846).

1832. *Traité de calcul différentiel et de calcul intégral.* Traduit de l'Allemand, et augmenté de plusieurs notes et solutions par Maurice Kartscher. Paris: Bachelier Père et fils. pp. xii + 411 + 2 plates.

    ◇ This is a translation of *Lehrbuch des hohern Kalkuls, fur Lehrer und Selbstlernende,* by Samuel Ferdinand Lubbe, published by G. Hayn, Berlin, 1825 (with 20 figures). Lubbe was a Privatdozent at the University of Berlin. (From the translator's Preface.) The notes and solutions are contained in an appendix.

Ludlam, William (1717–1788).

1809. *The Rudiments of Mathematics: Designed for the Use of Students at the Universities. Containing an Introduction to Algebra, Remarks on the First Six Books of Euclid, the Elements of Plane Trigonometry.* Fifth edition corrected and enlarged by M. Fryer. London: Printed for W. Baynes, et. al. pp. vi + 352 + plates.

Ludlow, Henry H. (1854–1926; USMA 1876).

1888. *Elements of Trigonomety.* New York: John Wiley and Sons. pp. iv + 192.

◇ Contains an order from the War Department, 9 Feb 1888, approving this book as a textbook for USMA.

○ Presented by Dept of Math (USMA) to Library Apr 26, 1895. Some marginal notes and corrections written in book.

n.d. *Exercises in Trigonometry.* West Point: Printed at West Point, USMA. pp. 41.

Ludlow, Henry H. (1854–1926) and Bass, Edgar W. (1843–1918).

1890. *Elements of Trigonometry, with Logarithmic and Other Tables.* New York: John Wiley and Sons. pp. iv + 174.

◇ This book was used as a text at USMA from 1888–1906.

○ Signed by Buell B. Bassette, 1890, 4th Class (freshman); contains numerous student notes.

1893. *Elements of Trigonometry with Logarithmic and Other Tables.* Third edition. New York: John Wiley and Sons. pp. [viii] + 174 + [vi] + 32 + 89.

◇ The tables carry the separate title "Logarithmic, Trigonometric and Other Mathematical Tables" [Second edition, second thousand, 1895].

○ Stamped "Textbook USMA 1892 to ?"

1896, [1888]. *Elements of Trigonometry with Logrithmic and Other Tables.* Third edition. New York: John Wiley & Sons. pp. iii + 174 + 16 (adds).

◇ This book was used as a text at USMA from 1888 to 1906.

○ Notes written in margin. Course outline inside front cover. Notes in blue – lower section (omitted material). Notes in red – upper section (omitted material).

1899. *Logarithmic, Trigonometric, and Other Mathematical Tables.* New York: John Wiley and Sons. pp. v + 31+ 89 pages of tables.

◇ This book was used as a text at USMA from 1892 until 1906.

○ Corrections and additions pasted into the text: "corrected copy" on front paper.

1900 [1888]. *Elements of Trigonometry with Logarithmic and Other Tables.* Third edition. New York: John Wiley and Sons. pp. [iii] + 174 + [ii] + 32 + 89 + 16.

○ Copy contains "Book Plate USMA: Gift of COL Perry Huston Eubank, USMA Class of 1937." Copy bound with tables. Also includes numerous students notes throughout.

Lull, Raymond (1235?–1315).

1598. *Raymundi Lullii opera eaquae ad adinventam ab ipso artem universalem, scientiarum artiumque omnium brevi compendio, firmaáque memoria apprehendendarum, locupletilsimaáque vel oratione ex tempore pertractandarum, pertinent. Ut et in eandem quorundam interpretum scripti commentarij; que omnia sequens indicabit pagina: & hoc demùm tempore coniunctim emendatiora locupletior aq non* *nihil edita sunt, accessit index cum capitum M, tùm rerum ac verborum locupletissimus.* Argentinae [Strasbourg]: Sumtibus Lazari Zetzneri. pp. [xxiv] + 992 + [xxxi].

o   Two copies present in the library.

Machin, John (?–1751). See Newton 1819.

Macfarlane, Alexander (1851–1913).
1906. *Vector Analysis and Quaternions.* Mathematical Monographs No. 8. Edited by Mansfield Merriman and Robert S. Woodward. Fourth edition. New York: John Wiley & Sons. pp. [x] + 50 + 18.

◇   The last 18 pages contain a short–title catalog of books published by John Wiley & Sons.

MacKay, Andrew (1760–1809).
1804. *A Collection of Mathematical Tables for the Use of Students in Universities and Academies, For the Practical Navigator, Geographer, and Surveyor, for Men of Business & c.* London: Printed for Longman, Hurst, Rees, and Orme; et. al. pp. xv + 260.

◇   Dedicated to George, Marquis of Huntly. Errata follow p. xv.

1811.   *The Description and Use of the Sliding Rule in Arithmetic, and in the Mensuration of Surfaces and Solids. Also, the Description of the Ship Carpenter's Sliding Rule, and its Use Applied to the Construction of Masts, Yards, &c. Together with the Description and Use of Gauging Rule, Gauging Rod, & Ullage Rule.* Second edition improved, enlarged, and illustrated with an accurate engraving of the different rules. Edinburgh: Printed for Oliphant, Waugh, and Innes. pp. [viii] + 138 + 1 plates.

◇   Errata on the verso of the title page.
◇   The preface is dated 1806.

1812.   *The Description and Use of the Sliding Gunter in Navigation.* Second edition improved and enlarged. Leith: Archibald Allardice. pp. viii + 168 + 1 plate.

◇   Frontispiece portrait of the author.

Maclaurin, Colin (1698–1746).
1720. *Geometria organica: sive descriptio linearum curvarum universalis.* Londini [London]: Impensis Gul. & Joh. Innys. pp. [xii] + 139 + 12 plates.

◇   Dedicated to Issac Newton. Errata following p. 139.

1742. *A Treatise on Fluxions in Two Books.* 2 vols. Edinburgh: Printed by T. W. and T. Ruddimans. pp. vi + 412; 413–762.

◇   One page of errata at the end of the second volume. Dedicated to the Duke of Argyle and Greenwich.
o   Some notes in the margins.

1749. *Traité des fluxions.* 2 vols. Traduit de l'Anglois par R. P. Pezenas. Paris: Charles–Antoine Jombert. pp. xvi + li + 344; viii + 322.

◇ At the end of each volume there are several unnumbered pages of errata.

◇ See Figure 19 for a photograph of the title page of this volume.

1750. *An Account of Sir Isaac Newton's Philisophical Discoveries, in Four Books. Published from the Author's Manuscript Papers, by Patrick Murdoch.* Second edition. London: Printed for A. Miller. pp. [x] + xxvi + 412 + 6 plates.

◇ Dedicated by Anne Maclaurin (the author's widow) to the "Duke." The six plates are arranged, I, II, IV, III, V, VI. Errata on p. [x]. There is a short biography of Maclaurin on pp. i–xxvi.

1779. *A Treatise of Algebra, in Three Parts, Containing I. The Fundamental Rules and Operations. II. The Composition and Resolution of Equations of all Degrees; and the Different Affections of Their Roots. III. The Application of Algebra and Geometry to each other. To which is Added, an Appendix Concerning the General Properties of Geometrical Lines.* Fourth edition. London: Printed for J. Nourse et al. pp. [xiv] + 504 + 12 plates.

◇ Dedicated to Thomas, Lord Archbishop of Canterbury by Anne Maclaurin.

◇ Contains "Appendix: de linearum geometricarum proprietatibus generalibus tractatus" and also "Appendix: Being a Treatise Concerning the General Properties of Geometrical Lines. Translated from the Latin by John Lawson, B. D."

○ Flyleaf signed "I. D. Campbell, 1820."

Macnie, John.
1876. *A Treatise on The Theory and Solution of Algebraical Equations.* New York: A. S. Barnes and Co. pp. x + 184.

○ Rebound in December 1883.

Magdeleine, Pierre de Sainte Marie, Dom. See Pierre de Sainte Marie Magdeleine, Dom.

Mahan, Dennis Hart (1802–1871; USMA 1824).
1867. *Descriptive Geometry, as Applied to the Drawing of Fortification and Stereotomy. For the Use of the Cadets of the U. S. Military Academy.* New York: John Wiley. pp. [iii] + 55 + 8 plates.

◇ Mahan served as Professor and Department Head, Department of Engineering, USMA (1832 – 1871).

○ The library has two copies.

1877. *Descriptive Geometry, as Applied to the Drawing of Fortification and Stereotomy. For the Use of the Cadets of the U. S. Military Academy.* New edition with additions. New York: John Wiley. pp. [iii] + 60 + 12 plates.

◇ Copyright 1864.

○ Signed by Greble.

1900. *Descriptive Geometry, as Applied to the Drawing of Fortification and Stereotomy. For the Use of the Cadets of the U. S. Military Academy.* New edition with additions. New York: John Wiley and Sons. pp. [iii] + 60 + 12 plates + 16 pages of advertisements of John Wiley books.

◇   Copyright 1891.

◇   Editions of this book were used as textbook at USMA starting in 1841.

Malcolm, Alexander (1687–ca 1749).

1730. *A New System of Arithmetick, Theorical and Practical. Wherein the Science of Numbers is Demonstrated in a Regular Course from its First Principles, thro' all the Parts and Branches thereof; Either known to the Ancients, or Owing to the Improvements of the Moderns. The Practice and Application to the Affairs of Life and Commerce being also Fully Explained: So as to make the Whole a Complete System of Theory, for the Purposes of Men of Science; And of Practice, for Men of Business.* London: for J. Osborne and T. Longman. pp. xx + 623.

◇   Dedicated to William Cruikshank, James Morrison, John Gordon, William Mowat, Hugh Hay, Alexander Livingston, Alexander Robertson, "And the Remnent Members of the Town–Council of Aberdeen."

◇   Errata follow p. 623.

Mannheim, Amédée (1831–1906).

1857. *Transformation des propriétés métriques des figures a l'aide de la théorie des polaires réciproques.* Paris: Mallet–Bachelier. pp. xix + 59.

1886. *Cours de géométrie descriptive de l'École polytechnique, comprenant les éléments de la géométrie cinématique.* Illustré de 256 figures dans le texte. Deuxième édition. Paris: Gauthier–Villars. pp. xix + 480.

Mannheim, Amédée. See also Poncelet 1862.

Manning, Henry P. (1859–1956) (Editor).

1910. *The Fourth Dimension Simply Explained: A Collection of Essays Selected from those Submitted in the Scientific American's Prize Competition with an Introduction and Editorial Notes.* New York: Munn and Company. pp. 250.

Mansfield, Jared (1759–1830).

1802. *Essays, Mathematical and Physical: Containing New Theories and Illustrations of some Very Important and Difficult Subjects of the Sciences. Never Before Published.* New Haven: William W. Morse. pp. viii + 274 + unnumbered pages of tables.

◇   Mansfield graduated from Yale in 1777 and was the first Head of the Department of Mathematics at the United States Military Academy, serving from 1802 until 1803. Later he served as Professor and Department Head of Natural and Experimental Philosophy, USMA, from 1812 to 1828.

◇   Errata follow the tables.

○ The library has three copies. One copy includes 13 plates, the others fewer.

○ One copy signed "Samuel S. Smith, March 13, 1817" on p. 1.

Marie, Maximilien (1819–1891).

1883. *Histoire des sciences mathématiques et physiques.* 12 vols. Paris: Gauthier–Villars. pp. [v] + 286; [v] + 315; [v] + 230; [v] + 246; [v] + 255; [v] + 359; [v] + 272; [v] + 259; [v] + 321; [v] + 229; [v] + 257; [v] + 258.

◇ A bibliographical dictionary. Each volume ends with a list of the scientists treated in that volume. The arrangement is roughly chronological.

Mariotte, Edmé (1620–1684).

1700. *Traité du mouvement des eaux et des autre corps fluides. Divisé en V. parties. Mis en lumiere par les soins de M. de la Hire.* Nouvelle édition corrigée. Paris: Jean Jombert. pp. [ix] + 390 + [18].

Marolois, Samuel (1572?–1627).

1628. *Oeuvres mathematiques de Samuel Marolois, traictant de la geometrie et fortification, reduictes en meilleur ordre, & corrigees d'un nombre infiny de fautes escoulees aux impressions precedentes: La geometrie par Theodore Verbeeck. Et la fortification par François van Schoten.* Amsterdam: Guillaume Iansson Caesius. pp. [viii] + 219 + 1 plate; [iv] + 248 + 16 plates + 1 table.

◇ See Figure 1 for a photograph of the title page of this volume.

◇ Two books bound together in one volume.

○ Note from COL Henry C. McClean pasted inside front cover.

Martin, Benjamin (1704–1782).

1736. *The Young Trigonometer's Compleat Guide. Being the Mystery and Rationale of Plain Trigonometry Made Clear and Easy in Two Parts.* 2 vols. London: Printed for J. Noon. pp. xvi + 328 + 3 plates; xii + 370 + 1 plate.

◇ There are numerous interesting diagrams in the text. Martin apparently had his own school in London (from the preface, Volume 1).

○ The set is marked #5.

1740. *Logarithmologia: or the Doctrine of Logarithms Common and Logistical, in Theory and Practice.* London: J. Hodges. pp. xii + 246 + 62 + 1 plate.

◇ The logarithm table in this book is copied from that of Sherwin, which is "in the main line of descent from Briggs".

○ Signature of Andrew Ellicott is on a front paper.

1759a; 1764. *A New and Comprehensive System of Mathematical Institutions, Agreeable to the Present State of the Newtonian Mathesis.* 2 vols. London: Owen. pp. viii + 410; 535 + 21 plates.

◇ Volume 1 is dedicated to George Augustus, Prince of Wales.

○ Volume 2 has the signature of Tho. Roberts on its title page.

1759b. *New Elements of Optics; or, the Theory of the Aberrations, Dissipation, and Colours of Light: of the General and Specific Refractive Powers and Densities of Mediums; The Properties of Single and Compound Lenses: and the Nature, Construction, and Use of Refracting and Reflecting Telescopes and Microscopes of*

Every Sort hitherto Published. London: Printed for the Author. pp. ix + [iii] + 120 + 4 plates + vi + [ii] + 131 + 8 plates.

◇ Errata on p. 120.

○ USMA bookplate reads "Quarters of the Superintendent". The title page, which has been repaired, carries the signature "Wm. Part[ridge] of U.S. Engineers".

1797. *Description and Use of the Pocket Case of Mathematical Instruments; Wherein are Particularly Explained the Nature and Use of all the Lines Contained on the Plain Scale, the Sector, the Gunter, and Proportional Compasses: also Their Practical Application, Exemplified in a Variety of Mathematical Problems; with Whole Illustrated with Copper–plate Figures.* New edition with corrections, etc, etc. by William Jones. London: Printed for W. and S. Jones. pp. 19 + 1 plate.

Mason, Charles (1730–1787). See Mayer 1787.

Mascheroni, Lorenzo (1750–1800).

1803. *Problêmes pour les arpenteurs avec différentes solutions.* Ouvrage traduit de l'Italien. Paris: Courcier. pp. 117 + 4 plates.

Maseres, Francis (1731–1824).

1758. *A Dissertation on the Use of the Negative Sign in Algebra: Containing a Demonstration of The Rules usually Given Concerning it; and Shewing How Quadratic And Cubic Equations may be Explained, without the Confideration of Negative Roots. To which is Added, as an Appendix, Mr. Machin's Quadrature of the Circle.* London: Printed by Samuel Richardson. pp. viii + 293 + 1 p. errata.

◇ Dedicated to Thomas Holles.

1760. *Elements of Plane Trigonometry. In which is Introduced a Dissertation on the Nature and Use of Logarithms.* London: T. Parker. pp. viii + 471.

◇ Errata follow p. 471. There are directions on p. iii for inserting thirteen plates.

1791; 1791; 1796; 1801; 1804; 1807. *Scriptores logarithmici, or a Collection of Several Curious Tracts on the Nature and Construction of Logarithms, Mentioned in Dr. Hutton's Historical Introduction to his New Edition of Sherwin's Mathematical Tables: Together with Some Tracts on the Binomial Theorem and Other Subjects Connected with the Doctrine of Logarithms.* 6 vols. London: Printed by J. White (vols. 1,2,3); Printed by Davis, Wilks, and Taylor (vol. 4); Printed by R. Wilks (vol. 5, 6). pp. xix + cxxi + 383; x + 591+ 1 Plate; cviii + 791; li + 695; clvi + 859; lxxxiv + 958.

◇ Errata for Volume 1 on p. xx; for volume 2 follow p. x; for volume 3 on pp. 787–791; for volume 6 on pp. 951–958. There are tables of contents for volumes 1–5 only. Five of the six volumes have prefaces (of the following lengths: 16 pp; 20 pp; 104 pp; 152 pp; 84 pp), and these form a significant work by themselves. Volume 1 contains Joannis Kepler's, *Mathematici chilias logarithmorum ad totidum numeros rotundes* (1624). Volume 2 contains Euclid Speidell's, *Logarithmotechnia: or the Making of Numbers called Logarithms to Twenty–five Places* (1688) in which a method for computing hyperbolic logarithms is given. Volume 6 contains John Napier's, *Mirifici logarithmorum canonis descriptio* (1614).

Maseres, Francis (Continued)

1795. *The Doctrine of Permutations and Combinations, being an Essential and Fundamental Part of the Doctrine of Chances; As it is delivered by Mr. James Bernoulli, in his Excellent Treatise on the Doctrine of Chances, Intitled, Ars Conjectandi, and by the Celebrated Dr. John Wallis, of Oxford, in a Tract Intitled from the Subject, and Published at the End of his Treatise on Algebra: In the Former of which Tracts is Contained, A Demonstration of Sir Isaac Newton's Famous Binomial Theorem, in the Cases of Integral Powers, and of the Reciprocals of Integral Powers, Together with some Other Useful Mathematical Tracts.* London: Francis Maseres. pp. xvi + 606.

⬦   This is essentially an anthology of the following works: *Artis Conjectandi Pars Secunda, continens Doctrinam de Permutationibus et Combinationibus* [by Jacques Bernoulli], pp. 1–34 (in Latin); *A translation of the Three First Chapters of the Second Part, or Book, of Mr. James Bernoulli's Excellent Treatise Intitled Artis Conjectandi; or "The Art of Forming Probable Conjectures Concerning Events that Depend on Chance." Published in a small Quarto Volume at Basil, or Basle, in Switzerland, in the Year 1713*, pp. 35–213; *A New and General Method of Finding the Sum of any Series of Powers of a Set of Quantities that are in Arithmetic Progression; Being the Tenth of the Late Learned Mr. Thomas Simpson's Mathematical Essays, published in the year 1740*, pp. 215–224; *An Investigation and Demonstration of Sir Isaac Newton's Binomial Theorem, in the Case of Integral and Affirmative Powers*, pp. 225–268; *A Discourse of Combinations, Alternations, and Aliquot Parts by John Wallis. Printed at London in the Year 1685, and Published with the Author's treatise of Algebra*, pp. 269–351; *An Appendix to the English Translation of Rhonius's German Treatise of Algebra, made by Mr. Thomas Brancker, and Published by him, with the Advise and Assistance of Dr. John Pell, at London, in the Year 1668; Containing a Table of Odd Numbers Less than One Hundred Thousand, shewing, first, which of Them are Incomposit, or Prime Numbers, and, Secondly, the Factors, or Co-efficients, by the Multiplication of which the Others are Produced; Supputated, or Computed, by the same Thomas Brancker*, pp. 353–416; *Of Rational Numbers that express the Sides of Right–angled Triangles* [by Francis Maseres], pp. 417–457; *Of the Differences of the Cubes of the Natural Numbers 1, 2, 3, 4, 5, 6, 7, &c.* [by Francis Maseres], pp. 459–504; *A General Method of Extracting the Roots of Numbers by Approximation; Invented by Monsieur de Lagny, a Member of the Royal Academy of Sciences at Paris, and Published in the Year 1697, in his Nouveaux Eléments d'Arithmétique et d'Algébre*, pp. 505–556; *Observations on Mr. Raphson's Method of Resolving Affected Equations of all Degrees by Approximation*, pp. 557–590; *A Table of the Square and Cube Roots of the Natural Numbers 1, 2, 3, 4, 5, &c to 180; Being Table XIX. of Mr. James Dodson's valuable Tables of Computation, intitled The Calculator, that were published in the Year 1747*, pp. 591–592; *A Table of the Square–Roots and Reciprocals of all Numbers, from 1 to 1000. Computed by Dr. Charles Hutton*, pp. 593–604; and *Dr. Hutton's Account of the foregoing Table of the Reciprocals and the Square–Roots of the Natural Numbers 1, 2, 3, 4, 5, 6, 7, &c, to 1000, given at the end of the Fourth Volume of his Collection of Mathematical Problems and Tracts, Intitled Miscellanea Mathematica, Published in Four Little Volumes, Duodecimo, in the year 1775*, pp. 605–606.

Maskelyne, Nevil (1731–1811).
1776; 1783. *Astronomical Observations made at the Royal Observatory at Green-wich, from the Year MDCCLXV to the Year MDCCLXXIV. Published by the President and Council of the Royal Society at the Public Expense in Obedience to his Majesty's Command.* 2 vols. London: Printed by William Richardson. pp. xii + 335 + 128 + xii + 52; 377 + 119 + 28.
◇   Dedicated to the King. The second volume is denoted, "Part of Vol. ii to be Completed hereafter."
○   Andrew Ellicott signed both volumes.
○   A second copy of Vol. 1 contains the bookplate of Francis Wollaston. The last four pages of this copy of Vol. 1 are missing.

Mason, Charles (1730–1787). See Mayer 1787.

Maupin, Georges (1867–?).
1898; 1902. *Opinions et curiosités touchant la mathématique d'après les ouvrages français des XVIe, XVIIe et XVIIIe siècles.* 2 vols. Paris: Georges Carré et C. Naud. pp. [vii] + 199 + 40; [vi] + 332.
◇   Volume 1 is dedicated to C.–A. Laisant. There is a reproduction of a portrait of Oronce Fine on p. [vii] of volume 1. Volume 2 is dedicated to M. and MM. A. Le Tallec.

Mayer and Choquet, Charles (1798–?).
1836. *Traité élémentaire d'algèbre.* Deuxième èdition. Paris: Bachelier. pp. xiv + 564.
◇   Errata follow p. xiv.

Mayer, Tobias (1723–1762).
1770. *Tabulae motuum solis et lunae, novae et correctae; auctore Tobia Mayer: quibus accedit methodus longitudinum promota, eodem auctore. Editae jussu prae fectorum rei longitudinariae.* Londini [London]: Typis Gulielmi et Johannis Richardson. pp. vii +1–89 in Latin + [3] + 91–136 in English translation of previous + 2 plates.
◇   Errata on p. [2]. Preface by Nevil Maskelyne.
○   Rebound and in a slipcase. Signed by Andrew Ellicott. Bound with Mayer 1787.

1787. *Mayer's Lunar Tables, Improved by Mr. Charles Mason. Published by Order of the Commissioners of Longitude.* London: Printed by William Richardson. pp. [iii] + 72.
◇   Errata on p. [iii].
○   Bound with Mayer 1770.

Mechain, Pierre–Francois–Andre (1744–1804), and Delambre, Jean Baptiste Joseph (1749–1822).
1806; 1807; 1810. *Base du système mètrique décimal, ou mesure de l'arc du méridien compris entre les parallèles de Dunkerque et Barcelone, exécutée en 1792*

et années suivantes. 3 vols. Paris: Baudouin. pp. 551 + 8 plates; xxiv (xvii–xxiv bound between xiii and ix) + 844 + 11 plates; 704 + 62 + 9 plates.

Medina Celi, duque de.

1708 *El architecto perfecto en el arte militar, dividio en cinco libros: el primero contienne, la fortification regular y irregular à la moderno: el II. la especulacion sobre cada una de sus partes: el III. la fabrica de quarteles, almazenes à prueva de bomba, y de toda fietre de muralles tanto en tierra firma como en el agua: el IV. la defensa y attaque de una plaza segun el nuevo modo de Guerrear: el V. la geometria, trigonometria, calculos, regla de proporcion, &c.* En Amberes: Henrico y Cornelio Verdussen, mercadores de libros. pp. [x] + 464 + [xiv] + 27 plates.

Merriman, Mansfield (1848–1925).

1906. *The Solution of Equations. Mathematical Monographs,* edited by Mansfield Merriman and Robert S. Woodard. No. 10. Fourth edition, enlarged. New York: John Wiley and Sons. pp. ii + 47 +18.

◇ This is a chapter from Merriman and Woodward 1896.

Merriman, Mansfield (1848–1925) and Woodward, Robert Simpson (1849–1924) (Editors).

1896. *Higher Mathematics, A Textbook for Classical and Engineering Colleges.* New York: John Wiley. pp. xi + 576 + 16.

◇ Eleven chapters by eleven authors.

Metius, Adriaen (1571–1635).

1626. *Adriani Metii Alcmariani arithmeticae libri duo: et geometriae lib. VI. Huic adiungitur trigonometriae planorum methodus succincta. Altera veró praeter alia, nova regulae proportionalis inventa proponit; et quaecunque loca adversus hostium insultus iuxta hoc seculo praxin (quam fortificationem vocant) munire solide docet.* Lugd. Batavorvm, Ex offician Elzeviriana. Published by Elsevier. pp. [xvi] + 118.

This work is followed by *Geometricae practicae Pars I. & II.— Quae rei cujus vis mensurabilis vim, proprietates & habitudines interpretatur & exercet. Authore Adriano Metio Alcmariano matheseos professore ordinario. Franekerae, excudebat Vldericus Balck, ordinum frisiae & eorundem academiae typographus.* Frankfurt: Balck. pp. 308 + 2 plates.

Finally, this is followed by another title page: *Geometriae practicae pars quinta. Continens problemata astronomica geometricè delineata, & Arithmeticè resoluta. Autore Adriano Metio Alcmariano mathes. professore ordinario. Franekeae, Ex officinâ Vlderici Dominici Balck, ordinum frisiae typographi.* Frankfurt: Balck. pp. 102 + 1 plate + [14].

◇ There is a 14 page "Index Capitum" at the end for all three works.

○ Given in memory of MAJ Donald T. MacLachlan (USMA 1907).

Meyer, Wilhelm Franz (1856-?) (Editor).

1898–1904. *Encyklopädie der Mathematischen Wissenschaften mit Einschluss ihrer Anwendungen. Herausgegeben im Auftrage der Akademieen der Wissenschaften zu Göttingen, Leipzig, München und Wien, sowie unter Mitwirkung Zahlreicher*

Fachgenossen. Erster Band in zwei Teilen. Arithmetik und Algebra. Erster Teil. Leipzig: B. G. Teubner. pp. xxxviii + 554.

1900–1904. *Encyklopädie der Mathematischen Wissenschaften mit Einschluss ihrer Anwendungen. Herausgegeben im Auftrage der Akademieen der Wissenschaften zu Göttingen, Leipzig, München und Wien, sowie unter Mitwirkung Zahlreicher Fachgenossen. Erster Band in zwei Teilen. Arithmetik und Algebra.* Zweiter Teil. Leipzig: B. G. Teubner. pp. x + 555–1197 + 2 pages of advertisements + 2 plates.

Michelson, Johann Andreas Christian (1749–1797). See Leonhard Euler 1788, 1788, 1791; Leonhard Euler 1790, 1790, 1793.

Michie, Peter S. (1838–1901; USMA 1863).
1887. *Elements of Analytical Mechanics.* Second edition. New York: John Wiley & Sons. pp. xiv + 292.
⋄ Michie served as Department Head, Natural and Experimental Philosophy, USMA (1871–1901).
○ A laudatory review is pasted inside the front cover. This was the copy of E. W. Bass, given to him by P. S. Michie, August 31, 1887.

1890. *Elements of Analytical Mechanics.* Fourth edition. New York: John Wiley & Sons. pp. xvi + 373.
○ Copy of Buell B. Bassette, dated West Point, August 29, 1891. Many marginal notes.

1893. *Elements of Analytical Mechanics.* Fourth edition. New York: John Wiley & Sons. pp. xvi + 373 + 16.
⋄ Short–title catalog of John Wiley & Sons, last 16 pages.

1900. *Elements of Analytical Mechanics.* Fourth edition. New York: John Wiley & Sons. pp. xvi + 373 + 16.
⋄ Short–title catalog of John Wiley & Sons, last 16 pages.
○ Some marginal notes.

1905. *Elements of Analytical Mechanics.* Fourth edition. New York: John Wiley & Sons. pp. xvi + 373 + 18.
⋄ Short–title catalog of John Wiley & Sons, last 18 pages.

1907. *Elements of Analytical Mechanics.* Fourth edition. New York: John Wiley & Sons. pp. xvi + 373.

Michie, Peter S. (1838–1901) See also Anonymous n.d.

Miles, Solomon Pearson (1791–1842).
1830. *Mathematical Tables; Comprising Logarithms of Numbers, Logarithmic Sines, Tangents and Secants, Natural Sines, Meridional Parts, Difference of Latitude and Departure, Astronomical Refractions, &c.* Stereotype edition. Boston: Carter and Hendee. pp. 10 + unnumbered pages of tables.

Miller, W. J. C. (1832–?) (Editor).

1894. *Mathematical Questions and Solutions from the "Educational Times", with Many Papers and Solutions in Addition to those Published in the "Educational Times" and an Appendix.* Volume 61. London: Francis Hodgson. pp. xxiii + 128 + 4.

o Frontispiece autograph "From E. Bass."

1896. *Mathematical Questions and Solutions from the "Educational Times", with Many Papers and Solutions in Addition to those Published In the "Educational Times", and an Appendix.* Volume 64. London: Francis Hodgson. pp. 4 + xxiii + 128 + 4.

Milner, Isaac (1750-1820).

1778. "Reflections on the Communication of Motion by Impact and Gravity."

◇ A paper, not a book, bound together with other papers in Anonymous n.d.

n.d. "Observations on the Limits of Algebraic Equations; and a General Demonstration of Descartes's Rule for Finding Their Number of Affirmative and Negative Roots."

◇ A paper, not a book, bound together with other papers in Anonymous n.d.

Moigno (L'Abbe), [François Napolean Marie] (1804–1884).

1840, 1844. *Leçons de calcul différentiel et de calcul intégral, rédigées d'après les méthodes et les ouvrages publiés ou inédits de M. A.–L. Cauchy.* 2 vols. Volume 1: Differential Calculus. Volume 2: Integral Calculus first part. Paris: Bachelier. pp. xxxv + 531; xlv + 783.

◇ Errata pages precede page 1 in each volume.

Moivre, Abraham de (1667–1754). See DeMoivre, Abraham.

Mole, John (1743-1827).

1788. *Elements of Algebra to which is Prefixed, a Choice Collection of Arithmetical Questions, with Their Solutions, Including Some New Improvements Worthy the Attention of Arithmeticians. The Principles of Algebra are Clearly Demonstrated, and Applied in the Resolution of a Great Variety of Problems on Different Parts of the Mathematicks and Natural Philosophy.* London: Printed for G. G. J. and J. Robinson. pp. 8 + [iv] + 320.

◇ Errata on p. [iv].

Monge, Gaspard (1746–1818).

1809. *Application de l'analyse a la géométrie, a l'usage de l'École impériale poly-technique.* Quatrième édition. Paris: Ve Bernard. pp iv + 444 + 5 plates.

o Thayer binding. The heading has been trimmed off plates 4 and 6.

1810. *Traité élémentaire de statique, a l'usage des écoles de la marine.* Cinquième édition, revue par M. Hachette. Paris: Courcier. pp. vi + 204 + 6 plates.

◇ Errata follow p. 204.
o Thayer binding.

1820. *Géométrie descriptive; augmentée d'une théorie des ombres et de la perspective, extraite des papiers de l'auteur, par M. Brisson.* Quatrième édition. Paris: Ve Courcier. pp. xx + 187 + 28 plates.

◇ Errata on p. 187. The author is widely known as the father of descriptive geometry. The introduction contains historical information.

1827. *Géométrie descriptive; augmentee d'une théorie des ombres et de la perspective, extraite des papiers de l'auteur, par M. Brisson.* Cinquième édition. Paris: Bachelier. pp. xx + 188 + 28 plates.

◇ The "Advertisement of the Editor" (pp. v–xiv) contains some history of descriptive geometry and some biographical information about Monge.

Monge, Gaspard. See also Dupin 1813, Hachette 1818, Heather 1851.

Montferrier, Alexandre André Victor Sarrazin de (1792–1863).
1835; 1836; 1840. *Dictionnaire des sciences mathématiques pures et appliquées, par une société d'anciens élèves de l'École polytechnique.* 3 vols. Paris: Dénain et Delamare; A. J. Dénain; Bureau de la Bibliothèque Scientifique. pp. viii + 384 + 34 plates; 620 + 24 plates; 490 + 22 plates.

◇ There are errata pages at the end of each volume. The dictionary runs A–Z in volumes 1 and 2. These volumes are under the direction of Montferrier. Then volume 3 is a "supplement containing several articles on geometry, trigonometry and astronomy," also with its topics arranged alphabetically.

Montucla, Jean Étienne (1725-1799).
1798a; 1798; May 1802; May 1802. *Histoire des mathématiques, dans laquelle on rend compte de leurs progrès depuis leur origine jusqu'à nos jours; où l'on expose le tableau et le développement des principales découvertes dans toutes les parties des mathématiques, les contestations qui se sont élevées entre les mathématiciens, et les principaux traits de la vie des plus célèbres.* 4 vols. Nouvelle édition, considérablement augmentée, et prolongée jusque vers l'époque actuelle. Paris: Henri Agasse. pp. viii + 739 + 13 plates; [iv] + 717 + 14 plates; viii + 832 + 17 plates; [iii] + 688 + 2 plates.

◇ Frontispiece portrait of Montucla in volume 1. Frontispiece portrait of Jerome de la Lande in volume 3. Volumes 3 & 4 are "completed and published" by Jerome de la Lande. Errata for volume 1 and 2, follows p. 717 of volume 2 and for the last two volumes is at the end of volume 4. The preface for volume 3, pp. v–viii, is by de la Lande.

◇ Portraits: Lande, Jerome de la; Montucla, Jean Étienne.

○ The library has two copies.

○ One copy has a Thayer binding.

1831. *Histoire des recherches sur la quadrature du cercle, avec une addition concernant les problèmes de la duplication du cube et de la trisection de l'angle.* Nouvelle édition revue et corrigée. Paris: Bachelier Père et Fils. pp. xvi + 300 + 4 plates.

◇ Errata on p. xvi. The first edition of this work (1754) was Montucla's first book; it established his scholarly reputation. This second edition which was prepared by Silvestre Francois Lacroix (1765–1843), "superseded completely the original one."

Montucla, Jean Étienne. See also Hutton 1803; Ozanam 1803.

Morgan, Charles.
1770. *Six Philosophical Dissertations on the Mechanical Powers, Elastic Bodies, Falling Bodies, the Cycloid, the Parabola, the Rain–Bow by Dr. Morgan. Published by Dr. Samuel Clarke, in his Notes upon Rohault's Physics.* Translated into English by John Clarke. Cambridge: Printed by Fletcher and Hodson. pp. iv + 50 + 3 plates.
　◇　List of subscribers on pp. i–iii. Errata on p. [v].
　○　Bound together with Cotes 1770.

Motte, Andrew (? –1734). See Newton 1819.

Moutard, Théodore Florentin (1827–1901). See Poncelet 1862.

Mudge, William (1762–1820) and Dalby, Isaac (1744–1824).
1799; 1801; 1811. *An Account of the Operations Carried on for Accomplishing A Trigonometrical Survey of England and Wales; from the Commencement, in the Year 1784, to the End of the Year 1796. Begun under the Direction of the Royal Society, and Continued by the Order of the Honourable Board of Ordinance. First Published in, and Now Revided from, The Philosophical Transactions. Volume I; An Account of the Operations carried on for Accomplishing a Trigonometrical Survey of England and Wales; from the Year 1797, to the End of the Year 1799. Volume II; An Account of the Measurement of an Arc of the Meridian, Extending from Dunnose in the Isle of Wight, to Clifton in Yorkshire, in the Course of the Operations Carried on for Accomplishing a Trigonometrical Survey of England, in the Years 1800, 1801, and 1802. Intended as a Second Part to volume II; An Account of the Trigonometrical Survey, Carried on by the Order of the Master–General of His Majesty's Ordnance, in the Years 1800, 1801, 1803, 1804, 1805, 1806, 1807, 1808, and 1809. Volume III.* London: Printed by W. Bulmer and Co. pp. xx + 437 + 2 fold–out tables + 22 plates; ix + 190 + 7 plates; xv + 382 + 14 plates.
　◇　Dedicated to the Master General and Principal Officers of His Magesty's Ordinance. Volume 1 is by Mudge and Dalby; volume 2 is by Mudge alone; and volume 3 is by Mudge and Thomas Colby.
　○　The library has two copies.
　○　One copy appears to have been the copy of John Bonnycastle; an unsuccessful atempt was made to erase his signature from the title page of Volume 1.

Mueller, Johann (1436–1476). See Regiomontanus 1496.

Muir, Thomas (1844-1934).
1882. *A Treatise on the Theory of Determinants with Graduated Sets of Exercises for Use in Colleges and Schools.* London: Macmillan and Co. pp. vii + 240 + 40 pages of advertisements.

Mulerius, Nicolaus (1564–1630). See Copernicus 1617.

Muller, John (1699–1784).
1757; 1757; 1756; 1757; 1757; 1757. *A System of Mathematics, Fortification and Artillery. In Six Volumes. Illustrated with above One Hundred Copper–plates. For the Use of The Royal Academy of Artillery at Woolwich.* 6 vols. Each volume has a distinct subtitle or title: Vol. 1 "Containing Algebra, Geometry and Conic Sections"; Vol. 2 "Containing Trigonometry, Surveying, Levelling, Mensuration, Laws of Motion, Mechanics, Projectiles, Gunnery, &c. Hydrostatics, Hydraulics, Pneumatics, Theory of Pumps"; the whole title of Vol. 3 is "A Treatise Containing the Elementary Part of Fortification, Regular and Irregular. With Remarks on the Construction of the most Celebrated Authors, particularly of Marshal de Vanban and Baron Coehorn, in which the Perfection and Imperfection of Their several Works are considered"; Vol. 4 "Containing Practical Fortifications; Theory and Dimensions of Walls, Arches and Timbers, Properties, Qualities and manner of using Materials; manner of Tracing a Fortress, and Estimate of the Works; Method of Building Aquatics, as Stone Bridges, Harbours, Quays, Warfs, Sluices and Aqueducts"; Vol. 5 "Containing General Constructions of Brass and Iron Guns and Carriages for Sea and Land; Mortars and Howites, their Beds and Carriages; Laboratory–Work; Theory of Powder applied to Fire–Arms, &c"; Vol. 6 "Containing 1. Attac from the. [sic] 2. Defense of every Beginning to the End; Part with the Requisites. 3. Manner to make and load Mines. Table of the proper Charges, &c". London: Vols. 1, 2, 4, 5, 6 are Printed for J. Millan; Vol. 3 is the Second edition and is Printed for J. Nourse. pp. liv + 188 + 12 plates; [ii] + 189–442 + 13 plates; xvi + 240 + 33 plates; xxiv + 304 + 26 plates; xvi + 309 + 28 plates; xvi + 247 + 25 plates.
⋄ Frontispiece in volume 1 is an illustration of cannons in a fort. Volume 1 is dedicated to Martin Folks. Volume 3 is dedicated to John, Duke of Montagu. Volume 4 is dedicated to George, Prince of Wales. Volume 5 is dedicated to Charles, Duke of Marlborough. Volume 6 is dedicated to William, Duke of Cumberland. Errata for volume 4 on p. xxiv, and for volume 5 on p. 309.

1765. *Elements of Mathematics Containing Geometry, Conic Sections, Trigonometry, Surveying, Levelling, Mensuration, Laws of Motion, Mechanics, Projectiles, Gunnery, &c. Hydrostatics, Hydraulics, Pneumatics, Theory of Pumps. To which is Prefixed, The First Principles of Algebra, by Way of Introduction. With an Addition of a New Treatise on Perspective.* 2 volumes bound together as one book. Third edition, improved. London: J. Millan. pp. xxxvi + 312 + 28 plates.
⋄ Dedicated to Ralph Willett. Errata on p. 312.
○ The library has two copies. One copy is from the Estate of Edwin N. Rich.

Murray, Daniel Alexander (1862–1934).
1898a. *An Elementary Course in the Integral Calculus.* Cornell Mathematical Series. New York: American Book Co. pp. xiv + 288.
⋄ Used as a text at USMA 1900–1936.
○ The library has four copies. One copy has a lesson outline for the USMA course enclosed in book. Problems solutions at the end of the book. Color annotations: red (omissions for upper sections) and blue (omissions for lower sections).

1898b. *Introductory Course In Differential Equations for Students in Classical and Engineering Colleges.* Second edition. New York: Longmans, Green & Co. pp. xvi + 236 + 3 pages of descriptions of math books published by Longmans, Green and Co.

1912. *Introductory Course in Differential Equations for Students in Classical and Engineering Colleges.* New York, London, Bombay and Calcutta: Longmans, Green and Co. pp. xvi + 236.

⋄ This book was used as text at USMA from 1912 to 1913.

⋄ Solutions to problems at end.

○ Numerous margin comments and inserts.

Murray, Daniel Alexander. See also Anonymous 1901.

Mydorge, Claude (1585–1647).

1641. *Claudii Mydorgii patricii Parisini prodromi catoptricorum et dioprtricorum: Sive conicorum operis ad abdita radii reflexi et refracti mysteria praeuii & facem praeferentis. Libri quatuor priores. D. A. L. G.* Parisiis [Paris]: I. Dedin. pp. [iv] + 308.

Napier, John (1550–1617).

1614. *Mirifici logarithmorum canonis descriptio, ejusque usus, in utraque trigonometria, ut etiam in omni logistica mathematica, amplissimi, facillimi, & expeditissimi explicatio.* Edinburgi [Edinburgh]: Ex officinâ Andreae Hart. pp. [vi] + 57 + unnumbered pages of tables.

⋄ Dedicated to King James.

○ This copy has been rebound and it had to be bound backwards because of the deterioration of the edges of the pages.

Napier, John. See also Briggs 1624a; Briggs 1624b; Briggs 1633; Maseres 1807.

N[audin].

1695. *L'ingenieur français, contenant la geometrie pratique sur le papier & sur le terrain, avec le toisé des travaux & des bois; La fortification reguliere & irreguliere; sa construction effective; l'attaque & la défense des places. Avec la methode de Monsieur de Vauban, & l'explication de son nouveau systeme.* Paris: Estienne Michallet. pp. [xxvii] + 312 + 24 plates.

⋄ The title page lists the author as "Par M. N*** Ingenieur ordinaire du Roy."

○ Donated in memory of MAJ Donald J. MacLachlan [USMA 1907].

Navier, Claude Louis (1785–1836).

1833; 1838. *Résumé des leçons données a l'École des ponts et chaussées, sur l'application de la mécanique a l'établissement des constructions et des machines.* 2 vols. Deuxième édition. Paris: Carilian–Goeury. xv + 448 + 5 plates; viii + 422 + 9 plates.

Newcomb, Simon (1835–1909).

1882. *Logarithmic and Other Mathematical Tables with Examples of Their Use and Hints on the Art of Computation.* Newcomb's Mathematical Course. New York: Henry Holt and Co. pp. vi + 80+ 104 pages of tables.

◇ This book was used as text at USMA from 1907 to 1936.

◦ The library has two copies.

◇ One copy has notes on lower sections in green, (the weakest students), middle sections in blue (the average students), all other sections (strongest students) in red. Copy used by "John R. Noges, USMA, 3rd Class, Room 233."

◦ Second copy is signed by Merrow E. Sorley, Cadet Sept 1 1890. Some notes are in the margins. Course outline and syllabus are given on back page.

1887. *Logarithmic and Other Mathematical Tables, with Examples of Their Use and Hints on The Art of Computation.* Newcomb's Mathematical Course. New York: Henry Holt and Company. pp. ii + 72 + 104.

◦ The library has two copies.

◦ One is signed "B. B. Bassette (Connecticut) United States Corps Cadet), 4th Class, 1889."

◦ The other copy has pp. ii + 80 + 104 pages of tables and is signed "Dennis Mahan Michie, 1888."

1903. *The Reminiscences of an Astronomer.* Boston and New York: Houghton, Mifflin, and Company. pp. x + 424.

◇ Contains an index.

◇ Frontispiece is a portrait of Newcomb.

Newton, Isaac (1642–1727).

1706. *Optice: sive de reflexionibus, refractionibus, inflexionibus & coloribus lucis libri tres. Authore Isaaco Newton, equite aurato. Latine reddidit Samuel Clark, A. M. Reverendo admodum Patri ac D^{no} Joanni Moore episcopo Norvicensi a sacris domesticis. Accendunt tractatus duo ejusdem authoris de speciebus & magnitudine figurarum curvilinearum, Latine scripti.* Londini [London]: Sam. Smith & Benj. Walford. pp. [xiii] + 348 + 12 plates + 24 + 43 + 6 plates.

◇ Translated by Samuel Clarke from the 1704 English original.

◇ The second and third paginations, which have separate title pages, are the "Enumeratio linearum tertii ordinis" and "Tractatus de quadratura curvarum" respectively. The last six plates are for the "Enumeration of Curves of the Third Order".

◇ Errata, corrigenda & addenda on pp. [vii]–[xiii].

◦ This copy is the rare first state of this work. In Query 20, on page 315, it declares that infinite space is the sensorium of God. In the second state, Newton added a 'tanquam' to refer to infinite space as if it were God's sensorium.

◦ Signed by Williams on the title page.

Newton, Isaac (Continued)

1729. *The Mathematical Principles of Natural Philosophy.* Volume 2. Translated by Andrew Motte. London: Printed for Benjamin Motte. pp. 393 + 13 page index + vii page appendix + 3 figures.

◇ This is the second volume (only) of the first English translation of the *Principia.* Volume 1 is not present.

◇ 16 plates and one table are bound in facing appropriate pages.

1736. *The Method of Fluxions and Infinite Series; with its Application to the Geometry of Curve–Lines, Translated from the Author's Latin Original not yet Made Publick. To which is Subjoin'd, A Perpetual Comment upon the whole Work, Consisting of Annotations, Illustrations, and Supplements, In order to Make this Treatise a Compleat Institution for the Use of Learners.* Translated by John Colson. London: Printed by Henry Woodfall. pp. xxiv + 140.

◇ Original Lation work entitled: *Artis analyticae specimina.*

1745. *Sir Isaac Newton's Two Treatises of the Quadrature of Curves and Analysis by Equations of an Infinite Number of Terms Explained: Containing the Treatises Themslves, Translated into English, with Large Commentary; in which the Demonstrations are Supplied where Wanting, the Doctrine Illustrated, and the Whole Accommodated to the Capacities of Beginners, for whom it is Chiefly Designed.* Translated by John Stewart. London: Printed by James Bettenham, at the expense of the Society for the Encouragement of Learning. pp. xxxii + 479.

◇ Dedicated to Thomas, Duke of Leeds.

○ This copy contains a few marginal notes.

1769. *Universal Arithmetic—Or a Treatise of Arithmetical Composition and Resolution. Translated by The late Mr. Ralphson; and Revised and Corrected by Mr. Cunn. To which is Added a Treatise upon the Measures of Ratios, By James Maguire. The whole illustrated and explained, In a Series of Notes by the Rev. Theaker Wilder.* London: Printed for W. Johnston. pp. viii + 536 + 63 + 8 plates.

◇ Errata (3 pages) follow p. 63.

◇ Dedicated to The Duke of Bedford by Theaker Wilder.

○ Holograph "Wm Partridge of US Enineers [sic]" on the title page.

1779; 1779; 1782; 1782; 1785. *Isaaci Newtoni opera quae exstant omnia.* 5 vols. Commentary by Samuel Horsley. Londini [London]: Excudebat Joannes Nichols. pp. xxii + 592 + 2 fold out tables; xxv+ [iii] + 460; [iii] + 437 + 13 plates + 48 + 3 plates; [iii] + 616 + 13 plates; [iii] + [iii] + 550 + 3 plates.

◇ Volume 1 is dedicated to the King.

◇ Corrigenda for volume 1 follow p. 592, for volume 2 on p. 460, for volume 3 follow p. 437; for volume 4 follow p. 616.

○ Signed "W. Courtenay" in each volume.

Newton, Isaac (Continued)

1819. *The Mathematical Principles of Natural Philosophy, by Sir Isaac Newton; Translated into English by Andrew Motte: to which are Added, Newton's System of the World; a Short Comment on, and Defence of, The Principia, by William Emerson; with the Laws of the Moon's Motion According to Gravity, by John Machin, a New Edition, with the Life of the Author; and a Portrait, Taken from the Bust in the Royal Observatory at Greenwich, Carefully Revised and Corrected by William Davis.* 3 vols. Translated by Andrew Motte. London: Printed for Sherwood, Neely, and Jones; and Davis and Dickson. pp. lx + 211 + 25 plates + a 4 page index; 321 + 19 plates + an 11 page index; vi + 231 + 10 plates.

◇ Frontispiece: reproduction of the bust of Newton at the Royal Observatory, Greenwich. Volume 1 is also dedicated by Andrew Motte to Hans Sloane, President of the Royal Society. Volume 1 contains: Newton's *Preface of 1686* (pp. ix–xi); an additional prefatory comment of Newton made after 1723 (pp. xi–xii); Roger Cotes' *Preface to the Second Edition* (pp. xiii–xxxi); a life of Newton (pp. xxxiii–lx). Volume 3 contains William Emerson's *Short Comment on Newton's Principia* (pp. 83–175 with 5 plates); Emerson's *Defense of Newton* (pp. 177–192); John Machin's *Laws of Moon's Motion* (pp. 193–231 with 3 plates).

○ Signature of "Mahan" is on the frontispiece.

1822. *Philosophiae naturalis principia mathematica. Auctore Isaaco Newtono, eq. aurato. Perpetuis commentariis illustrata, communi studio PP. Thomae Le Seur et Francisci Jacquier ex Gallicana Minimorum Familia, matheseos professorum. Editio nova, summa cura recensita.* 4 vols. Glasguae [Glasgow]: Ex prelo academico, typis Andreae et Joannis M. Duncan. pp. xxxi + 431; vi + 320; xxxvi + 344; vi + 203.

◇ Contains Newton's preface of May 8, 1686, his preface of March 28, 1713, Roger Cotes' preface of May 12, 1713, Newton's preface of January 12, 1725–26, and Edmund Halley's poem. This is a four volume reissue of the three volume Jesuit edition of 1739–42.

1888. *A Catalogue of the Portsmouth Collection of Books and Papers Written by or Belonging to Sir Isaac Newton, the Scientific Portion of Which has been Presented by the Earl of Portsmouth to the University of Cambridge. Drawn up by the Syndicate Appointed the 6th November, 1872.* Cambridge: At the University Press. pp. xxx + 56.

Newton, Isaac. See also Anonymous 1725b; Barrow 1735; Ball 1893; Collins 1725; Ditton 1726; Domcke 1730; 'sGravesande 1784; Gray 1907; Maseres 1795; Pemberton 1728; Vince 1781; Whiston 1710.

Nicholson, Peter (1765–1844).

1817. *An Introduction to the Method of Increments Expressed by a New Form of Notation; Shewing more Intimately its Relation to the Fluxional Analysis.* London: Scientific Press, Printed and published by Davis and Dickson. pp. viii + xxxvii + 130.

◇ The preface is followed by a long Introduction, "... the Decomposition of Powers."

1822. *The Rudiments of Practical Perspective, in which the Representation of Objects is Described by Two Easy Methods, One Depending on the Plan of the Object, the Other on its Dimensions and Position, each Method being Entirely Free from the Usual Complication of Lines, and from the Difficulties Arising from Remote Vanishing Points.* London: Printed for J. Taylor. pp. xviii + 122 + 38 plates.

◇   An advertisement for five other works by the author precedes the plates.

Nicollet, Joseph Nicolas (1786–1843).

1830.   *Cours de mathématiques a la usage de la marine.   Géométrie, trigonométrie, applications diverses.*   Tome second. Paris: Bachelier, successeur de Mme Ve Courcier; Brussels: Librairie Parisienne. pp. xvi + 320 + 10 plates.

◇   The library has only the second volume. Errata on p. 16.

Niewenglowski, Boleslas (1846–?).

1894; 1895; 1896.   *Cours de géométrie analytique l'usage des élèves de la classe de mathématiques spéciales et des candidats aux écoles du gouvernement.*   3 vols. Paris: Gauthier–Villars et Fils. pp. [vi] + [483]; [292]; [572].

◇   Subtitles for the volumes are: volume 1. conic sections; volume 2. construction of plane curves – relative complements to conics; volume 3. geometry in space, with a note on the transformations in geometry by Emile Borel.

1902.   *Cours d'algèbre a l'usage des élèves de la classe de mathématiques spéciales et des candidats à l'École normale supérieure et à l'École polytechnique.*   2 vols. Cinquième édition, entièrement refondue. Paris: Armand Colin. pp. [vi] + [387]; [488].

Niewenglowski, Boleslas (1846-?) and Gerard, Louis.

1898; 1899.   *Cours de géométrie élémentaire a l'usage des élèves de mathématiques élémentaires, de mathématiques spéciales; des candidats aux ecoles du gouvernement et des candidats a l'agrégation.*   2 vols. Paris: Georges Carré et C. Naud. pp. [iii] + [xii] + [362].

◇   Volume 1 is plane geometry and volume 2 is geometry in space.

Norris, John Saurin (1795–1873).

1854.   *A Sketch of the Life of Benjamin Banneker; From Notes Taken in 1836.* [Baltimore]: Printed for the Maryland Historical Society. pp. 20.

◇   A paper read by J. Saurin Norris before the Maryland Historical Society, October 5, 1854.

Ocagne, Maurice Philbert d' (1862–1938).

1896.   *Cours de géométrie descriptive et de géométrie infinitésimale.*   Encyclopédie des travaux publics. Paris: Gauthier–Villars et Fils. pp. xi + 4238.

◇   One Erratum on p. 428.

○   Signed by the author.

1905. *Le calcul simplifie par les procedes mécaniques et graphiques. Histoire et déscription sommaire des instruments et machines a calculer, tables, abaquès et nomogrammes.* Paris: Gauthier–Villars. pp. viii + 228.

Oldenburg, Henry (ca 1618–1677). See Collins 1725.

Olivier, Théodore (1793–1853).
1847. *Additions au cours de géométrie descriptive. Demonstration nouvelle des propriétés principales des sections coniques.* 2 vols. Paris: Carilian–Goeury et Vor Dalmont. pp. xvi + 94; 15 plates.
 ◇ Volume 1 is the text and volume 2 is the "Atlas". Errata follow p. 94.
 ◦ Bound together with parts 1 and 2 of Olivier 1852 and 1853.

1852. *Cours de géométrie descriptive. Première partie. Du point, de la droite et du plan.* Deuxième édition. Paris: Carilian–Goeury et Vor Dalamont. pp. xii + 136; 43 plates.
 ◇ Seven works in descriptive geometry by Olivier are listed on p. iv. Volume 1 contains the text, and volume 2 is the "Atlas," or plates. The bastaro title (pp. [i]–[ii]) is missing. First edition was in 1843.
 ◦ Bound together with Olivier 1847 and 1853.

1853. *Cours de géométrie descriptive. Deuxième partie. Des courbes et des surfaces courbes et en particulier des sections coniques et des surfaces du second ordre.* Deuxième édition. Paris: Carilian–Goeury et Vor Dalmont. pp. viii + 413; 54 plates numbered 44–97.
 ◇ There are no pages 281–286. There is no gap in the text, only in the pagination. Plates are in the 2nd Volume.
 ◦ Bound with Olivier 1847 and 1852.

Olney, Edward (1827–1887).
1872. *A Treatise on Special or Elementary Geometry. School Edition; Elements of Trigonometry, Plane and Spherical; Introduction to the Table of Logarithms.* Stoddard's Mathematical Series. 3 vols. New York: Sheldon and Company. pp. ix + 239 + 113 + 88.
 ◇ These three books are bound as one book.
 ◦ This copy owned by Edgar Bass.

Otho, Lucius Valentinus (ca. 1550–1605). See Rhäticus 1596.

Oyon, J. B.
1824; 1825. *Tables d'multiplication a l'usage de MM. les préfets et sous-préfets, les directeurs des contributions directes, les géometres en chef du cadastre, les officiers du génie, les ingénieurs du Corps Royal des Ponts et Chaussées, les arpenteurs de l'administration des forêts, les architectes, commerçans, banquiers, agens de change, etc., etc.* 2 vols. Troisieme édition. Approuvées par son excellence le ministre des finances. Paris: Hre. Pèrèz et Compagnie. Volume 1: instruction on

the use of the tables pp. 5–12; table pp. 2–500; errata on one page following p. 500. Volume 2: errata page; pp. 501–1001.

Ozanam, Jacques (1640–1717).

1697. *Cours de mathematique, qui comprend toutes les parties les plus utiles & les plus necessaires à un homme de guerre, & à tous ceux qui se veulent perfectionner dans cette science.* In volumes 3, 4 and 5, the title changes slightly to *toutes les parties de cette science les plus utiles & les plus necessaires à un homme de guerre, & à tous ceux qui se veulent perfectionner dans les mathématiques* 5 vols. Nouvelle edition, revue et corrigèe. Paris: Jean Jombert. pp. [xv] + 80 + 4 plates + 295 + 17 plates; [viii] + 93 + [iii] + 153 + 7 plates + [xiv] unnumbered pages of tables; [xxiv] + 215 + 24 plates + [viii]; [xxiv] + 185 + 28 plates + [vii] + 80 + 36 plates + [viii]; [xvi] + 191 + 14 plates + 131 + 30 plates + [x].

⬦   See Figure 8 for a photograph of the frontispiece in volume 1.

○   Volumes 1, 2, 3 and 5 each have a bookplate inside the front cover, "William Byrd of Westover in Virginia." Erased signature inside of front cover of each volume.

1803. *Recreations in Mathematics and Natural Philosophy: Containing Amusing Dissertations and Enquiries Concerning a Variety of Subjects the Most Remarkable and Proper to Excite Curiosity and Attention to the Whole Range of the Mathematical and Philosophical Sciences: The Whole Treated in a Pleasing and Easy Manner, and Adapted to the Comprehension of All who are the Least Initiated in Those Sciences: First Composed by M. Ozanam, Lately Recomposed and Greatly Enlarged, in a new Edition, by the Celebrated M. Montucla and now Translated into English, and Improved with Many Additions and Observations by Charles Hutton.* 4 vols. London: Printed for G. Kearsley. pp. xv + xvi + xxi + 447 + 21 plates; xiii + 464 + 34 plates; ix + 501 + 34 plates; xii + 516 + 8 plates.

⬦   In volume 1, errata and addenda follow p. xv.

⬦   Volume 1 also contains short "Life and Writings of Montucla," pp. viii– xii, "Life and Writings of Oznam," pp. xiii–xv, and Montucla's preface, pp. i–xvi.

1814. *Recreations in Mathematics and Natural Philosophy: Containing Amusing Dissertations and Enquiries Concerning a Variety of Subjects the Most Remarkable and Proper to Excite Curiosity and Attention to the Whole Range of the Mathematical and Philosophical Sciences: The Whole Treated in a Pleasing and Easy Manner, and Adapted to the Comprehension of All who are the Least Initiated in Those Sciences: First Composed by M. Ozanam, Lately Recomposed and Greatly Enlarged, in a new Edition, by the Celebrated M. Montucla and now Translated into English, and Improved with Many Additions and Observations by Charles Hutton.* 4 vols. London: Printed for Longman, Hurst, Rees, Orme, and Brown. pp. xxxv + 379 + 22 plates; xv + 389 + 34 plates; xi + 422 + 34 plates; xii + 440 + 8 plates.

⬦   This book has the same "Advertisement", dated 1801, as the 1803 edition, but lacks Montucla's preface and the short biography of the life of Montucla.

Ozanam, Jacques. See also Hutton 1803.

Pappus (fl, 300-350).
1660. *Pappi Alexandrini mathematicae collectiones a Federico Commandino urbin-ate in Latinum conversae, & commentarijs illustratae. In hac nostra editione ab innumeris, quibus scatebant mendis, & praecipuè in Graeco contextu diligenter vin-dicate, et serenissimo principi Leopoldo Gulielmo archiducci Austriae, &c. dictatae.* Bononiae [Bologna]: H.H. de Duccijs. pp. [xii] + 490.
 ◇ Dedicated to Leopold Gulielmo, Archduke of Austria. Frontispiece probably depicts Leopold Gulielmo. Contains books III-VIII of the eight books forming the original Greek text.

Pappus. See also Apollonius 1795; Chasles 1860; Euclid 1860; Wallis 1699.

Parkinson, Thomas (1745–1830).
1785. *A System of Mechanics, being the Substance of Lectures upon that Branch of Natural Philosophy.* Cambridge: Printed by J. Archdeacon. pp. [viii] + 255 + 22 plates.
 ◇ Dedicated to the "tutors, and other members of the university." Errata follow p. 255. Bound together with Parkinson 1789. The table of contents is on p. [iii] of Parkinson 1789.

1789. *A System of Mechanics and Hydrostatics, being the Substance of Lectures upon Those Branches of Natural Philosophy.* Cambridge: Printed by J. Archdea-con. pp. [iv] + 192 + 10 plates.
 ◇ Errata on p. iv. Bound together with Parkinson 1785. The table of contents of Parkinson 1785 is combined with the table of contents of this book.

Pascal, Blaise (1623–1662).
1819. *Oeuvres de Blaise Pascal.*   5 vols. Nouvelle édition. Paris: Lefèvre. pp. [viii] + ccxxviii + 488; xliv + 551; [iv] + 619; [iv] + 404; [vi] + 435.
 ◇ Volume 1 has a frontispiece portrait of Pascal. There is a biographical sketch of Pascal at the beginning of volume 2. The mathematics is in volume 4. Two plates deal with his arithmetic machine. A plate in volume 5 shows Pascal's triangle.

Pasley, Charles William (1780–1861).
1822. *A Complete Course of Practical Geometry Including Conic Sections, and Plan Drawing; Treated on a Principle of Peculiar Perspicuity. Orginally Published as the First Volume of a Course of Military Instruction.* Second edition, much enlarged. London: T. Egerton. pp. xlvii + 608.
 ◇ Errata follow p. xlvii.

Patoun, Archibald (1706–1775).
1765. *A Complete Treatise of Practical Navigation Demonstrated from its First Principles: With all the Necessary Tables. To which are Added the Useful Theo-rems of Mensuration, Surveying, and Gauging; with Their Application to Practice.*

Seventh edition, revised and corrected by the author with large additions and alterations. London: Printed for W. Mount and T. Page, et al. pp. xii + 525 + 1
plate (facing p. 262).

⋄ Dedicated to Sir John Ligonier. The first two sections (72 pp.) deal with geometry and trigonometry.

○ This copy belonged to Jared Mansfield in 1790.

Patterson, Robert. See Brewster 1814.

Peacock, George (1791–1858).
1820. *A Collection of Examples of the Applications of the Differential and Integral
Calculus.* Cambridge: J. Smith. pp. viii + 506 + 5 plates.
⋄ Addition and corrections on pp. 503–506.

Peacock, George. See also Herschel 1820.

Peck, William Guy (1820–1892; USMA 1844).
1871. *Practical Treatise on the Differential and Integral Calculus with Some of its
Applications to Mechanics and Astronomy.* New York: A. S. Barnes. pp. 208.
⋄ Peck served as an Associate Professor of Mathematics at USMA from 1852 to
1856.

1873. *A Treatise on Analytical Geometry, with Applications to Lines and Surfaces
of The First and Second Orders.* New York and Chicago: A. S. Barnes and Co. pp.
i–viii, 9–212.
○ This copy has been rebound.

1887. *Elementary Treatise on Determinants.* New York: A. S. Barnes and American Book Company. pp. [iv] + 47, pp. [vi] + 69.
⋄ This book was used as a text at USMA in 1899.

Peck, William Guy (1820–1892). See Davies 1855.

Peirce, Benjamin (1809–1880).
1836. *First Part of an Elementary Treatise on Spherical Trigonometry.* Boston:
James Munroe and Company. pp. iv + 71 + 1 plate.
⋄ Begins with Napier's rules and then treats the solution of spherical right and
oblique triangles. There is no preface or other indication of the origin or purpose
of the book.
○ The library has two copies.
○ One copy was donated by LT Norvin N. Van Nostrand. Contains notes by a
student.

1840. *An Elementary Treatise on Plane & Spherical Trigonometry with Their Applications to Navigation, Surveying, Heights, & Distances, and Particulary Adapted
to Explaining the Construction of Bowditch's Navigator, and the Nautical Almanac.*
Boston: James Munroe and Co. pp. iv + 428 + 5 plates.
⋄ Errata on p. iv.

1843, [1837]. *An Elementary Treatise on Algebra to which are Added Exponential Equations and Logarithms.* Boston: James Munroe and Co. pp. iv + 284.

1852. *An Elementary Treatise on Curves, Functions, and Forces.* 2 vols. Boston and Cambridge: James Munroe and Co. pp. v + 301 + 14 plates; vii + 290 + 1 plate.

⋄ Corrections follow p. v of volume 2.

1855. *A System of Analytical Mechanics.* Boston: Little, Brown, and Company. pp. xxxix + 496 + 1 plate.

⋄ Dedicated to Nathaniel Bowditch. List of subscribers is on pp. ix–xii. Errata listed on pp. 483–486. This is the first "system" of Peirce's "Physical and Celestial Mechanics," the other three being *Celestial Mechanics*, *Potential Physics*, and *Analytical Morphology.*

1870. *Linear Associative Algebra.* Washington City: Unknown. pp. 153.

⋄ Lithograph of the handwritten manuscript. "Read before the National Academy of Sciences" is on the title page.

Peirce, Benjamin Osgood (1854–1914).
1886. *Elements of the Theory of the Newtonian Potential Function.* Boston: Ginn and Company. pp. x + 163.

Peirce, Benjamin Osgood. See also Byerly 1889.

Peirce, James Mills (1834–1906).
1857. *A Text–Book of Analytic Geometry on the Basis of Professor Peirce's Treatise.* Cambridge: John Bartlett. pp. vii + 228 + 6 plates.

⋄ Errata follow p. vi.

∘ A small amount of writing in the margins. Book flyleaf signed "William P. Blake. Harvard College". On the title page is inscribed the couplet "None knew him eir to love him./ None named him eir to praise." Also on the title page is the autograph "E. L. Zalinski U. S. A."

Pell, John (1611-1685). See Maseres 1795.

Pemberton, Henry (1694–1771).
1728. *A View of Sir Isaac Newton's Philosophy.* London: Printed by S. Palmer. pp. [l]+ 407 + 12 plates.

⋄ See Figure 14 for a photograph of the title page of this volume.

⋄ Dedicated to Sir Robert Walpole. Errata on p. [xxxiv]. There is "A poem on Sir Isaac Newton" by R. Glover, pp. [xiii]–[xxvii].

Péquégnot (1819–1878).
1872a. *Leçons de perspective.* Paris: l'auteur. pp. [iv] + 56.
∘ "Van Nostrand, publisher and Importer" stamped on cover.

1872b. *Leçons de perspective.* Paris: l'auteur. pp. [ii] + 82 plates.

Perkins, George (1844–1926).

1842. *A Treatise on Algebra, Embracing, Besides the Elementary Principles, All the Higher Parts Usually Taught in Colleges; Containing Moreover, The New Method of Cubic and Higher Equations, as well as The Development and Application of the More Recently Discovered Theorem of Sturm.* Utica: Saxton and Miles O. Hutchinson. pp. viii + 360.

◇   The author claims this is the first "American school book" containing Sturm's Theorem.

Peslouïan, Charles Lucas (1878–?).

1906. *N. -H. Abel, sa vie et son oeuvre.* Paris: Gauthier–Villars. pp. xiii + 168.

◇   Frontispiece is the Gobitz portrait of Abel painted in Paris in 1826.

Peyrard, François (1760–1822). See Archimedes 1807; Euclid 1814a.

Pfaff, Johann Friedrich (1765–1825).

1788. *Versuch einer neuen Summationsmethode nebst andern damit zusammenhängenden analytischen Bemerkungen.* Berlin: Christian Friedrich Himbrugh. pp. [x] + 120.

Pflieger, Wilhelm.

1901. *Elementaire planimetrie.* Sammlung Schubert II. Leipzig: G. J. Göschensche. pp. [vii]+[430].

Phillips, Andrew W. (1844–1915) and Fisher, Irving (1867–1947).

1896. *Elements of Geometry.* Phillips–Loomis Mathematical Series. New York/Cincinnati/Chicago: American Book Company. pp. viii + 540 with approximately 20 pages of advertising notes in cover.

◇   Errata are taped into book. Corresponding to most of the illustrations of three dimensional geometric objects, there are photographs of models (made at Yale University) of these objects. This the first in the Phillips–Loomis Mathematical Series which was "founded on the works of the late Professor Elias Loomis."

○   The library has two copies.

○   One copy was owned by Arthur R. Hercz.

1898. *Elements of Geometry.* Phillips–Loomis Mathematical Series. New York: Harper. pp. viii + 540.

◇   This book was used as a text at USMA from 1902 until 1943.

Picard, Charles Emile (1856–1941).

1891; 1893; 1896. *Traité d'analyse.* 3 vols. Cours de la faculté des sciences de Paris. Paris: Gauthier–Villars, et Fils. pp. xii + 457; xiv + 512; xiv + 568.

◇   Volume 1 contains simple and multiple integrals, Laplace's equation and its applications, development of series, geometric applications of infinitesimal calculus.

◇   Volume 2 contains harmonic functions and analytic functions, an introduction to the theory of differential equations, Abelian integrals and Riemann surfaces.

◇   Volume 3 contains singularities of integrals of differential equations, curves defined by differential equations; study of the case where the variable remains real, linear equations–analogies between algebraic equations and linear equations.

Pickering, John (1777–1846).

1838. *Eulogy on Nathaniel Bowditch, LL. D., President of the American Academy of Arts and Sciences; Including an Analysis of his Scientific Publications. Delivered before the Academy, May 29, 1838.* Boston: Charles C. Little and James Brown. pp. [iv] + 101.

○   "From the Author" inscribed on the title page.

Pierre de Sainte Marie Magdeleine, Dom.

1665. *Traitté [Traité] d'horlogiographie, contenant plusieurs manières de constru-ire, sur toutes surfaces, toutes sortes de lignes horaires: & autres cercles de la sphère. Avec quelques instrumens pour la mesme pratique, & pour connoistre les heures durant la nuict [nuit]: & l'heure du flus & reflus de la mer. Plus la méthod de couper, en pierre ou en bois, les corps réguliers & autres polyèdres, par le cube & par le cylindre.* Reveu, corrigé & augmenté en cette troisiéme edition, de plusieurs propositions & figures. Paris: Jean Dupuis. pp. [xvi] + 312 + 72 plates.

◇   Dedicated to Salomon Phelippeaux. Frontispiece.

○   This copy contains an old label: "C. A. Force, Bookbinder, 216 William St., N. York". Several plates are out of order; #34 is inverted.

Pike, Nicolas (1743–1819).

1788. *A New and Complete System of Arithmetic, Composed for the Use of the Citizens of the United States.* Newbury–Port, [Massachusetts]: John Mycall. pp. 512.

◇   Dedicated to James Bowdoin, Governor of MA.

○   Title page signed several times by "Daniel Dearborn, Concord, NH," once by "Shillings," and also "Trueworth Gore Dearborn's property." Pages 19–24 are missing.

Pillet, Jules–J. (1842–1912).

1875–1876. *Notes prises par les elèves du cours de géométrie descriptive et de stéréotomie.* Cours préparatoires. Paris: Ecole Nationale des Ponts et Chaussées. pp. 322 + 3 plates.

◇   Dactotile.

1887. *Traité de géométrie descriptive ligne droite et plan – polyèdres – surfaces. Texte et dessins.* Cours de science appliquées aux arts. Paris: Ch. Delagrave; Leipzig: H. Lesoudier. pp. [ii] + [272].

◇   Dedicated to Ossian Bonnet.

Plauzoles, Charles.

1814. *Tables de logarithmes des nombres depuis 1 jusqu'a 21750; des sinus, cosinus, tangentes et cotangentes pour chaque minute du quart de cercle; suivies d'une table centésimale donnant les logarithmes de ces mêmes lignes pour chaque cent-millième du quadrans, depuis 0,00000 jusqu'à 0,03000, et pour chaque dixmillième depuis 0,0300 jusqu'à 0,5000.* Édition stéréotype. Paris: Didot. pp. 22 + unnumbered pages of tables.

Playfair, John (1748–1819).

1778. "On the Arithmetic of Impossible Quantities."

◇ A paper, not a book, bound together with other papers in Anonymous n.d.

1822. *The Works of John Playfair with a Memoir of the Author.* 4 vols. Edinburgh: Printed for Archibald Constable & Co. pp. cv + 514 + 4 plates; [iv] + 445; [iii] + 474 + 4 plates; [vi] + 535.

Playfair, John. See Euclid 1819b, 1861.

Poincaré, Jules Henri 1854–1912).

n.d. *La Science et l'hypothèse.* Bibliothèque de philosophie scientifique. Paris: Ernest Flammarion. pp. [ii] + 284.

Poincaré, Jules Henri (1854–1912) and Quiquet, Albert.

1912. *Calcul des probabilitiés.* Cours de la facultý des sciences de Paris. Deuxième édition. Paris: Gauthier–Villars. pp. [ii] + 333 + 1.

Poisson, Simeon–Denis (1781–1840).

1833. *Traité de mécanique.* 2 vols. Seconde édition, considérablement augmentée. Paris: Bachelier. pp. xxx + 696 + 4 plates; xxxv + 782 + 3 plates.

◇ Errata for volume 1 follow p. xxx, and in volume 2 follow p. xxxv.

1835. *Théorie mathématique de la chaleur.* Paris: Bachelier. pp. [iv] + 532 + 1 plate.

◇ Errata on p. [iv].

○ Bound together with Poisson 1837.

1837. *Théorie mathématique de la chaleur; mémoire et notes formant un supplément a l'ouvrage publié sous ce titre.* Paris: Bachelier. pp. 72.

○ Bound together with Poisson 1835.

○ The library has two copies.

○ In one copy there is a second copy of the plate.

○ In the second copy, there is "Extrait du catalogue des livres de fonds et d'assortiment de Bachelier," dated October 1848, and which consists of an alphabetized list of scientific books published by Bachelier (pp. 1–19) and the tables of contents of the *Journal de Mathématiques Purés et Appliquées*, 1836–1847, the *Comptes Rendus*, and of the first 31 volumes of the *Journal de École Polytechnique* (pp. 20–27).

Pommies, L.
1808. *Manuel de l'ingénieur du cadastre, précédé d'un traité de trigonométrie rectiligne, et des instructions publièes pour l'exécution des arpentages parcellaires, approuvées par le Ministre des finances.* Paris: L'Imprimerie Impériale. pp. xxxvi + xlvi + 213 + [iv] + 2 fold out tables + 11 plates.

◇ Dedicated to Hennet, Imperial Commissioner of the Survey.
◇ The treatice on trigonometry is by A. A. L. Reynaud.
○ Thayer binding.

Poncelet, Jean Victor (1788–1867).
1822. *Traité des propriétés projectives des figures; ouvrage utile a ceux qui s'occupent des applications de la géométrie descriptive et d'opérations géométriques sur le terrain.* Paris: Bachelier. pp. xlvi + 426 + 12 plates.

◇ Errata follow p. 426. The Preface, p. v–vi, contains Poncelet's personal history of this work beginning with his capture during Napolean's Russian campaign of the winter of 1813.

1839. *Introduction a la mécanique industrielle, physique ou expérimentale.* Deuxieme edition. Metz: Thiel and Paris: Leneveu. pp. xvi + 719 + 3 plates.

◇ Errata on pp. 709–711.

1862; 1864. *Applications d'analyse et de géométrie, qui ont servi de principal fondement au traité des propriétés projectives des figures.* Avec additions par M. M. Mannheim et Moutard. 2 vols. Paris: volume 1 Mallet–Bachelier; volume 2 Gauthier–Villars, successeur de Mallet–Bachelier. pp. xiii + 563 + 4 fold–out sheets; vii + 602.

1874. *Cours de mécanique applquée aux machines.* publié par M. X. Kretz. Paris: Gauthier–Villars. pp. v + 520 + 6 plates.

◇ The preface by Kretz describes some of the origins of this work in Poncelet's courses.

Pond, J. see Laplace 1809.

Potier.
1817. *Traité de géométrie descriptive.* Paris: Didot. pp. 96 + 2 plates.
○ Thayer binding.

Potts, Robert (1805–1885). See Euclid 1845.

Powell, Baden (1796–1860).
1842. *History of Natural Philosophy from the Earliest Periods To The Present Time.* London: Longman, Brown, Green, and Longmans. pp. xvi + 396.

◇ There is a "Chronology of Physical and Mathematical Science" on pp. xi–xvi.

Prestet, Jean (1648–1690).
1689; 1700. *Nouveaux elemens des mathematiques ou principe generaux de toute les sciences qui ont les grandeurs pour objet.* 2 vols. Premier volume. Qui comprend la science des nombres & l'algébre, ou l'art de comparer toute sorte de grandeurs par le moyen des chiffres & des lettres. Et où tout est démontré dans un ordre naturel & facile, & les choses expliquées plus à fond, & poussées plus loin que l'on n'a fait jusqu'ici. Second volume. Que comprend un corps d'analyse, ou l'art de résoudre les questions qu'on propose sur toutes les diverses grandeurs. Et où tout est expliqué dans un ordre naturel & facile, & les choses traitées bien plue à fond, & poussées plus loin que l'on n'a fait jusqu'ici. Seconde edition, plus ample et mieux digerée. Paris: Volume 1 André Pralard, and volume 2 Nicolas Pepie. pp. [xxxii] + 588 + 2 plates; [xii] + 452 + 2 plates.
◇   Volume 1 is dedicated to the King. First edition ca. 1675.

Proclus (410–485).
1792. *The Philosophical and Mathematical Commentaries of Proclus, on the First Six Books of Euclid's Elements. To which are Added, a History of the Restoration of Platonic Theology, by the Latter Platonists: and a Translation from the Greek of Proclus's Theological Elements.* 2 vols. London: Printed for the Author. pp. cxxx + 183; [iv] + 444.
◇   Thomas Taylor is the translator and the "Author."

Proclus. See also Euclid.

Prony, Gaspard Clair François Marie Riche de. (1755–1839).
1791. *Exposition d'une méthode pour construire les équations indéterminées qui se rapportent aux sections coniques. A l'usage de l'École des ponts et chaussées.* Paris: Didot. pp. viii + 28 + 2 plates.
◇   Errata on p. 26.
○   Plates have been cut and bound as 4 leaves. Volume is rudely trimmed.

1800 (an VIII). *Mécanique philosophique, ou analyse raisonnée des diverses parties de la science de l'équilibre et du mouvement.* Paris: C.en Bernard. pp. vii + 477.
◇   Errata (two pages) follows p. 477. The left pages consist of paragraphs of exposition, numbered consecutively to 519; the right pages are divided into four columns headed notation, definitions, theorems, and problems.

Prus, C.
1846. *Tables relatives au tracé des courbes de raccordement.* Angers: Cornilleau et Maige. pp. xxiii + 477.

Pruvost, E.
1893. *Leçons de géométrie analytique à l'usage des élèves de la classe de mathématiques spéciales et des candidats à l'École normale supérieure et à l'École polytechnique.* 2 vols. Paris: Paul Dupont. pp. 703; 437.
◇   Volume 2 is designated "Gémétrie dans l'espace, Quatrième édition."

Ptolemy, Claudius (ca. AD 150).

1528. *Claudii Ptolemaei pheludiensis Alexandrini almagestum seu magnae constructionis mathematicae opus plane divinum Latina donatum lingua ab Georgio Trapezuntio usq veq vaq. doctissimo. Per Lucam Gavricum Neapolit. divinae matheseos professorem egregium in alma urbe Veneta orbis regina recognitum. Anno salvatis M D XXVIII labente. Nequispiam alius Calcographus/Venetiis aut usquá locorum Venetae detionis impune almagestum hunc imprimat per Decennium/Senatus Veneti Decreto cautum est.* In urbe Veneta [Venice]: Luceantonii Iunta. pp. [xii] + 143 + [1].

1538. ΚΛ—ΠΤΟΛΕΜΑΙΟΥ. ΜΕΓΑΛΗΣ ΣΥΝΤΑΞΕΩΣ ΒΙΒΛ. ΙΓ ΘΕΩΝΟΣ ΑΛΕΞΑΝ ΔΡΕΩΣ ΕΙΣ ΤΑ ΑΥΤΑ ΥΓΟΜΝΗΜΑΤΩΝ ΒΙΒΛ. ΙΑ. *Claudii Ptolemaei magnae constructionis, id est perfectae coelestium motuum pertractationis, lib. XIII. Theonis Alexandrini in eosdem commentariorum lib. XI.* Basileae apud Ioannem VValdervm, Cum Privilegio Caesareo ad Quinquennium. Basileae [Basel]: J. Walderum. pp. [xvi] + 327.

◇ This work is in Greek, with the exception of the dedication.

○ Stamped "De La Lande." on the inside front cover. Some notes in French on the front endcovers.

1795. *Beobachtung und Beschreibung der Gestirne und der Bewegung der himmlischen Sphäre: mit Erläuterungen, Vergleichungen der neuern Beobachtungen und [e]inem stereographischen Entwurf der beyden Halbkugeln des gestirnten Himmels für die Zeit des Ptolemäus von J. E. Bode.* Berlin: F. Nicolai. pp. viii + 260 + 1 plate.

Ptolemy, Claudius. See also Flamsteed 1725; Regiomontanus 1496; Wallis 1699.

Puissant, Louis (1769–1843).

1807. *Traité de topographie, d'arpentage et de nivellement.* Paris: Courcier. pp. xx + 331 + 20 unnumbered pages of tables + 6 plates.

◇ Dedicated to Prince Alexander.

◇ Bound together with Puissant 1810.

◇ Index pp. 325–331.

○ Thayer binding.

1809. *Recueil de diverses propositions de géométrie, résolues ou démontrées par l'analyse algébrique; précédé d'un précis du levé des plans.* Seconde édition, considérablement augmentée. Paris: Ve Bernard. pp. xx + 442 + 6 plates.

◇ Errata follow p. 442.

1810. *Supplément au second livre du traité de topographie, contenant la théorie des projections des cartes.* Paris: Courcier. pp. iv + 143 + 1 fold out table + 1 plate.

◇ Bound together with Puissant 1807.

◇ Errata follow p. 143.

○ Thayer binding.

1812. *Mémoire sur la projection de Cassini, pour servir de supplément a sa théorie des projections des cartes géographiques.* Paris: Ve Courcier. pp. 43.
  ◇ Erratum follows p. 43.
  ○ Copy is very brittle.

Puissant, Louis. See also Allaize 1813.

Pund, Otto (1867–?).
1899. *Algebra mit Einschluss der elementaren Zahlentheorie.* Sammlung Schubert VI. Leipzig: G. J. Göschensche Verlagshandlung. pp. [vii] + 345.

Pyne, George (1800–1884).
1851. *A Rudimentary and Practical Treatise on Perspective For Beginners; Simplified for the Use of Juvenile Students and Amateurs in Architecture, Painting, etc.; also Used for Schools and Private Instructors.* Third edition, revised and enlarged. London: John Weale. pp. viii + 165 + 134 plates inserted in the text.

Quinby, Issac Ferdinand (1821–1891; USMA 1843) (Editor).
1868. *A New Treatise on the Elements of the Differential and Integral Calculus.* Robinson's Mathematical Series. New York: Ivison, Phinney, Blakeman and Company. pp. 472.

Raphson, Joseph (? –1715). See Maseres 1795.

Ray, Joseph (1807–1855).
1908 [1880]. *Ray's New Higher Arithmetic.* Eclectic Educational Series, A Revised Edition of the Higher Arithmetic. New York: American Book Co. pp. viii + 408.
  ○ USMA special collection book plate: "In memory of BG Fenton, Class of 1904". Inscription inside front cover: ref Stilwell and McNair and problem on page 406.
  ○ Earlier editions of this book used in the entrance exam for West Point, 1900.

Ray, Joseph. See also Howison 1869.

Raymond, William G. (1859–1926).
1896. *A Text Book of Plane Surveying.* New York: American Book Co. pp. ii + 485.
  ◇ This book contains both color and black and white plates. [I–VI].
  ○ Signed by LT Castle.

Rebière, Alphonse (1842–1901).
1897. *Les femmes dans la science notes recueillies.* Deuxième edition très augmentée et ornée de portraits et d'autographes. Paris: Nony & Cie. pp. ix + 361.
  ◇ Contains portraits of 26 women scientists, including Agnesi, Chatelet, Germain, Kowalevski, Ladd–Franklin, Scott, and Somerville.

Reech, Ferdinand (1805– ?).
1852. *Cours de méchanique d'après la nature généralement flexible et élastique des corps, comprenant la statique et la dynamique avec la théorie des vitesses virtuelles celle des forces vives et celle des forces de réaction, la théorie des mouvements relatifs et le théorème de Newton sur la similitude des mouvements.* Paris: Carilian–Goeury et Vor Dalmont. pp. [vi] + x + 275.

Regiomontanus (Johannes Müller) (1436–1476).
1496. *Epytoma Joānis de mōte regio in almagestii ptolomei.* Veneto [Venice]: J. Hamman–Hertzog. pp. [214].
◇   All figures are in the margins. There is a handsome full page illustration on p. [6] of Ptolemy and Regiomontanus seated beneath an armillary sphere.
○   Title page signed "Bibliotheca Colbertinae". There are some annotations in the margins of the first few pages.

Renwick, James (1790?–1863).
1832. *The Elements of Mechanics.* Philadelphia: Carey & Lea. pp. xxxii + 508 + 24 pages of advertisements.
◇   Page [ix] lists 26 authors consulted in preparing the work.

1840. *Applications of the Science of Mechanics to Practical Purposes.* New York: Harper & Brothers. pp. xv + 327 + 8.
◇   The last eight pages list "valuable works" published by Harper & Brothers. The book contains no plates, but has 97 detailed figures.

Regnier. See Burg 1847.

Reye, Theodor (1838–1919).
1898. *Lectures on The Geometry of Position Part I.* Translated by Thomas Holgate. New York: The MacMillan Co. pp. xix + 248.

Reynaud, Antoine André Louis (1777–1844).
1804. *Traité d'arithmétique, à l'usage des ingenieurs du cadastre, et des èléves qui se destinent, à l'École polytechnique, à la marine, à l'artillérie et an commerce.* Paris: Publié pour l'auteur, et. al. pp. lii + 377 + 7 pages of tables.

1805. *Trigomométrie analytique, précédée de la théorie des logarithmes, et suivie des tables de logarithmes des nombres et des lignes trigonométriques, à l'usage des ingénieurs du cadastre, et des eléves qui se destinent à l'École polytechnique.* Paris: Courcier. pp. 106 unnumbered pages of text + 2 plates + unnumbered pages of tables.
○   Thayer binding (but no border).

1819. *Traité d'application de l'algèbre a la géométrie, et de trigonométrie, a l'usage des élèves qui se destinent a l'École royale polytechnique.* Paris: Ve Courcier. pp. xii + 340 + 2 fold out tables + 10 plates.
○   The library has three copies. One copy has torn front cover.
○   One copy has been rebound.

1821. *Traité d'algèbre a l'usage des élèves qui se destinent a l'École royale poly-technique et des élèves de l'École spéciale militarie.* Cinquieme édition. Paris: Ve Courcier. pp. xvi + 336 + 1 plate.

Reynaud, Antoine André Louis (1777–1844) and Duhamel, Jean Marie Constant (1797–1872).

1823. *Problèmes et développemens sur diverses parties des mathématiques. Par M. Reynaud et M. Duhamel.* Paris: Bachelier. pp. viii + 399 + 2 pages of errata + 11 plates.

Reynaud, Antoine André Louis. See also Pommies 1808; Bezout 1777, 1771, 1812, 1799, 1799, 1800.

Reynaud, Emmanuele.

1855. *Résolution des questions relatives a l'épreuve pratique d'après le programme officiel du 20 Avril 1853.* Licence ès sciences mathématiques. Paris: Victor Dalmont. pp. viii + 120.

Reyneau, Charles René (1656–1728).

1708. *Analyse demontrée ou la methode de resoudre les problêmes des mathematiques, et d'apprendre facilement ces sciences; Expliquée & démontrée dans le premier volume, & appliquée, dans le second, à découvrir les proprietés des figures de la geometrie simple & composée; à resoudre les problêmes de ces sciences & les problêmes des sciences physico-mathematiques, en employant le calcul ordinaire de l'algebre, le calcul differentiel & le calcul integrale. Ces derniers calculs y sont aussi expliqués & démontrés.* 2 vols. Paris: Jacque Quillau. pp. [vi] + xxiv + 486 + 1 fold out page; xxviii + 487–914 + 5 plates.

◇ Volume 2 has a somewhat different title. *Usage de l'analyse, ou la maniere de l'appliquer à découvrir les proprietés des figures de la geometrie simple & composée; à resoudre les problêmes de ces sciences & les problêmes des sciences physico-mathematiques, en employant le calcul ordinaire de l' algebre, le calcul differentiel & le calcul integrale. Ces derniers calculs y sont aussi expliqués & démontrés.* Presented by the Minister of Public Instruction of France. Dedicated to the Duke of Bourgogne. Volume 1 errata follow p. 486 and volume 2 errata follow the plates.

Rhäticus, Georg Joachim (1514–1576).

1596. *Opus palatinum de triangulis a Georgio Ioachimo Rhetico coeptum. Neostandii in Palatinatu.* [Prague]: Excudebat Matthaeus Harnisius. pp. [xix] + 104 + 140 + 341 + [i] + 121 + 554 + 181.

◇ Contains (1) *De fabrica canonis doctrinae triagulorum;* (2) *De triangulis globi cum angulo recto;* (3) *L. Valentini Othonis parthenopolitani. De triangulis globi sine angulo recto libri quinque. Quibus tria meteo roscopia nomerorum accesserunt;* (4) *L. Valentini Othonis parthenopolitani. Meteoroscopium numerorum primum. Monstrans proportronem sigulorum parallelorum ad aequatorem vel meridianum;* (5) *Georgii Ioachimi Rhaetiei. Magnus canon doctrinae triangulorum ad decades secundorum scrupulorum et ad partes 10000000000; (6) Tertia series magni canonis doctrinae triangulorum in quo triquerti cum angulo recto in planite minus latus includentium angulum rectum ponitur partium 10000000.*

◇ Place of publication and printer from colophon.

Richard, Claude (1589–1664). See Apollonius 1655.

Risner, Friedrich (?–1580). See Alhazen 1572.

Ritt, Georges (1801–1864).
1847. *Problèmes d'algèbre et exercices de calcul algébrique avec les solutions.* Ouvrage autorisé par le conseil royal de l'université. Troisième édition. Paris: L. Hachette et Cie. pp. 399.
○ Autographed in the front by A. Northrop Bellinger.

Rivard, Dominique François (1697–1778).
1816. *Traité de la sphère et du calendrier.* Septième édition, revue et augmentée par M. Puissant. Paris: Ve Courcier. pp. xii + 13–263 + 1 fold out table (facing p. 250) + 3 plates.
◇ Errata on p. 263. Contains an explanation of the French Revolutionary Calendar.
○ Thayer binding.

Robertson, John (1712–1776).
1775. *A Treatise of such Mathematical Instruments, As are Usually put into a Portable Case. Shewing some of Their Uses in Arithmetic, Architecture, Geometry, Surveying, Trigonometry, Geography, Spherics, Perspective, &c. with An Appendix; Containing the Description and Use of the Gunners Callipers. And the Description of, and Precepts for the Delineation of, Ship–guns and Sea Mortars. To this Treatise, is Prefixed a Brief Account of Authors, who have Wrote on the Proportional Compasses and Sector.* Third edition, with many additions. London: J. Nourse. pp. 1 plate + xxiv + 2333 + 3 + xi plates.
◇ Errata on p. xxiv. Dedicated to Sir James Burrow, Vice President of the Royal Society.
◇ Frontispiece is a plate showing several proportional compasses.
○ Special collection book plate: from the estate of Edwin N. Rich. Book plate: C.F. Northumberland.

1805. *The Elements of Navigation; Containing The Theory and Practice: With the Necessary Tables and Comendiums for Finding the Latitude and Longitude at Sea. To which is Added, a Treatise of Marine Fortification. Composed for the Use of the Royal Mathematical School at Christ's Hospital, the Royal Academy at Portsmouth, and the Gentlemen of the Navy.* 2 vols. Seventh edition, with additions, carefully revised and corrected by Lieut. Lawrence Gwynne. London: Printed for F. Wingrave; et al. pp. xlviii + 422 + 6 plates; 392 + 76 + 16 plates.
◇ Dedicated to Sir John William Anderson, James Palmer, and the rest of the Governors of Christ's Hospital, London. Two volumes bound together as one.

Roberval, Giles Personne de (1602–1675).
1736. *Ouvrages de mathematique de M. de Roberval.* Amsterdam: Pierre Mortier. pp. 399 + 26 plates.

Robins, Benjamin (1707–1751).
1761. *Mathematical Tracts of the Late Benjamin Robins, Esq; Fellow of the Royal Society, and Engineer General to the Honourable the East India Company.* 2 vols. London: James Wilson. Printed by J. Nourse. pp. xlviii + 341 + 2 plates; [viii] + 7–380.
◇ Volume 1, which deals with gunnery, was first published in 1742; the second, on fluxions, in 1735. The preface, pp. iii–xlvi of the first volume, by James Wilson, is dated 1761 and is largely a biography of Robins.
○ Copy owned by Wm. Partridge, U.S. Engineers.

Robinson, Horatio Nelson (1806–1867).
1862. *Conic Sections and Analytical Geometry; Theoretically and Practically Illustrated.* Robinson's Mathematical Series. New York: Ivison, Phinney and Co. pp. viii + 280 + 70 pages of tables.
◇ The tables at the end of the book are printed on a specially colored (yellow) paper; see also Babbage 1827.

Robinson, Horatio Nelson. See also Quinby 1868.

Robinson, James Watts (1827–1918; USMA 1852).
1896. *Robinsonian Unique Calculator, An Accountants' and Teachers' Assistant.* Dorchester, MA: J. Robinson. pp. [viii] + unnumbered pages of tables.

Rodriguez de Quijano y Arroquia, Angel.
1850. *Complemento á la geometria descriptiva. Empleo de un solo plano de proyeccion valiéndose del sistema de acotaciones para servir de aplicacion de los principios generales de la riencia á las superficies irregulares, y como preliminar á la topografia y á la desenfilada de las obras de fortificaion.* Madrid: Boix mayor y Compañia. pp. ii + 19 plates.

1869. *Tratado sobre las escalas gráficas en general y sus aplicaciones al dibujo geometrico. Atlas.* Madrid: no publisher listed. pp. 38 plates.

Roguet, Christoph Marie Michel (1809–?).
1860. *Leçons de géométrie analytique a deux et à trois dimensions, a l'usage des candidats a l'École polytechnique et a l'École normale, précédées d'une introduction renfermant les premières notions sur les courbes usuelles; à l'usage des candidats au baccalauréat ès sciences; ouvrage entirèrement conforme aux programmes officiels de l'enseignement scientifique des lycées.* Deuxième édition, revue et augmentée. Paris: Dalmont et Dunot. pp. xv + 502.

Rohault, Jacques (1618–1672). See Watts 1716.

Ronayne, Phillip (1683–1755).
1727. *A Treatise of Algebra in Two Books: The First Treating of the Arithmetical, and the Second of the Geometrical Part.* Second edition with additions. London: William and John Innys. pp. [viii] + v + 461.
⋄ Errata follow p. 5.

Ross, Edward C. (1801–1851; USMA 1821).
1831. *Elements of Algebra Translated from the French of M. Bourdon, for the Use of the Cadets of the U.S. Military Academy.* Translator LT Edward C. Ross. New York: E. B. Clayton. pp. vii + 389.
⋄ Errata page at end. This book was the foundation for Charles Davies's *Bourdon's Algebra.*
⋄ This book was used as a textbook at USMA from 1831 until 1833.
⋄ Ross served as an Assistant Professor in Mathematics Department, USMA, from 1825 to 1833.
○ Ten copies of this book are in the collection.
○ One copy has a title page signed by H.S. Kendrick (USMA 1835; Prof. of Chem., USMA). Otherwise a clean copy.

Ross, Edward C. (1806–1851; USMA 1821). See also Bourdon 1831a; Davies (Algebra).

Rossi, Gaetanuo (1780–1855).
1804. *Soluzione Esatta e Regolare Del Difficilissimmo Problema della Quadratura Del Circolo; Produzione Sintetica ed Analitica di Gaetano Rossi Di Catanzaro, Nel Regno di Napoli; Arricchita di Molti Corollarj, non che di Moltre altre Importanti Dimostrazioni Concernenti al Soggetto, e Dedicata a' Degnissimi e Sempre Lodevolissimi Soscrittori di Essa.*   Londra [London]: W. Spilsbury. pp. xx + 105 + 2 plates.
⋄ Portrait of Gaetanuo Rossi is the frontispiece.

Rouché, Eugène (1832–1910) and Comberousse, Charles de (1826–1897).
1879.   *Traité de géométrie, conforme aux programmes officiels, renfermant un très–grand nombre d'exercises et plusieurs appendices consacrés a l'exposition des principales méthodes de la géométrie moderne. Première partie. Géométrie plane.* Quatrième édition, revue et augmentée. Paris: Gauthier–Villars. pp. xxii + 549.

Rowe, John (1735–c. 1776).
1800. *An Introduction to the Doctrine of Fluxions.* Fourth edition, with additions and alterations. To which is Added "An Essay on the Theory." The whole revised, carefully corrected, and prepared for the press by the late William Davis. London: Printed for Anne Davis. pp. vi + 7–197.
⋄ Advertisement of other books "printed for Anne Davis" on three unnumbered pages at the end of the volume, before the plates.

Rowe, John. See also West 1763.

Royal Military Academy, Woolwich
1853; 1852. *An Elementary Course of Mathematics, Prepared for the Use of the Royal Military Academy. By Order of the Master–General and Board of Ordinance.* 2 vols. London: John Weale. pp. viii + 455; viii + 387.
◇ Dedicated to Lieutenant–General Lord Raglan. In volume one, the sections on arithmetic and algebra, as well as the sections on the differential and integral calculus were written by William Rutherford (1798–1871); applications of algebra to geometry, trigonometry, mensuration, and coordinate geometry were written by Stephen Fenwick. The portion of the second volume on geometry was written by Thomas Stephens Davies (1795–1851), while that on the conic sections is by Fenwick.

Rüdiger, Christian Friedrich. See Bicquilley 1788.

Russell, Bertrand A. W. (1872–1970).
1897. *An Essay on the Foundations of Geometry.* Cambridge: University Press. pp. xvi + 201.

1903. *The Principles of Mathematics.* vol. 1. Cambridge: University Press. pp. xxix + 534.
◇ No more volumes published. The material intended for the second volume was later developed into an independent work: *Principia mathematica* by Russell and A. N. Whitehead.

Russell, Bertrand A. W. (1872–1970) See also Whitehead 1910.

Rutherforth, Thomas (1712–1771).
1748. *A System of Natural Philosophy, being a Course of Lectures in Mechanics, Optics, Hydrostatics, and Astronomy; which are Read in St. Johns College Cambridge.* 2 vols. Cambridge: Printed by J. Bentham. pp. [xxiv] + 496 + 17 plates; [iii] + 497–1105 + 13 plates.
◇ Dedicated to Peter Burrell Junior. An index (6 pages) appears at the end of volume 2.
◇ The author's last name is also spelled "Rutherford."

Rutherford, William (1798–1871). See Royal Military Academy, Woolwich 1853; 1852; and Hutton 1843.

Ryan, James (died 1839, age 43).
1826. *An Elementary Treatise on Algebra, Theoretical and Practical, Adapted to the Instruction of Youth in Schools and Colleges.* Second edition, revised and corrected. New York: Collins and Hannau. pp. xi + 383.
◇ The Appendix, mentioned in the title, is contained on pp. 363–383.

1828. *The Differential and Integral Calculus.* New York: White, Gallagher and White. pp. viii + 328.

◇ This was the first Calculus book prepared by an American.

Saint Vincent, Gregorius (1584–1667).

1647. *P. Gregorii a S^{to} Vincentio opus geometricum quadraturae circuli et sectionum coni decem libris comprehensum.* Antverpiae [Antwerp]: Apud Ioannem et Iacobum Mevrsios. pp. [lii] + 1226 + 3 pp. errata.

○ See Figure 2 for a photograph of the frontispiece of this volume.

○ Many notes on separate slips in the volume. See Figure 5 for a photograph of one of them.

○ A few marginal notes. See Figure 4 for a photograph of one of them.

○ Copy signed by François de Sluse. See Figure 3 for a photograph of this signature.

1668. *R. P. Gregorii à S^{to} Vincentio ex Socieate Iesu opus geometricum posthumum ad mesolabium per rationum proportionalium noves proprietates. Finem operis mors authoris antevertit.* Gandavi [Ghent]: Typis Balduini Manilii. pp. [xix] + 297.

◇ Frontispiece portrait of Saint Vincent. Errata follow p. 297.

Salmon, George (1819–1904).

1879. *A Treatise On The Higher Plane Curves* [Intended as a sequel to A Treatise on Conic Sections]. Third edition. Dublin: Hadges, Foster, and Friggis. pp. xix + 395 + 1.

1900. *A Treatise on Conic Sections Containing an Account of Some of the Most Important Modern Algebraic and Geometric Methods.* London: Longmans, Green, and Co. pp. xv + 3999 + 40.

Saunderson, Nicholas (1682–1739).

1740. *The Elements of Algebra, In Ten Books, To which is Prefixed an Account of the Author's Life and Character, Collected from his Oldest and most Intimate Acquaintance.* 2 vols. *Volume the First, Containing the First Five Books. To which are Prefixed I. The Life and Character of the Author. II. His Palpable Arithmetic Decyphered* and *Volume the Second, Containing the Five Last Books.* Cambridge: University Press. pp. [xxv] + xxvi + [iv] + 360 + plate facing p. xxiv; 363–748 + [xxxii] + 8 plates.

◇ See Figure 16 for a photograph of the frontispiece, which is a portrait of the author.

◇ Dedicated to John Earl of Radnor. Following the dedication is a second title page: *The Elements of Algebra, in Ten Books: To which is prefixed, An Account of the Author's Life and Character, Collected from his oldest and most intimate Acquaintance,* 1740. There are two lists of subscribers pp. [xvi]–[xxiii], and p. [iv]. Errata are on p. [xxiv] and p. [iv]. The table of contents for the whole work comes at the end of volume II, pp. [i]–[xxxii]. Volume II also contains printed versions of correspondence between the author and Abraham DeMoivre, pp. 743–748.

1756. *The Method of Fluxions Applied to a Select Number of Useful Problems: Together with the Demonstration of Mr. Cotes's Forms of Fluents in the Second Part of his Logometria; The Analysis of the Problems in his Scholium Generale; and an Explanation of the Principal Propositions of Sir Issac Newton's Philosophy.* London: Printed for A. Millar, et al. pp. xxiv + 309.

⋄ Errata page follows p. 309. An advertizement for two other books of Saunderson's follows the errata page. The introduction consists of the following: "Of the Composition and Resolution of Forces;" "Of the Descent of Heavy Bodies;" and "Of Powers and their Indexes."

1778. *Select Parts of Saunderson's Elements of Algebra For The Use of Students at the Universities.* Fourth edition, revised and corrected. London: W. Bowyer and J. Nichols, et al. pp. iv + 412 + 2 plates (facing pp. 351 and 365).

⋄ The editor of this work is unknown.

Sauri (1741–1785).

1774. *Cours complet de mathématiques.* 5 vols. Paris: de Ruault. pp. xxxii + 515 + 6 plates; 351 + 8 plates; xxxvi + 318 + 6 plates; 503 + 4 plates; 656 + 8 plates.

⋄ Volume 1. calculations pp. 1–340, geometry pp. 341–508; volume 2. geometry of curves pp. 1–349; volume 3. differential calculus pp. 1–318; volume 4. integral calculus pp. 1–500; volume 5. integral calculus pp. 1–199, calculus of variations pp. 200–278, problems of physics and mathematics pp. 279–652. Dedicated to Count DeNoailles.

1835. *Institutions mathématiques servant d'introduction a un cours de philosophie a l'usage des universités de France.* Sixième édition, revue et corrigée. Paris: Bachelier. pp. xvi + 480 + 5 plates.

Schlömilch, Oskar (1823–1901).

1847, 1848. *Handbuch der Differenzial– und Integralrechnung.* 2 vols. The first of these volumes has the subtitle *Erster Theil. Differenzialrechnung,* the second *Dritte Lieferung. Integralrechnung. Erste Hälfte.* Greifswald: Ferd. Otte. pp. [214] + 1 plate; [327] + 2 plates.

○ Signed by E. Bass.

Schönberg, Curt Friedrich von.

1773. *Abhandlung von den Tangenten, Quadraturen und Rectificationen der Kegelschnitte, nebst einigen andern, diese Linien betreffenden Aufgaben.* Leipzig: Siegfried Lebrecht Crusius. pp. [viii] + 198 + 4 plates.

⋄ Dedicated to Georg Heinrich Bork.

Schooler, Samuel (1827–1873).

1853. *Elements of Descriptive Geometry, The Point, The Straight Line, and The Plane.* Richmond, Virginia: J. W. Randolph. pp. ix + 112.

⋄ Dedicated to Lewis Minor Coleman, Principal of Hanover Academy, Virginia. There is a brief history of descriptive geometry (pp. 9–12) and a "list of the chief works on descriptive geometry" (pp. 13–15). Errata on p. [x].

Schooten, Frans van (1615–1660). See Vieta 1646; Descartes 1649; Marolois 1628.

Schott, Gaspard (1608–1666). See Guericke 1672.

Schubert, Hermann Cäsat (1848–1911).
1898. *Mathematical Essays and Recreations.* Translated by Thomas J. McCormack. Chicago: Open Court Publishing Company. pp. [vii] + 147.
◇ Translated from the German. Includes essays on magic squares, the fourth dimension, and squaring the circle.

Schulze, Johann Carl (1749–1790).
1778. *Neue und erweiterte Sammlung logarithmischer, trigonometrischer und anderer zum Gebrauch der Mathematik unentbehrlicher Tafeln.* 2 vols. Berlin: August Mylius. pp. viii + [xii] + 260 + unnumbered pages of tables; 319.
◇ The preface (pp. i–viii) and the introductions to both volumes (pp. [i]–[xii] in Volume I) are in German and in French.
○ Some writing on an endpaper.

Scott, Charlotte Angas (1858–1931).
1894. *An Introductory Account of Certain Modern Ideas and Methods in Plane Analytical Geometry.* London: Macmillan and Co. pp. xii + 288.
○ Front matter bound out of order.

Scott, Robert Forsyth (1849–1910).
1880. *A Treatise on the Theory of Determinants and Their Applications in Analysis and Geometry.* Cambridge: University Press. pp. xi + 251 + 32 of advertisements.

Sedillot, Louis Pierre Eugéne Amèlie (1808–1875).
1845–1849; 1849. *Matériaux pour servir a l'historire comparée des sciences mathématiques chez les Grecs et les orientaux.* 2 vols. Paris: Didot Frères. pp. [iv] + vi + 420; [iv] + xvi + pages numbered 467–771 + 4 tables + 8 plates + 2 maps.
◇ The "Orient" here refers to the Arabic world.

Serret, Joseph–Alfred (1818–1885).
1849. *Cours de algèbre supérieure, professé a la faculté des sciences de Paris.* Paris: Bachelier. pp. xi + 400 + 1 plate.

1850. *Traité de trigonométrie.* Paris: Bachelier. pp. viii + 215 + 2 plates.

1862. *Traité de trigonométrie.* Troisième édition, revue et augmenté. Paris: Mallet–Bachelier. pp. [xi] + 298 + 1 plate.

1877; 1879. *Cours d'algèbre supérieure.* 2 vols. Quatrième édition. Paris: Gauthier–Villars. pp. [xiii] + [647]; xii + 694.

1879. *Cours de calcul différentiel et intégral.* 2 vols. Deuxième édition. Paris: Gauthier–Villars. pp. xii + 617; xii + 734.

○ A small number of marginal notes in Volume 1.

Serret, Paul Joseph (1827–1898).
1869. *Géométrie de direction. Application des coordonnées polyédriques. Propriété de dix points de l'ellipsoïde, de neuf points d'une courbe gauche du quatriéme ordre, de huit points d'une cubique gauche.* Paris: Gauthier–Villars. pp. [xx] + 523.

Servois, François Joseph (1767–1847).
1803. *Solutions peu connues de différens problèmes de géométrie–pratique; pour servir de supplément aux traités connus de cette science.* Metz: Devilly; Paris: Courcier. pp. [ii] + 79 + 28 + 1 plate.

◇ Errata on the last page.

1814. *Essai sur un nouveau mode d'exposition des principes du calcul différentiel; suivi de quelques réflexions relatives aux divers points de vue sous lesquels cette branche d'analise a été envisagée jusqu'ici, et, en général, à l'application des systèmes métaphysiques aux sciences exactes.* Nismes [Nîmes]: P. Blachier–Belle. pp. 79.

◇ Errata on p. 80.
◇ On p. 8 he defines a function to be "distributive" if $P(x+y+...) = Px+Py+...$. This is the origin of our phrase, "The distributive law". "Commutative" is defined on the same page.

Servois, François Joseph. See also Brianchon 1818a.

Sganzin, Joseph Mathiu (1750–1837). See Anonymous 1833.

'sGravesande. See Gravesande, Willem Jacob van, 's, listed alphabetically under G.

Sharp, Abraham (1651–1742). See Sherwin 1761.

Sherwin, Henry.
1761. *Sherwin's Mathematical Tables, Contriv'd after a most Comprehensive Method: Containing, Dr. Wallis's Account of Logarithms, Dr. Halley's and Mr. Sharp's Way of Constructing Them, with Dr. Newton's Contraction of Brigg's Logarithms, viz. a Table of Logarithms of the Numbers from 1 to 101000 ... and Tables of Natural and Logarithmic sines, tangents, secants, and versed sines, to Every Minute of the Quadrant ... Carefully Revised and Corrected by William Gardner.* Fourth edition. London: Printed for W. and J. Mount, T. Page and Son. pp. [8] + 46 + 64 + [306] + 2 plates.

Sherwin, Henry. See also Maseres 1791.

Simon, Max (1844–1918).
1900a. *Analytische Geometrie der Ebene.* Sammlung Schubert VIII. Leipzig: G. J. Göschensche Verlagshandlung. pp. [vii] + 372.

1900b. *Analytische Geometrie des Raumes. I. Teil: Gerade, Ebene, Kugel.* Sammlung Schubert IX. Leipzig: G. J. Göschensche Verlagshandlung. pp. 152.

Simonin.
1792. *Traité élémentaire de la coupe des pierres ou art du trait.* Paris: Basset; Jean. pp. [vi] + 74 + 49 plates.

Simpson, Thomas (1710–1761).
1737. *A New Treatise of Fluxions: Wherein the Direct and Inverse Method are Demonstrated after a New, Clear, and Concise Manner, with Their Application to Physics and Astronomy: also the Doctrine of Infinite Series and Reverting Series Universally, are Amply Explained, Fluxionary and Exponential Equations Solved: Together with a Variety of New and Curious Problems.* London: Printed by Tho. Gardner and the Author. pp. iv + 5–216.
◇ Errata on p. 164 and on p. 216. No table of contents. Pages 165–216 contain a "Supplement."

1740. *Essays On Several Curious and Useful Subjects, in Speculative and Mix'd Mathematicks. Illustrated by a Variety of Examples.* London: Printed by H. Woodfall, jun[ior] for J. Nourse. pp. viii + 142.
◇ Errata follow p. 142.
○ Bound together with copies of Simpson 1743 and Simpson 1757.

1743. *Mathematical Dissertations on a Variety of Physical and Analytical Subjects.* London: T. Woodward. pp. [viii] + 168.
◇ Three interesting essays in this book are "A Mathematical Dissertation on the Figure of the Earth," "A General Investigation of the Attraction at the Surfaces of Bodies nearly spherical," and "To determine the Length of a Degree of the Meridian, and the Meridional Parts answering to any given Latitude, according to the true spherodical Figure of the Earth."
○ A second copy is bound together with Simpson 1740.

1757. *Miscellaneous Tracts on Some Curious, and Very Interesting Subjects in Mechanics, Physical–Astronomy, and Speculative Mathematics; wherein, the Precession of the Equinox, the Nutation of the the Earth's Axis, and the Motion of the Moon in Her Orbit, are Determined.* London: Printed for J. Nourse. pp. [vii] + 179 + 3 plates.
◇ Dedicated to the Earl of MacClesfield, President of the Royal Society.
◇ Errata on p. 179.
◇ Following p. 179 is a list of nine works by Simpson published by J. Nourse. In this collection, there are papers on the theory of errors, fluents and fluxions, isoperimetric problems, the solution of equations using surds, in addition to works on astronomy and mechanics. There are a total of 18 papers listed.
○ A second copy is bound together with Simpson 1740.

Simpson, Thomas (Continued).

1799. *Trigonometry, Plane and Spherical; with the Construction and Application of Logarithms.* Fifth edition. London: F. Wingrave. pp. 248 + 4 pages of advertisements.

1810. *Trigonometry, Plane and Spherical; with the Construction and Application of Logarithms with an Appendix on Spherical Projections.* Philadelphia: Kimber and Conrad. pp. [iii] + 125 + 7 plates.

◇ This book was used as a text at USMA beginning in 1816.

◦ The library has two copies.

◦ One copy in the library lacks at least two plates. Some notes in the margins and on the title page. The last, and only dated, signature on the title page is "Henry D. Burke 1820".

◦ The other copy contains a two page list of other math books published by Kimber and Conrad. Written on the title page "John M. Gough, Cadet at the U.S. Mil. Academy at March 1, 1815" and "G. W. Whistler," the famous painter's father and graduate of USMA in 1819. Artist James Whistler attended USMA in 1851, but did not graduate. Other signatures and annotations throughout the book.

1821. *Elements of Geometry; with Their Application to the Mensuration of Superfices and Solids, to the Determination of the Maxima and Minima of Geometrical Quantities, and to the Construction of a Great Variety of Geometrical Problems.* New edition carefully corrected–with additional notes and an appendix, containing a description of the analytical and synthetical modes of reasoning made use of by mathematicians–an account of the lost writings of Euclid and Apollonius, and of the several attempts of the moderns to restore them. London: Printed for J. Collingwood. pp. viii + 283.

◇ Dedicated to Charles Frederick, Surveyor–General of His Majesty's Ordinance, dated March 3, 1760. There is an "advertisement" to his new edition signed by M. Fryer and dated October, 1821 on pp. vii–viii. Between pages 278 and 279 are bound pages ix–[xii]. "A brief account of the lost writings of Euclid, Apollonius, and others on the geometrical analysis of the ancients, and of several attempts of the moderns to restore them" is on pp. 268–283.

1823. *The Doctrine and Application of Fluxions. Containing (Besides what is Common on the Subject) a Number of New Improvements in the Theory; and the Solution of a Variety of New and Very Interesting Problems, in the Different Branches of the Mathematics.* 2 vols. New edition carefully revised, and adapted, by copious appendixes, to the present advanced state of science. By a Graduate of the University of Cambridge. London: Printed for J. Collingwood and G. and W. B. Whittaker. pp. xii + 344; 371.

◇ Dedicated to the Right Honourable George Earl of MacClesfield. Appendix to volume 1 (pp. 275–344) written after Simpson's death. Simpson acknowledges his indebtedness to *An Explanation of Fluxions in a Short Essay on the Theory* "printed for W. Innys: wrote by a worthy friend of mine (who was too modest to put his name to that, his first attempt) whose manner of determining the fluxion of a rectangle, I have, in particular, followed." This proof of the product rule, undoubtedly a reaction to criticisms of Berkeley, simply omits the second order differential.

Simpson, Thomas. See also Maseres 1795; Euclid 1810; Leybourn 1817.

Simson, Robert (1687–1768).
1792. *Elements of the Conic Sections.* Second edition, translated from the Latin original. Edinburgh: Printed for James Dickson and William Creech. pp. vii + 284 + 14 plates + [2].
⋄ Book was used as textbook at USMA until 1818.
○ Clean copy.

1804. *Elements of the Conic Sections. By the Late Dr. Robert Simson, Professor of Mathematics in the University of Glasgow. Translated from the Latin Original, for the Use of Students of Mathematics.* New York: Printed for W. Falconer. pp. [iii] + 278 + 15 plates.
○ Copy of William S. Mailland. Copy is misbound. Pages 275–278 and Plate 14 are bound between pages 174 and 175.

Simson, Robert. See also Apollonius 1749; Euclid 1756, Euclid 1810; Euclid 1811; Euclid 1814b; Euclid 1845; Chasles 1860; Keith 1814.

Sleeman, Thomas (fl. 1780–1806).
1805. *A Practical Treatise on the Use of Portable Mathematical Instruments, in Various Parts of the Mathematics to which is Added, a Complete System of Land–Surveying.* London: Printed by W. Spilsbury. pp. viii + 273 + 5 plates.
⋄ Frontispiece is an illustration of a Gunter's Quadrant and a Sector. Dedicated to John Rook Grosett.

de Sluse, René François (1622–1685). See Collins 1725; Saint Vincent 1647; Guericke 1672.

Smith, Charles (1844–1916).
1893a. *A Treatise on Algebra.* Fourth edition. London: Macmillan and Co. pp. xviii + 646.
⋄ First edition was published in 1888.
⋄ This book was used as textbook at USMA from 1900 to 1906. See Anonymous 1900.
○ Clean copy.

1893b. *Solutions of the Examples in a Treatise on Algebra.* Fourth edition. London and New York: Macmillan & Co. pp. 315.
⋄ First edition was published in 1888.
⋄ This book was used as textbook supplement at USMA from 1900 to 1906.
○ Copy signed "E.W. Bass".

1894. *An Elementary Treatise on Conic Sections.* London: Macmillan and Co. pp. xi + 352.
⋄ From the preface, this is the second edition, 1883. The first edition was published in 1882.
○ "Textbook U.S.M.A., 1899 to 18__" stamped in the gutter of the title page.

Smith, Charles (Continued)

1895a. *Solutions of the Examples in A Treatise on Algebra.* Fourth edition. London: Macmillan. pp. [vi] + 314 + 64 of advertisements.

○ Signed "Peirce, W.W." and "Fenton." A student has underlined "The average student ought to find this book easy and pleasant reading."

1895b. *An Elementary Treatise on Solid Geometry.* Fifth edition. New York: Macmillan. pp. xv + 242.

◇ This book was used as a text at USMA from 1900 until 1906.

○ The library has two copies.

○ One copy has a book plate: USMA received stamp, USMA library 16 Oct 1895. Stamped: "Textbook USMA 1899."

○ The other copy is annotated with red and blue to show what sections to keep or omit. Margin notes and inserts throughout the book.

1898a. *Solutions of the Examples in an Elementary Treatise on Conic Sections.* London: Macmillan & Co. pp. 268.

◇ This book was used at USMA from 1899 to 1914.

◇ Book of solutions to problems from Smith's conic sections book (See Smith 1904).

1898b. *A Treatise on Algebra.* London: Macmillan & Co. pp. xviii + 646.

○ Copy used by Fulton Q. C. Gardner. He first signed it Nov 3, 1900 & last on "Christmas Day, 1901. Almost through C. Smith for Good." On p. 22, there is a posted in sheet related to p. 16 and which contains the words "commutative law", "associative law" and "distributive law".

1898c. *Arithmetic for Schools.* New York: MacMillan & Co. pp. x + 329.

1899a. *An Elementary Treatise on Solid Geometry.* Seventh edition. London: Macmillan. pp. xv + 242.

○ Signed "Fulton Q. C. Gardner". A few printed slips posted in this copy.

1899b. *Examples from Conic Sections and Solid Geometry.* West Point, NY: Press of USMA. pp. 187.

◇ This text was written and printed at West Point.

○ The library has two copies.

○ One copy was rebound in 1929.

1899c. *An Elementary Treatise on Conic Sections.* London: Macmillan. pp. xi + 352.

○ Signed "Fulton Q. C. Gardner, USMA, 1904". There are a number of handwritten and "mimeographed" sheets pasted in. The notes in the text are those of a student.

1904. *An Elementary Treatise on Conic Sections.* London: Macmillan & Co. pp. [xi] + [352].

◇ This book was used as a text at USMA from 1899 to 1914.

○ Margin notes and inserts throughout the book. Color coded with blue and red marks to show what to keep or omit.

Smith, Charles (Continued)

1905. *A Treatise on Algebra.* London: Macmillan & Co., Limited. pp. xviii + 646.
⋄ This book was used as a text at USMA from 1900 to 1906.
○ A few annotations that are highlighted in different colors. Several margin notes.

1906. *An Elementary Treatise on Conic Sections.* London: Macmillan. pp. xi + 352.
⋄ First edition 1882.
○ Signed "Wm Cooper Foote." Contains handwritten and "mimeograhphed" notes pasted in. Lessons covered in 1909–10 written on front endpapers.

1910. *An Elementary Treatise on Conic Sections by the Methods of Co–Ordinate Geometry.* New Edition revised and enlarged. London: Macmillan. pp. x + 449.

1914. *An Elementary Treatise on Conic Sections by the Methods of Co–Ordinate Geometry.* New edition revised and enlarged. London: Macmillan. pp. x + 449.
○ Signed "M. C. Grenata, 1916". Printed slips pasted in.

1916. *An Elementary Treatise of Conic Sections by the Methods of Co–ordinate Geometry.* New edition revised and enlarged. London: Macmillan & Co. pp. x + 449.
⋄ This book was used as a text at USMA from 1912 to 1919.
○ Annotated for sections (lower and middle) as to what to omit from book.

Smith, Charles (1844–1916). See also Anonymous 1900; Steese 1905.

Smith, Charles (1844-1916) and Harington, Charles L.
1898. *Arithmetic for Schools.* Rewritten and revised. New York: MacMillan Co. pp. x + 329 + 9 pages of advertising of math books.

Smith, David Eugene (1860–1944). See Fink 1903 and Klein 1897.

Smith, Edmund Dickerson (1854–1900; USMA 1879).
1888. *Exercises in Geometry.* West Point, NY: US Military Academy Press. pp. 49.
⋄ These exercises are arranged to correspond with the propositions in "Davies Legendre."
○ The library has three copies of this book.

1896. *Exercises in Geometry.* West Point: United States Military Academy Printing Office. pp. 53.
○ There are two copies of this book in the library. One copy is signed on endpaper "Chas. P. Echols Assoc. Professor Maths." Copy specially bound with a sheet of graph paper between each pair of pages, but no annotations have been added.

Smith, Percey Franklyn (1867–1956) and Gale, Arthur Sullivan.
[**1904**]. *The Elements of Analytic Geometry.* Boston: Ginn and Company. pp. xii + 424 + 2 plates illustrating plaster models of surfaces.
○ The library has two copies.

○ One copy is clean.

○ The second copy has a book plate, "gift of Arthur R. Hercz." There are notes in the margins and some corrections are pasted in. There is a course outline and syllabus for a USMA course on the back cover and examination problems inserted in the back.

Smith, Richard Somers (1813–1877; USMA 1834).
1864. *A Manual of Linear Perspective: Perspective of Form, Shade and Shadow and Reflection.* New York: John Wiley. pp. [i–ii] + i–viii + 1–80 + 18 plates throughout text.

○ This copy contains an inscription honoring Professor Charles Davies. Approximately 30 blank pages in rear.

Smith, Robert (1689–1768).
1738. *A Compleat System of Opticks in Four Books, viz. A Popular, a Mathematical, a Mechanical, and a Philosophical Treatise. To which are Added Remarks upon the Whole.* 2 vols. Cambridge: Printed for the author. pp. xiv + 1 plate + 280 + 43 plates; 1 plate + 282–455 + 171 + [13] + 82 plates.

◇ Dedicated to Edward Walpole. Subscribers listed on pp. vii–x.

◇ In Volume 2, there are 171 pages of notes on the preceding work, and there is a 9 page index.

◇ Two pages of errata follow the index.

1749. *Harmonics or the Philosophy of Musical Sound.* Cambridge: J. Bentham, Printer to the University. pp. xv + 292 + 25 plates + an unpaginated index (11 pp).

◇ Dedicated to William, Duke of Cumberland. Errata follow the index.

Smith, Robert. See also Cotes 1722a.

Smyth, William (1797–1868?).
1859 (1853). *Elements of the Differential and Integral Calculus.* Second edition. Portland, Maine: Sanborn & Carter. pp. 240.

Snel, Willebrord (1580–1626). See Lansberge 1616.

Souciet, Etienne (1671–1744).
1729; 1732; 1732. *Observations mathématiques, astronomiques, geographiques, chronologiques, et physiques, tirées des anciens livres chinois; ou faites nouvellement aux Indes et a la Chine, par les Peres de la Compagnie de Jesus.* 3 vols. Paris: Chez Rollin. pp. [viii] + xxv + [vi] + 178; xxix + 294; 373.

◇ Subtitles are : Tome 2 *Contenant une histoire de l'astronomie Chinoise, avec des dissèrtations par le P. Gaubil* and Tome 3 *Contenant un traité de l'astronomie Chinoise par le P. Gaubil.*

◇ Dedicated to the King. Errata follow p. 294 of volume 2 and p. 373 of volume 3.

Spence, William.
1820. *Mathematical Essays by the Late William Spence, Esq. Edited by John F.
W. Herschel, Esq. with a Brief Memoir of the Author.* London: Printed by J.
Moyes for Oliver, Boyd and G. & W.B. Whittaker. pp. xxxii + 295.

Spidel, Euclid. See Maseres 1791.

Spiller, W. H. See Sturm 1835.

Spottiswoode, William (1825–1883).
1851. *Elementary Theorems Relating to Determinants.* London: Longman, Brown,
Green, and Longman. pp. viii + 63.
  o   The library has two copies.

Stainville, Nicolas Dominique Marie Janot de (1783–1828)
1815. *Mélanges d'analyse algébrique et de géométrie.*   Paris: Ve Courcier. pp. ii
+ vii + 680 + 3 plates.
  ◇   Dedicated to Delambre. Errata on p. vii and the following page. On pp. 339–
341 he gives the now standard proof of the irrationality of *e*, and attributes it to
Fourier.

Steese, James Gordon (1882–1958; USMA 1907).
1905. *Notes and Problems on Solid Geometry* [Text of C. Smith]. West Point:
United States Military Academy Printing Office, Class of 1908, by the New Era
Printing Co. pp. v + 3–67.
  o   The library has two copies with variant title pages.
  o   One copy has [iv] + 67, the other v + 3–67.

1906. *Problems and Solution in Analytical Mechanics.* [West Point, NY]: The
classes of 1909 and 1910, United States Military Academy. pp. xiv + 137 (includes
22 plates).
  ◇   There are 44 illustrations, photographs, and line drawings.
  ◇   There are two copies in the library.

Steiner, Jakob (1796–1863).
1832. *Systematische Entwickelung der Abhängigkeit geometrischer Gestalten von
einander, mit Berücksichtigung der Arbeiten alter und neuer Geometer über Poris-
men, Projections–Methoden, Geometrie der Lage, Transversalen, Dualität und Re-
ciprocität, etc.* Erster Theil. Berlin: G. Fincke. pp. xvi + 322 + 4 plates.
  ◇   The library has only Volume 1.

Sterry, Consider (1761–1817) and Sterry, John (1766–1823).
1790. *The American Youth: Being a New and Complete Course of Introductory Mathematics: Designed for the Use of Private Students.* Providence: Bennet
Wheeler. pp. xv + 387.
  ◇   Errata follow p. xv. There was never a volume 2.

Stevin, Simon (1548–1620).

1634. *Les oeuvres mathematiques de Simon Stevin de Bruges. Ou sont inserées les mémoires mathematiques.* Le tout reveu, corrigé, & augmenté par Albert Girard. Leyde [Leiden]: Bonaventure & Abraham Elsevier. pp. [vi]+ 224 + 678.

◇ Contains a translation of the first four books of Diophantus by Stevin and the next two by Albert Girard.

◇ Dedicated to les Estats Generann de Pais Bas Unis and to Prince d'Aurenge by Albert Girard.

Stewart, John (?–1766). See Newton 1745.

Stirling, James (1692–1770). [Jacob Stirling]

1753. *Methodus differentialis: sive tractatus de summatione et interpolatione serierum infinitarum.* Londini [London]: Impensis Ric. Manby. pp. [vi] + 153.

◇ One page of errata follows page 153. First edition was 1730.

Stokes, George Gabriel (1819–1903).

1880; 1883; 1901; 1904; 1905. *Mathematical and Physical Papers. Reprinted from the Original Journals and Transactions, with Additional Notes by the Author.* 5 vols. Cambridge: University Press. pp. x + 328; viii + 366; viii + 415; viii + 378; xxv + 370.

◇ The frontispiece of volume 4 is a portrait of Stokes taken from an oil paining made in 1874 by Lowes Dickinson. The frontispiece of volume 5 is a portrait of Stokes, 1892.

Stone, Edmund (1700?–1768).

1730. *The Method of Fluxions Both Direct and Inverse. The Former being a Translation from the Celebrated Marquis De L'Hospital's Analyse des Infinements Petits: and the Latter Supply'd by the Translator.* London: Printed for William Innys. pp. xix + iv + 238 + 212.

◇ "Part II" is also called an "Appendix of the Inverse Method of Fluxions" and comprises the last 212 pages. There is an errata page for the Appendix at the end of the Appendix. This work is a translation, accompanied by a change to Newton's fluxional notation, of the first calculus book, that of L'Hospital in 1696.

○ A few marginal notes.

Stone, Edmund (1700?–1768). See also L'Hospital 1723; 1781; Barrow 1735; Euclid 1763.

Strauch, Aegidius (1632-1682).

1700. *Aegidii Strauchii, Tabellen der Sinuum, Tangentium, Logarithmorum, und zu der Gantzen Mathesi denen zu mehrer Vollkommenheit beygefüget Leonh. Christ. Sturmii Prof. Publ. Matth. In Wolffenbüttel Sehr nützliches Handbüchlein vor alle Baumeister, bestehend in sonderbahren gantz neuen Tabellen zu der Architectur, Fortification, Geometrie, und dem Proportional-Circul. Gedruckt in Amsterdam: Zufinden bey Christ: Henr. Schumachernl.* This is followed by *Vade*

*mecum architectonicum bestehend in neu ausgerechneten Tabellen zu der Civil-
und Militar-Baukunst dergleichen so vollständig/ ordentlich/ beqvehm und leicht
zugebrauchen nochniemahl gesehen worden/ Samt einem Anhang. 1. Einem sehr
vollständigen Catalogo aller guten Bücher/ so von beyderley Baukunft handeln. 2.
Einer Tabelle zu denen Wasserwägen. 3. Einer Tabelle die Winckel ohne Instru-
ment zumessen und abzustecken. 4. Aller Tabellen, welche zu auftragung der Linien
des proportional-Circuls dienen nicht nur denen Lernenden/ sondern auch denen
Practicirenden selbst zu grossen Nützen gerechnet und abgefasset. von Leonh. Chr.
Sturm.* Amsterdam. pp. 46 + 561 pp of tables + 102. The last 102 pages are
Sturm.

◇ Volume donated to "The Military Engineering Research Collection" in memory
of MAJ Gen Julian L. Schley, USMA 1903.

Strong, Theodore (1790–1869).

1859. *A Treatise on Elementary and Higher Algebra.* New York: Pratt and Oakley.
pp. ix + 551.

Strong, Wendell Melville (1871–1942).

1897. *Key to Phillips & Fisher's Elements of Geometry Including the Abridged
Edition.* New York: American Book Company; Harper & Brothers. pp. v + 240.

Stuart, Sidney E. (1857–1899; USMA 1880)

1888. *Formulas, Compiled Especially for the Use of Instructors and Cadets at
the United States Military Academy.* West Point: United States Military Academy
Press. pp. 30.

◇ Formulas on trigonometry, analytic geometry, calculus and algebra.
○ The library has two copies.
○ One copy in a slipcase marked "Stuart's Formulas. West Point, 1888."

Sturm, Charles–François (1803–1855).

1835. *On the Solution of Numerical Equations.* Translated from the "Mémoires
présentés par divers Savans à l'Académie Royale des Sciences de l'Institute de
France" by W. H. Spiller. London: John Souter. pp. iv + 51.

Sturm, Johann Christophorus (1635–1703).

1709. *Mathesis juvenilis: Or a Course of Mathematicks for Young Students and
Such as have not Arriv'd to a Great Perfection in those Studies. Containing Plain
and Easie Treatises, by Way of Question and Answer in the Following Sciences,
viz. Arithmetick, Algebra, Geometry, Trigonometry, Navigation, Surveying, Fortification, Architecture, Staticks, Mechanicks, Opticks, Catoptricks, Dioptricks, Astronomy, and the Use of the Globes, Chronology, Dialling, &c.* Translator George
Vaux. London: Printed for Dan. Midwinter. pp. [xiv] + 402 + 39 plates.

◇ Volume 1 (only) of 3.
◇ Dedicated, by the translator, to John Harris. Errata on p. [xiv]. The last leaf
is misnumbered 385–386.

1724. *Mathesis enucleata: or, the Elements of Mathematicks to which is Annexed, an Introduction to Specious Analysis, or Algebra.* London: Midwinter. Second edition / corrected and much amended by E. Stone. pp. xvi + 300 + 30 plates.

⬦ First published in Nurnberg in 1689.

⬦ Errata on p. xvi.

⬦ *An Introduction to Specious Analysis; or, the New Geometry, Chiefly According to the Method of Des Cartes; But much Facilitated by Later Inventions, &c. By J. Chris. Sturmius* is bound together with this book, pp. 213–300.

○ Title page is missing from this book.

Sturm, Leonhard Christoph (1669–1719). See Strauch 1700.

Tannery, Jules (1848–1910).

1886. *Introduction a la théorie des fonctions d'une variable.* Paris: A. Hermann. pp. [xii] + 401.

1904. *Leçons d'arithmétique théorique et pratique.* Cours complet de mathématiques élémentaires publié sous la direction de M. Darboux. Troisième édition. Paris: Armand Colin. pp. [xii] + 509 + 3 tables at end.

1906. *Leçons d'algèbre et d'analyse a l'usage des élèves des classes de mathématiques spéciales.* 2 vols. Paris: Gauthier–Villars. pp. [vii] + [423]; 633.

Tannery, Paul (1843–1904). See Fermat 1891; Heiberg and Zeuthen 1912.

Tartaglia, Niccolò (c.1500–1557).

1592. *Tutte l'opere d'arithemetica del famosissimo Nicolo Tartaglia. Nelle quali in XVII. libri con varie prove, & ragioni, mostrasi ogni prattica naturale, & artificiale; i modi, & le regole da gli antichi, & moderni usate nell'arte mercantile; & oue interuiene calcolo, pesi, denari, tariffe, calmeri, baratti, cambi di banchieri, e di fiere, saldi, sconti, giuochi, traffico di compagnie, compre, vendite, portar mercantie da un paese all'altro, conuertir monete, congiungimento di metalli, & opere de zecchieri. Sopre le qual cose tutte, formansi bellissimi quesiti, & si sciolgono le difficolta, con vgual chiarezza, & diligenza, per utile rileuato de i mercanti, & teforieri, à capitani, e matematici, & astrologhi, &c.* In Venetia [Venice]: All'Insegna del Leone. pp. [viii] + 199 (+verso) + [vi] + 273 (&verso).

⬦ Only the recto pages are numbered.

⬦ Part one dedicated to Leonardo Neri, part two to Giouanni Arcieri. Portrait of Tartaglia (facing p. 1).

Taylor, Brook (1685–1731).

1811. *New Principles of Linear Perspective: Or the Art of Designing on a Plane, the Representations of all Sorts of Objects, in a More General and Simple Method, than has been Hitherto Done.* Fourth edition, revised. London: Printed for J. Taylor. pp. xxvi + 70 + 13 plates.

⬦ This is a corrected version of the first edition (1719) to which two proofs by John Colson (1749) have been added.

⬦ Frontispeice is a portrait of Brook Taylor.

Taylor, Michael (1756–1789).
1780. *A Sexagesimal Table, Exhibiting, at Sight, the Result of Any Proportion, where the Terms do not Exceed Sixty Minutes. Also Tables of the Equation of Second Difference, and Tables for Turning the Lower Denominations of English Money, Weights, and Measures into Sexagesimals of the Higher, and Vice Versa. And the Sexigesimal Table Turned into Seconds as far as the 1000th Column, Being a very Useful Millesimal Table of Proportional Parts. With Precepts and Examples. Useful for Astronomers, Mathematicians, Navigators, and Persons in Trade.* London: Printed by William Richardson, Published by order of The Commissioners of Longitude. pp. xlv+ 1 fold-out table + 316.
◇ Errata follow p. xlv. This is Taylor's most important table. Taylor worked on the staff of the *Nautical Almanac* and was also a pupil of Hutton.
○ The signature of Andrew Ellicott is on one of the front papers.

1792. *Tables of Logarithms of All Numbers, from 1 to 101000; and of the Sines & Tangents to every Second of the Quadrant.* London: Christopher Buckton. pp. [xvi] + 64 + unnumbered pages of tables.
◇ Dedicated, by Nevil Maskelyne, to the Commissioners of Longitude. Errata on p. 64 and at the end of the tables. See also Maskelyne 1776.
○ This book was owned by Andrew Ellicott (but he is not in the "list of subscribers").

Taylor, Thomas (1758–1835).
1816. *Theoretic Arithmetic, in Three Books; Containing the Substance of all that has been Written on this Subject by Theo of Smyrna, Nicomachus, Iamblichus, and Boetius.—Together with some Remarkable Particulars Respecting Perfect, Amicable, and Other Numbers, which are not to be Found in the Writings of any Ancient or Modern Mathematicians. Likewise, a Specimen of the Manner in which the Pythagoreans Philosophized about Numbers; and a Development of Their Mystical and Theological Arithmetic.* London: Printed for the author by J. Valpy. pp. xliii + 252.
◇ Errata on p. 252. This is a most unusual book, being a cross between philosophy of arithmetic and numerology.

Taylor, Thomas. See also Proclus 1792.

Tempelhoff, George Friedrich von (1737–1807).
1770. *Anfangsgründe der Analysis des Unendlichen. Erster Theil, welcher die Differential–Rechnung enthält; zum Gebrauch der Königlichen Preussischen Artillerie entworfen.* Berlin und Stralsund: Gottlieb August Lange. pp. [xvi] + 622 + 2 pages of errata.
◇ Dedicated to Carl Friedrich von Ritscher, his Royal Majesty of Prussia.

Thénot, Jean Pierre (1803–1857).
1826. *Essai de perspective pratique pour rectifier ses compositions et dessiner d'après nature dédié a feu Jean–Thomas Thibault par son élève J. P. Thénot.* Paris: L'Auteur, Panckoucke, Goujon et Mille Formentin. pp. [iv] + 84 + 47 plates interspersed in the text.

⋄   The plates are quite interesting. Since this is a practical work, the plates are part of real drawings, many with people, buildings, and trees, rather than the abstract constructions which are typical of books on descriptive geometry.

⋄   Frontispiece is a plate of a large building and a tower shown in perspective.

Theodosius of Bithynia (Born second half of second century B.C.). See Euclid 1803.

Theon of Alexandria (fl. second half of fourth century A.D.). See Ptolemy 1538.

Théveneau, Charles M. Simon (1759–1821). See Clairaut 1801.

Thévenot, Melchisédec (1620–1692); Boivin, Jean (1665–1726); La Hire, Philippe de (1640–1718).
1693. *Veterum mathematicorum Athenaei, Bitonis, Apollodori, Heronis, Philonis, et aliorum opera, Graece et Latine pleraque nunc primum edita. Ex manuscriptis codicibus bibliotheca regia.* Parisiis [Paris]: Ex typographia regia. pp. xvi + 365.
⋄   Printed in Greek and Latin. There is a Latin index and a Greek index at the end of the book.

Thibault, Jean–Thomas.
1827. *Application de la perspective linéaire aux arts du dessin.* Ouvrage posthum. Paris: Mesdames Thibault, Jules Renouard, Bance aîné, Carilian–Goeury. pp. xv + 168 + frontispiece + 53 plates.
⋄   Prepared by Chapuis, who was Thibault's student.
⋄   Contains a 12 page historical introduction. The frontispiece is a portrait of Thibault.

Thibault, Jean–Thomas. See also Thénot 1826.

Thomson, James B. (1808–1883)
1848, [1846]. *Practical Arithmetic, Uniting the Inductive with the Synthetic Mode of Instruction. Also, Illustrating the Principles of Cancelation. For Schools and Academies.* Day and Thomson's Series. Twenty-fifth edition, revised and enlarged. New York: Mark H. Newman and Co. pp. xii + 366 + 6 pages of "recommendations".
○   Stamped inside cover "W.F. Steele"; signed on title page by Fred Steele.

Thomson, William (1824–1907).
1882. *Mathematical and Physical Papers Collected from Different Scientific Periodicals from May, 1841, to the Present Time.* 6 vols. Cambridge: University Press. pp. xiii + 558 + 2 plates; xi + 407 + 20 plates; ix + 3 errata + 529; xv + 1 errata

+ 563; xvi + 602; viii + 378.
- o The library has two sets of volumes 1–3.

Tillett, Francis.
1824. *A New Key to the Exact Sciences: or, a New and Practical Theory by which Mathematical Problems or Algebraic Equations of Almost every Description can be Solved with Accuracy, and with Greater Facility and Simplicity than They can be by any Method that has yet been Given by any Other Author: In which are also Introduced, a Variety of Useful and Interesting Problems that have never Before been Proposed, and which it is Believed Cannot be Solved by any Methods or Rules Except Those here Laid Down.* Winchester, [Virginia]: Printed for the author, by S. H. Davis. pp. vi + [7]–64.

Tisserand, François (1845–1896).
1877. *Recueil complémentaire d'exercises sur le calcul infinitésimal.*   Paris: Gauthier–Villars. pp. xiv + 388.

Todhunter, Isaac (1820–1884).
1861. *A History of the Progress of the Calculus of Variations during the Nineteenth Century.* Cambridge: Macmillan and Co. pp. xii + 532 + 1 plate + book advertisements.
- o The library has two copies. One copy is signed "From the Author."

1873. *A History of the Mathematical Theories of Attraction and the Figure of the Earth, from the Time of Newton to that of Laplace in Two Volumes.* Volume 2. London: Macmillan. pp. 508.
- ◇ The library has volume 2 only.
- o Up to page three has been replaced with photocopies.

1882. *An Elementary Treatise on the Theory of Equations, with a Collection of Examples.* Fifth edition. London: Macmillan and Co. pp. vii + 328.

1901. *Spherical Trigonometry for the Use of Colleges and Schools.* Revised by J.G. (John Gaston) Leathem (1871–?). London: Macmillan and Co. pp. xii + 275

Todhunter, Isaac. See also Boole 1865; Euclid 1869.

Toplis, John. See Laplace 1814b.

Toussaint, Claude Jacques (1781–?).
1811; 1812. *Traité de géométrie et l'architecture théorique et pratique, simplifié, ouvrage classique présenté à son Exc. Mgneur. le ministre de l'interieur.*   2 vols. Paris: Published by the author. pp. 24 + 134 + 7 + [iii] + 105 + 4 + 6; unpaginated.
- ◇ Volume 1 contains "theory" and "practice" and volume 2 contains plates.
- o Thayer Binding. The title page is missing from volume 2.

Townsend, Richard (1821–1884).
1863; 1865. *Chapters on the Modern Geometry of the Point, Line, and Circle;*
Being the substance of Lectures Delivered in the University of Dublin to the Can-
didates for Honors of the First Year in Arts. 2 vols. Dublin: Hodges, Smith and
Co. pp. xx + 300; xx + 400.
⋄ The chapters are numbered consecutively through both volumes.

Trail, William (1746–1831).
1789. *Elements of Algebra For the Use of Students in Universities.* Third edition.
Edinburgh: Creech and Elliot. pp. viii + 261 + 50.
⋄ There are six appendices at the end of this book; they touch on a wide variety
of topics from logarithms to continued fractions.

Tralles, Johann Georg (1763–1822).
1786. *Physikalisches Taschenbuch für Freunde der Naturlehre und Künstler.* Göttingen: Johann Christian Dieterich. pp. [lxvix] + 60–270 + 1 plate.
⋄ Frontispiece portrait of Euler.

Trembley, Jean (1749–1811).
1783. *Essai de trigonométrie sphérique, contenant diverses applications de cette*
science à l'astronomie. Neuchatel: Samuel Fauche Pere & Fils. pp. 270.
⋄ Two pages of errata follow p. 270.

Tresca, Henri Edouard (1814–1885).
1852. *Traité élémentaire de géométrie descriptive.* Rédigé conformément au
dernier programme d'admission à cette école d'après les ouvrages et les leçons de
Th. Olivier. Paris: L. Hachette et Cie. pp. viii + 223 + viii + 63.
⋄ The text and the plates are bound as one volume.

Tresse, A. (1808–?), and Thybaut, A. (1870–?).
1904. *De géométrie analytique a l'usage des candidats a l'École centrale des arts et*
manufactures, aux Écoles des mines, à l'École des ponts et chaussées, et des élèves
de première année de mathématiques spéciales. Paris: Armand Colin. pp. [III] +
[549].

Tripon, J. –B.
1848a. *Études de projections, d'ombres et de lavis a l'usage de toutes les écoles,*
des architectes et des mécaniciens. Ouvrage divisé en quatre parties: Première
partie–Études des projections orthogonales. Deuxième partie–Études des projections
obliques. Troisième partie–Études des ombres. Quatriéme partie–Cours élémentaire
de lavis appliqué à l'enseignement due dessin, des machines, de architecture, etc.
Paris: Carilian–Goeury et Vor Dalmont. pp. 140.

1848b. *Études de projections, d'ombres et de lavis a l'usage des toutes les écoles,*
des architectes et des mécaniciens. Ouvrage divisé en quatre parties: Première
partie–Études des projections orthogonales. Deuxième partie–Études des projections

obliques. Troisième partie–Études des ombres. Quatriéme partie–Cours élémentaire de lavis appliqué à l'enseignement due dessin, des machines, de architecture, etc. Atlas. Paris: Carilian–Goeury et Vor Dalmont. pp. 21 plates.

◇ Plate "21" is marked "Frontispiece" for "Cours élèmentaire de lavis appliqué a l'étude de projections orthoganales, obliques et a l'étude des ombres". It is unclear if this should be separate work.

Ulūgh Beg (1393–1449). See Flamsteed 1725.

Vallée, Louis Leger (1784–1864).
1819. *Traité de la géométrie descriptive.*   Paris: Ve Courcier. pp. xx + 355 + 8 + 40 plates.
◇ Frontispiece is an engraving of a portrait of Gaspard Monge. See Figure 24 for a photograph of this frontispiece.
◇ Dedicated to Monge. The plates were engraved by Ambroise Tardieu, and form essentially a second volume bound together with the text.

Van Amringe, J. Howard (1835–1915). See Davies (Geometry) [1882]a; Davies (Surveying) 1883.

Vastel, Louis Guillaume François (1746–1819). See Bernoulli, Jakob 1801.

Vaux, George. See Sturm 1709.

Vega, Baron von Georg (1756–1802).
1857. *Logarithmic Tables of Numbers and Trigonometrical Functions.* Stereotyped. Translated by W. L. F. Fischer. London: Williams and Norgate. pp. xxvii + 575 pages of tables.
○ This particular copy was owned by the U.S. Navy Department, Office of Ordinance and Hydrography at one time.

1859. *Logarithmic Tables of Numbers and Trigonometrical Functions.* Stereotyped. Translated by W. L. F. Fischer. Berlin: B. Weidmanns. pp. xxvii + 575 pages of tables.

1883. *Logarithmic Tables and Trigonometrical Functions.* Sixty-seventh edition. 2 vols. Berlin: Weidmanns. pp. xxviii + 575 pages of tables.

Venable, Charles Scott (1827–1900).
1877. *Elements of Geometry after Legendre, with a Selection of Geometrical Exercises, and Hints for the Solutions of the Same.* New York: University Publishing. pp. vi + 7–366.
○ Rebound in 1938. Signature on page before title page.

Venturoli, Giuseppe (1768–1846).
1823. *Elements of Practical Mechanics, to which is Added a Treatise upon the Principle of Virtual Velocities, and its Use in Mechanics.* Translated from the

Italian by David Cresswell. Cambridge: J. Smith. pp. [ii] + iii + iv + 238 + 3 plates.

  ○ Book is rebound.

Verbeeck, Theodore. See Marolois 1628.

Vieille, Jules Marie Louis (1814–1880).
1854. *Théorie général des approximations numériques, suivie d'une application a la résolution des équations numériques. A l'usage des candidats aux écoles spéciales du gouvernement.* Seconde édition, revue, corrigée et augmentée. Paris: Mallet–Bachelier. pp. xii + 200.

Viète, François [Franciscus Vieta] (1540–1603).
1646. *Francisci Vietae opera mathematica, in unum volumen congesta, ac recognita, operâ atque studio Francisci à Schooten Leydensis matheseos professoris.* Lugduni Batavorum [Leiden]: Ex Officinâ Bonaventurae & Abrahami: Elzeviriorum. pp. [xii] + 554 + 1 plate.

  ◇ There is a short biographical sketch of Viète on pp. [vii]–[ix]. Errata are on p. 554. The contents are listed on p. [xii].

Viète, François (1540–1603). See also Apollonius 1764; Apollonius 1795.

Vigenere, Blaise de (1523–1596).
1586. *Traicté [Traité] des chiffres, ou secretes manieres d'escrire.* Paris: Abel L'Angelier. pp. 343 (recto only).

  ○ Many pages have been carefully repaired; a few marginal notes.

Vignola, Giocomo Barocchio (1507–1573).
1611. *Le due regole della prospettiva pratica di M. Iacomo Barozzi da Vignola con i commentarij del R. P. M. Egnatio Danti dell'ordine de Predicatori Mathematico dello Studio di Bologna.* In Rome: Nella Stamparia Camerale. pp. viii + 145 + [v].

  ○ Glued in plates on many pages (but other figures are printed). Pages 95–96 discuss drawing anamorphoses.

Villeneuve, Bardet de. See Bardet de Villeneuve, P. P. A.

Vince, Samuel (1749–1821).
1781. *The Elements of the Conic Sections, as Preparatory to the Reading of Sir I. Newton's Principia.* Cambridge: Printed by J. Archdeacon. pp. [xii] + 55 + 3 plates.

  ◇ Errata are on p. [xii] and following p. 55. A list of subscribers is given on pp. [iii]–[xi].

1812. *The Principles of Fluxions: Designed for the Use of Students in the University.* The First American edition, corrected and enlarged. Philadelphia: Kimber and Conrad. pp. [4] + 256.

  o   Title page is signed "C. S. Patterson, Professor, Halle 1st July 1820."

1821. *A Treatise on Plane and Spherical Trigonometry; With an Introduction, Explaining the Nature and Use of Logarithms Adapted to the Use of Students in Philosophy.* Fourth edition, corrected. Cambridge: J. Smith. pp. 158 + 2 plates.

Vincent, Alexandre Joseph Hidulphe [Auguste Jules H.](1797–1868).
1826. *Cours de géométrie élémentaire, a l'usage des élèves qui se destinent a l'École polytechnique ou aux Écoles militares.* Reims: Delaunois; Paris: Bachelier. pp. vi + 355 + 4 plates.

  ◇   Errata are on p. vi.
  o   Title page has author's name written is as "A[uguste] J[ules] H." Paste in label on bottom of title page "A. J. Kilian. libraire."

Vitello [Witelo] (ca. 1230–ca. 1275). See Alhazen 1572.

Viviani, Vincenzo (1622–1703).
1659. *De maximis et minimis geometrica divinatio in quintum conicorum Apollonii Pergaei adhuc desideratum.* Florentiae [Florence]: Apud Ioseph Cocchini. pp. [xvi] + 154 + [iv] + 154 + 2 plates.

  ◇   Volume 1 is dedicated to Ferdinando II, and volume 2 is dedicated to Prince Leopold.
  o   Two volumes bound as one.

1701. *De locis solidis secunda divinatio geometrica in quinque libros iniuria temporum amissos Aristae senioris geometrae autore Vincentio Viviani magni Galilae i novissimo discipulo regiae celsitud. Cosmi III. M. D. Etruriae mathematico primario a Ludovico Magno inter octo exteros Regiae Academ. Scientiar. Socios Adscripto et Regalis Societatis Londini Sodali opus conicum continens elementa tractatuum ejusdem Viviani, quibus tunc ipse multa, maxima, & abdita in mathesi theoremata demonstrare cogitaverat. Elaboratum anno 1646. Impressum Florentiae ab Hippolyto Navesi anno 1673. Addendis auctum, & in lucem prolatum anno 1701. At, si extabit umquam ab autore completum, uti est in animo, solus Deus scit.* Florentiae [Florence]: Typis Regiae Celsitudinis Apud Petrum Antonium Brigonci. pp. [xxvi] + 2 plates + 164 + 128.

  ◇   See Figure 9 for a photograph of the frontispiece portrait of Galileo.
  ◇   Errata are on pp. 117–118.

Vlacq, Adriaan (1600–1666 or 1667).
1633. *Trigonometria artificialis: sive magnvs canon triangulorum logarithmicus, ad radium 100000,00000, & ad dena scrupula secunda, ab Adriano Vlacco Goudano constructus. Cui accedunt Henrici Briggii geometriae professoris in Academiâ Oxoniensi P. M. chiliades logarithmorum viginti pro numeris naturali serie crescentibus ab unitate ad 20000. Quorum ope triangula plana & sphaerica, inter alia nova eximiaque compendia è geometricis fundamentis petita, folâ additione, subtractione, & bipartitione, exquisitissimè dimetiuntur.* Govdae [Gouda]: Excudebat Petrus Rammasenius. pp. [vi] + 52 + unnumbered pages of tables.

⋄ Dedicated to Carolo Ludovico.

○ Bookplate (Avito Candore) with crest of Lindsay. Signature "Johannis Valentini Scheidraftig 1671" on flyleaf.

1681. *Tabulae sinuum, tangentium, et secantium, et logarithmi sinuum, tangentium, & numerorum ab unitate ad 10000. Cum methodo facillimâ, illarum ope, resolvendi omnia triangula rectilinea & sphaerica, & plurimas quaestiones astronomicas. Ab A. Vlacq. Editio ultima emendata & aucta.* Amstelaedami [Amsterdam]: Apud Henricum & Viduam Theodori Boom. pp. 48 + unnumbered pages of tables.

Voiron.
1810. *Histoire de l'astronomie, depuis 1781 jusqu'a 1811, pour servir de suite a l'histoire de l'astronomie de Bailly .* Paris: Courcier, pp [vi] + ix + 383.
⋄ Thayer binding. Dedicated to Laplace. Errata follow p. ix.

Walker, John. See Euclid 1827.

Walker, Richard.
1801. "On the Production of Artificial Cold by Means of Muriate of Lime."
⋄ This is a paper, not a book, bound with other papers in Anonymous n.d.

Walker, Timothy (1806–1856).
1829. *Elements of Geometry with Practical Applications, for the Use of Schools.* Second edition, improved. Boston: Richardson, Lord, and Holbrook. pp. xvi + 129 + 6 plates.
⋄ The introduction on pp. [vii]–xv is historical.

Wallace, James (?–1851).
1812. *A New Treatise on the Use of the Globes and Practical Astronomy or a Comprehensive View of the System of the World in Four Parts.* New York: Smith and Forman. pp. viii + 512.
⋄ Errata on p. 512.

Wallace, James. See also Brown 1826.

Wallis, John (1616–1703).
1685; 1684; 1684; 1684; 1685. *A Treatise of Algebra both Historical and Practical. Shewing, the Original, Progress, and Advancement thereof, from Time to Time; and by what Steps it hath Attained to the Heighth at which now it is. With Some Additional Treatises, I. Of the* Cono-Cuneus; *being a Body Representing in Part a* Conus, *in Part a* Cuneus. *II. Of* Angular Sections; *and Other Things Relating thereunto, and to* Trigonometry. *III. Of the* Angle of Contact; *with Other Things Appertaining to the* Composition of Magnitudes, *the* Inceptives of Magnitudes, *and the* Composition of Motions, *with the Results thereof. IV. Of* Combinations, *Alternations, and* Aliquot Parts. London: Printed by John Playford. pp. [xx] + 374 +

1 plate; [iv] + 17 + 7 plates; [i] + 68 + 2 plates; [i] + 71–105; [i] + 109–176.
◇ Separate title pages and dates are included for the four additional treatises.
◇ At the conclusion of his *Combinations, Alternations, and Aliquot Parts*, Wallis has *Additions and Emendations*, pp. 153–176. Errata for the whole book, up to the *Additions and Emendations*, are on pp. 175–176 (mislabeled 75 and 76).
o Notes in the margins.
o There is no frontispiece portrait of Wallis as in other copies.
o Bound with Caswell 1685.

1695. *Opera mathematica. Volumen primum.* Oxoniae [Oxford]: E Theatro Sheldoniano. pp. [xv] + 1064.
◇ Frontispiece portrait of Wallis by David Loggan (1635–1700?). See Figure 7 for a photograph of the frontispiece from this volume.
◇ Errata on p. [vii].

1693. *De algebra tractatus; Historicus & practicus. Anno 1685 Anglice editus; Nunc actus latine. Cum variis appendicibus; partim prius editis Anglice, partim nunc primum editis. Operum mathematicorum volumen alterum.* Oxoniae [Oxford]: E Theatro Sheldoniano. pp. [xvi] + 879 + 1 plate.
◇ Contains a work on trigonometry by John Caswell. Errata follow p. 879.

1699. *Operum mathematicorum volumen tertium. Quo continentur Claudii Ptolemaei, Porphyrii, Manuelis Bryennii, Harmonica: Archimedis, Arcenarius, & dimensio circuli; cum Eutocii commentario: Aristarchi Samii, De magnitudinibus & distantiis solis & lunae, liber: Pappi Alexandrini, Libri secundi collectaneorum, hactenus desiderati, fragmentum: Graece & Latine edita, cum notis. Accedunt epistolae nonnullae, rem mathematicam spectantes; et opuscula quaedam miscellanea.* Oxoniae [Oxford]: E Theatro Sheldoniano. pp. [vii] + [xvi] + 708 + [xiv] + 445.
◇ Frontispiece portrait of Wallis. See Figure 6 for a photograph of the frontispiece from this volume. Errata on p. [vii].
o There are two copies of volume three (this volume) in the library.

Wallis, John (1616–1703). See also Collins 1725; Maseres 1795; Sherwin 1761.

Ward, John, of Chester .
1730. *Posthumous Works.* Published by a particular Friend of the Author's from the Original Manuscript, and Revised by Mr. George Gordon. London: Printed for A. Bettesworth. pp. viii + 493 + 1 plate.

1740. *The Young Mathematician's Guide: Being a Plain and Easy Introduction to the Mathematics in Five Parts.* Seventh edition, carefully corrected. London: printed for S. Birt, et. al. pp. [viii] + 480.
◇ The "Five Parts" of the title are: "Arthmetick, Algebra, the Elements of Geometry, Conick Sections, and the Arithmetick of Infinites"; in addition, there is an Appendix on "Practical Gauging," and a Supplement containing a history of logarithms and a brief introduction to trigonometry.

Waring, Edward (1736–1798).
1762. *Miscellanea analytica, de aequationibus algebraicis, et curvarum proprietatibus.* Cantabrigiae [Cambridge]: Typis academicis excudebat J. Bentham. pp. [iv]

+ iv + [viii] + 162 + 3 plates.
◇ There is an extra leaf 27 which folds out. Dedicated to Thomas Holles.
◇ Corrigenda follow the plates.

1770. *Meditationes algebraicae.* Cantabrigiae [Cambridge]: Typis academicis excudebat J. Archdeacon. pp. [iii] + viii + 219.
◇ Dedicated to August–Henry Fitzroy. Corrigenda follow p. 219.

Warren, Samuel Edward (1831–1909).

1860. *General Problems from the Orthographic Projections of Descriptive Geometry; with Their Applications to Oblique — including Isometrical — Projections, Graphical Constructions in Spherical Trigonometry, Topographical Projection ("One Plane Descriptive"), and Graphic Transformations.* New York: John Wiley. pp. 412 with 36 plates.

Waters, William George (1844-1928).

1898. *Jerome Cardan, a Biographical Study.* London: Lawrence & Bullen. pp. vi + 301.
◇ Frontispiece portrait of Cardan.

Watts, Thomas.

1716. *A Treatise of Mechanicks: or, the Science of the Effects of Powers or Moving Forces, as Apply'd to Machines, Demonstrated from its First Principles. Done out of French [of Rohault]. In this Translation are Inserted Particular References to the Several Propositions of Euclid, on which the Demonstrations are Built: with Considerable Additions, whereby the Whole is More Compleat and Universal, and Particularly the Rising and Falling of the Quicksilver in the Weather-glass Explain'd and Accounted for. Useful for all Artificers, as well as Natural Philosophers and Mathematicians.* From the French of Rohault. London: for Edward Symon. pp. x + 160 + 4 plates + 4.

Waud, Samuel Wilkes.

1835. *A Treatise on Algebraical Geometry.* Library of Useful Knowledge. London: Baldwin and Cradock. pp. xxiv + 260.

Weber, Heinrich (1842–1913).

1898a; 1899. *Lehrbuch der Algebra.* Zweite Auflage. 2 vols. Braunschweig: Friedrich Vieweg und Sohn. pp. [xv] + 703; [xvi] + 855.

1898b. *Traité d'algèbre supérieure. Principes. Racines des équations. Grandeurs algébriques. Théorie de Galois.* Traduit de l'Allemand sur la deuxième édition par J. Griess. Paris: Gauthier–Villars. pp. xi + 764.
◇ This is a translation of Volume I of the first edition of Weber 1898a only.

Weidler, Johann Friedrich (1691–1755).

1741. *Io. Friderici Weidleri historia astronomiae siva de ortu et progressu astronomiae. Liber singularis.* Vitembergae [Wittemberg]: sumtibvs Gottlieb Heinrici

Schwartzii. pp. [xxiv] + 624 + [40] + 12 plates.

o   Autograph of Georgius Mauritius Lowitz 1751. Numerous blank pages are
bound and spaced evenly into the book, and some pages contain written notes.

Weisbach, Julius (1806–1871).
1899. *Mechanics of Engineering. Theoretical Mechanics with an Introduction to
the Calculus. Designed as a Text–Book for Technical Schools and Colleges, and for
the Use of Engineers, Architects, etc.* Translated from the Fourth Augmented and
Improved German Edition by Eckley B. Coxe. Ninth American edition. New York:
D. Van Nostrand Co. pp. xxix + 1112 + 73.

Weld, Laenas Gifford (1862–?).
1896. *A Short Course in the Theory of Determinants.* Second edition. New York:
MacMillan. pp. xiii + 238.

Wells, Webster (1851–1916).
1904. *Advanced Course in Algebra.* Boston: D. C. Heath and Co. pp. viii + 581
+ 41.
o   The library has two copies, different printings but same preface date.
o   Inside the cover of one copy is the lesson plan for upper sections Jan–Jun 1936.
Special collections book plate: Gift BG Fraser and is signed by Harvey R. Fraser
Company I. There are notes and corrections throughout the text.
o   The second copy is signed "Robert F. Tate, Company I–1, West Point, NY,
Sept 2, 1924" inside the front cover. There are notations throughout the text.

n.d. *Exercises in Advanced Course in Algebra.* no publisher listed. pp. 91.
◇   USMA Crest on page 2.

Wendell, Abraham (1796–1817; USMA 1815)
1814. [Algebra Notebook, 17 June 1814]. West Point. pp. 39.
◇   This copy book deals with algebra (pp. 3–23), equations (pp. 25–33), geometry (pp. 36–39). Topics include "Sir Isaac Newton's Rule for raising a Binomial
to any power whatsoever" (pp. 13–14), "Infinite Series" (pp. 18-20), "Arithmetical
Proportion" (p. 20), "Application of Arithmetical Progression to Military Affairs"
(pp. 21-22), "Of Computing Shot and Shells in a Finished Pile" (pp. 22–23), "Quadratic Equations" (pp. 30–31), and "Resolution of Cubic and Higher Equations"
(pp. 31–32).
o   Rebound in 1983. "Sarah Wendell Account book" is at the end of this volume.

Wentworth, George A. (1835–1906).
1887. *Elements of Algebra.* Boston: Ginn and Company. pp. xv + 510 + answers
insert 64 pages.
o   Book plate: Gift of Mr. William Post, autographed by Benjamin Arnolu, 1888.

**Wentworth, George A. (1835–1906) and Smith, David Eugene (1860–
1944).**
1913 [1911, 1899, 1888]. *Plane Geometry.* Wentworth–Smith Mathematical

Series. Boston: Ginn & Company. pp. iv + 287.

○ Stamped H. F. Foster.

West, John (1756–1817).

1820. *The Elements of the Conic Sections: for the Use of Students in the Universities.* Revised and improved. New York: Printed for F. Nichols. pp. [iv] + 64 + 10 plates.

West, William (?–1760).

1763. *Mathematics.* Second edition, with additions. London: J. Kippax. pp. [iv] + 84 + 11 plates.

◇ The book consists of two chapters, the first on fluxious and the second containing miscellaneous problems with their solutions. Edited by John Rowe.

○ This copy belonged to Andrew Ellicott. His signature is on a front paper (1779).

Whewell, William (1794–1866).

1819. *An Elementary Treatise on Mechanics. Vol. I. Containing Statics and Part of Dynamics.* Cambridge: J. Smith. pp. xxii + 348 + 15 plates.

1823. *A Treatise on Dynamics. Containing a Considerable Collection of Mechanical Problems.* Cambridge: J. Smith. pp. xvi + 1 page errata + 403 + vi plates.

1833. *Analytical Statics. A Supplement to the fourth edition of an Elementary Treatise on Mechancis.* Cambridge: Pitt Press. pp. viii + 3 plates + 152.

1838. *The Mechanical Euclid, Containing the Elements of Mechanics and Hydrostatics Demonstrated after the Manner of the Elements of Geometry; and Including the Propositions Fixed upon by the University of Cambridge as Requisite for the Degree of B.A. To which are Added Remarks on Mathematical Reasoning and on the Logic of Induction.* Third edition. Cambridge: J. & J. J. Deighton. pp. ix + 187 + 12 page list of books.

◇ In 1838, Professor Church requested that this volume be purchased.

1847. *The Philosophy of the Inductive Sciences, Founded upon Their History.* 2 vols. A new edition, with corrections and additions, and an appendix, containing philosophical essays previously published. London: John W. Parker. pp. xxxiv + 708; xiv + 679 + 1 plate.

◇ A large portion of Volume I deals with the philosophy of mathematics. Volume II contains considerable biographic information.

Whewell, William. See also Anonymous 1828.

Whiston, William (1667–1752).

1716. *Sir Isaac Newton's Mathematick Philosophy More Easily Demonstrated: with Dr. Haley's Account of Comets Illustrated. Being Forty Lectures Read in the Publick Schools at Cambridge.* In this English Edition the Whole is Corrected and Improved by the Author. London: Printed for J. Senex. pp. [iv] + 443 + 8 plates.

⋄   Errata start on p. 443. No table of contents.
○   The library has two copies.

Whitehead, Alfred North (1861–1947).
1898. *A Treatise on Universal Algebra with Applications.* Volume 1. Cambridge: University Press. pp. xxvi + 586.

Whitehead, Alfred North (1861–1947) and Russell, Bertrand (1872–1970).
1910; 1912; 1913. *Principia Mathematica.* 3 vols. Cambridge: Cambridge University Press. pp. xiii + 666; xxxiv + 772; viii + 491.
⋄   This is a first edition. Errata for volume 1 follow p. xiii. Errata to volume 2 and 3 follow p. viii of volume 3.

Whitworth, William Allen (1840–1905).
1886. *Choice and Chance, Elementary Treatise on Permutations, Combinations, and Probability, with 640 Exercises.* Fourth edition, enlarged. Cambridge: Deighton, Bell and Co. pp. viii + 299 + 16 plates.

Wiener, Christian (1826–1896).
1884, 1887. *Lehrbuch der darstellenden Geometrie.* 2 vols. Volume one has the subtitle *Geschichte der darstellenden Geometrie, ebenflächige Gebilde, krumme Linien (Erster Teil), projektive Geometrie,* the second *Krumme Linien (Zweiter Teil) und krumme Flächen, Beleuchtungslehre, Perspektive.* Leipzig: B. G. Teubner. pp. [xx] + [477]; [xxx] + [649].

Wildbrett, Adolf.
1915. *Analytische Geometrie und Elemente der Differentialrechnung. Lehrbuch mit Aufgabensammlung für die Oberklasse von Gymnasien und Realgymnasien.* Nürnberg: Friedr. Kornschen. pp. [viii] + [124].
○   Numorous translations of words throughout the book in pencil. Signed "Pfaff" on title page.

Wilkins, John (1614–1672).
1680. *Mathematical Magick: or, the Wonders that may be Performed by Mechanichal Geometry. In Two Books Concerning Mechanical Powers [and] Motions. Being one of the most Easie, Pleasant, Useful (and yet most neglected) Part of Mathematicks. Not before Treated of in this Language.* London: Printed for Edw. Gellibrand. pp. [xvi] + 295.
⋄   Dedicated to the Prince Elector Palatine.
○   The library has two copies.

Williams, John D. See Young 1833b.

Wilson, Edwin Bidwell (1879–1964).
1902. *Vector Analysis, a Text-Book for the Use of Students of Mathematics and*

Physics Founded upon the Lectures of J. Willard Gibbs. New York: Charles Scribner's Sons. xviii + 436.

◇ Based on the lectures of J. Willard Gibbs at Yale, 1899–1900.

Wilson, George Cox.

1782. *A New System of Geometry.* Little Chelsea. pp. 33 of drawings (9 are watercolor).

○ Manuscript in a protective case.

Witelo [Vitello] (ca. 1230–ca. 1275). See Alhazen 1572.

Wolff, Christian, Freiherr von (1679–1754).

1739. *A Treatise of Algebra; with the Application of it to a Variety of Problems in Arithmetic, to Geometry, Trigonometry, and Conic Sections. With the Several Methods of Solving and Constructing Equations of the Higher Kind. To which is Prefix'd, What he Refers to in his Three Preliminary Treatises.* Translated from the Latin by John Hanna. London: Printed for A. Bettesworth and C. Hitch. xii + 340 + 8 plates.

◇ Dedicated to William Jones by John Hanna. Errata on p. xii.

1763. *Der Anfangsgründe aller mathematischen Wissenschaften.* Each of the five volumes in this 'Neue, verbesserte und vermehrte Auflage' has a separate subtitle. After the main title they continue as follows:

(1) *Erster Theil, welcher einen Unterricht von der mathematischen Lehrart, die Rechnenkunst, Geometrie und Trigonometrie in sich enthält, und zu mehrerem Ausnehmen der Mathematik so wohl aus hohen als niederen Schulen ausgesetzt worden.*

(2) *Anderter Theil, welcher die Baukunst, Artillerie und Fortification in sich enthält, und zu mehrerem Ausnehmen der Mathematik so wohl aus hohen als niederen Schulen ausgesetzt worden.*

(3) *Dritter Theil, welcher die Mechanik, Hydrostatik, Aerometrie und Hydraulik in sich enthält, und zu mehrerem Ausnehmen der Mathematik so wohl aus hohen als niederen Schulen ausgesetzt worden.*

(4) *Vierter Theil, welcher die Optik, Katoptric, Dioptric, und die Perspectiv in sich enthält, und zu mehrerem Ausnehmen der Mathematik so wahl aus hohen als niederen Schulen ausgesetzt worden.*

(5) *Fünfter Theil, welcher die Sphärische Trigonometrie, Astronomie, Geographie, Chronologie und Gnomonik in sich enthält, und zu mehrerem Ausnehmen der Mathematik so wohl aus hohen als niederen Schulen ausgesetzt worden.*

Wien [Vienna]: Johann Thomas Trattnern. pp. [xviii] + 254 + 29 plates; 386 + 50 plates; 186 + 12 plates + 123 + 7 plates; 416 + 13 plates.

○ Volumes 3 and 4 are bound together.

Wood, James (1760–1839).
1824. *The Principles of Mechanics: Designed for the Use of the Students in the University.* Seventh edition. Cambridge: J. Smith. pp. iv + 211.

Woodhouse, Robert (1773–1827).
1801. "On the Necessary Truth of Certain Conclusions Obtained by Means of Imaginary Quantities."
⋄   This is a paper, not a book, bound with other papers in Anonymous n.d.

1803. *The Principles of Analytical Calculation.* Cambridge: Cambridge University Press. pp. xxxiv + 219.
⋄   The preface (pp. i–xxxiv) is an historical essay on much of eighteenth century fluxions.

1810. *A Treatise on Isoperimetrical Problems and the Calculus of Variations.* Cambridge: J. Smith. pp. x + 154.
∘   Copy contains several marginal notes.

1813. *A Treatise on Plane and Spherical Trigonometry.* Second edition, corrected, altered, and enlarged. Cambridge: Printed by J. Smith, Printer to the University. pp. v + 240.
⋄   The figures are set in the text. There are some historical notes in the preface.

Wright, John Martin Frederick [I. M. F.]
1825. *Solutions of the Cambridge Problems from 1800 to 1820.* 2 vols. London: Black, Young, and Young. pp. xix + 653 + 4 plates; xv + 747 + 5 + 6 plates.
⋄   Dedicated to the "Tutors of the several colleges'. Corrections for both volumes follow p. 747 in volume 2. There are indexes to the problems wherein the problems are grouped by type or subject: in volume 1, pp. ix–xix; and in volume 2, pp. iii–xv.

Wright, William James (1831–?).
1875. *Tracts Relating to the Modern Higher Mathematics. Tract No. 1. Determinants.* London: C. F. Hodgson. pp. viii + 72.
∘   Corrections are in pencil and notes are in pencil. Bound with Wright 1877 and 1879.

1877. *Tracts Relating to the Higher Modern Mathematics. Tract No. 2. Trilinear Coordinates.* London: C. F. Hodgson and Son. pp. vi + 77.
∘   Bound with Wright 1875 and 1879.

1879. *Tracts Relating to the Modern Higher Mathematics. Tract No. 3. Invariants.* London: C. F. Hodgson and Son. pp. viii + 75.
⋄   Bound with Wright 1875 and 1877.

Wrigley, Alfred.
1875. *A Collection of Examples and Problems in Pure and Mixed Mathematics, with Answers and Occasional Hints.* Ninth edition, corrected. London: Longmans, Green, and Co. pp. viii + 310 + 4.

1883. *A Collection of Examples and Problems in Pure and Mixed Mathematics, with Answers and Occasional Hints.* Tenth edition, completing twenty thousand copies. London: Longmans, Green, Reader and Dyer. pp. viii + 310 + 4.
○ Rebound in 1952. Belonged to A. J. Perkins. Signed by A. J. Perkins and Harry Buesav.

Wronski, Hoëné de (1776–1853).
1811. *Introduction a la philosophie des mathématiques, et technie de l'algorithmie.* Paris: Courcier. pp. [vii] + 269 + 1 fold out table + vi.
◇ Dedicated to the Emperor Alexander the First of Russia. Errata follow p. 269.
○ The fold out table, the "Architecture" of mathematics, is regretably incomplete.

1812. *Réfutation de la théorie des fonctions analytiques de Lagrange.*   Paris: Blankenstein. pp. 135.
◇ Dedicated to the Imperial Institute of France
○ Thayer binding.

Young, Charles A. (1834–1908).
1889. *A Text–Book of General Astronomy for Colleges and Scientific Schools.* Boston: Ginn and Company. pp. viii + 551 + advertisements.

Young, Gordon Russell. (1891–1963; USMA 1913).
1912. *Notes on Gordon's Mechanics.* [West Point, N.Y.]: The class of 1914, USMA. pp. [v] + 81.
◇ The Gordon in the title is William Brandon Gordon (USMA 1877).
○ Gift of Mrs. Gordon Russell Young, the author's wife (handwritten note on the verso of the title page). Dedicated to "goats," the cadets in the lower portion of their class in academic standing.

Young, John Radford (1799–1885).
1831a. *Elements of the Differential Calculus; Comprehending the General Theory of Curve Surfaces, and of Curves of Double Curvature. Intended for the Use of Mathematical Students in Schools and Universities.* London: John Souter. pp. xxiv + 2 pages of errata + 252.
○ Bound with Young 1831b.

1831b. *The Elements of the Integral Calculus; with its Applications to Geometry and to the Summation of Infinite Series. Intended for the Use of Mathematical Students in Schools and Universities.* London: John Souter. pp. xi + 1 page of errata + 306.
◇ This work is the third (*The Elements of Analytic Geometry* (not at USMA) and Young 1831a were the previous volumes) volume of a "pretty comprehensive view of the principles of modern analytical science." (p. [iii]) A fourth volume encompassing finite differences, partial differential equations, and definite integrals was planned.
○ Bound with Young 1831a.

1833a. *Elements of Geometry, with Notes.* Revised and corrected, with additions, by M. Floy, Jun. A. B. Philadelphia: Carey, Lea, and Blanchard. pp. vii + 216.
◇ Stereotyped.

1833b. *The Elements of Analytical Geometry; Comprehending the Doctrine of the Conic Section, and the General Theory of Curves and Surfaces of the Second Order. Intended for the Use of Mathematical Students in Schools and Universities.* Revised and corrected by John D. Williams. Philadelphia: Carey, Lea and Blanchard. pp. 288.

◇   An American edition of an English work.

1833c. *Elements of Plane and Spherical Trigonometry, With its Applications to the Principles of Navigation and Nautical Astronomy; with the Logarithmic and Trigonometrical Tables. To which are Added Some Original Researches in Spherical Geometry; by T. S. Davies.* Revised and corrected by J. D. Williams. Philadelphia: Carey, Lea, and Blanchard. pp. 148 + xxiii + 200.

◇   The tables, which follow p. 148, in essence form a second work, with title page reading "Mathematical Tables; Comprehending the Logarithms of all Numbers from 1 to 36,000; also the Natural and Logarithmic Sines and tangents; Computed to Seven Places of Decimals, and Arranged on an Improved Plan; with Several Other Tables, Useful in naviation and natical Astronomy, and in Other Departments of practical Mathematics; Revided and Corrected by J. D. Williams."

Young, Robert.

1788. *An Essay on the Powers and Mechanism of Nature; Intended, by a Deeper Analysis of Physical Priniples, to Extend, Improve, and More Firmly Establish, the Grand Superstructure of the Newtonian System.* London: for the author by Fry and Couchman. pp. xxix + 336 + 2 plates.

Zeuthen, Hieronymus Georg (1839–1920). See Heiberg 1912.

Zoretti, Ludovic (1880–?).

1914. *Leçons de mathématiques générales, avec une préface de P. Appell.* Paris: Gauthier–Villars. pp. xvi + 753.

◇   Book is in three parts. Part I is geometry and analytical geometry. Part II is theory of functions, derivatives and application. Part III is integral calculus and applications.

Zucchetti, Ferdinando.

1878. *Statica Grafica sua Teoria ed Applicazioni con tavole illustrative.* Turino [Turin]: Augusto Federico Negro. pp. [ii] + 33 plates.

◇   Text for this atlas is not in the library.

Appendix 1: "Catalog of 1803"

Inventory of Books, Maps and Charts, belonging to the Military Academy at West Point. –

Architecture Hydraulic, par M Bellidor. 3 vol.

Adam's Geometrical and Graphical Essays.–

Architecture des Fortiesses.

Atlas of the fortified Harbours of France.–

Barrow's Euclid.–

Considerations D'arcou.

Clairac Engineer.

Charts of the Sea Coast of France. 6 plates.

Chaptals practical Astronomy

Description of the Atlas of the fortified Harbours of France. by Jeffreys.

Desaguliers–

Essai de Tactique.

Emerson's Algebra.

Ewing's Practical Astronomy.–

Fortification perpendiculaire. par M. Montalawbest. 10 vol.

Fortification de Bousmard. 3 vol. in 1.–

Ferguson's Astronomy.

Ferguson's Lectures.

Fields of Mars. 2 vol.

Fortification de Belair.

Fortification perpendiculaire, par M Plusiers.

Ferguson's Philosophy.

Ferguson's Tables & Tracts.

Fenn's Mathematics.

Franklin's Miscellanies.–

Gravesend's Philosophy. 2 vol.

Gregory's Geometry.–

Hutton's course of Mathematics 34 Nos. 1st vol. and 33 Nos. 2nd volume.

Hutton's Logarithms.

Holliday's Gunnery.

Hauxley's Navigation.

Jones's FieldWorks.

Keils' Astronomy.

Keils' Trigonometry.

Lochee's elements of fortification.

L'artillerie Raisonues.

L'Blond.

Muller's Works.

Muller's attack and defense.

Martin's Philosophy.

Martin's Optics.

Newtons Algebra.

Nicholson's practical Astronomy.

Oeuvres de Vauban.

Oeuvres de Bellidor.

Plates to Bousmard's Fortification.

Palladio's Architecture.

Plairfair's Euclid.

Patoun's Navigation.

Preceptor 2 vol.

Robertson's Navigation.

Robius' Gunnery.

Robertson's Mensuration.

Rohaults' Phisics.–

Science des Ingenieurs par Bellidor.

Struensee's field Fortification.–

Stone's Euclid.

Simpson's mathematical Tracts.

Simpson's Cyphers.

Solar Tables.

Smith's Harmonics.–

Trait des project.

Tables du Tir.

Tielkes Engineer by Hugill. 2 vol.

Walkers Geography.

Whiston's Astronomy.

Whiston's Philosophy.

Ward's Mathematics.

Winkler's natural philosophy.

Military Journal. 2 vol.

West Point. 11th July 1803. Rec'd of Colonel Jonathan Williams,

the whole of the Books, contained in the foregoing catalogue.

[Signed] Wm. A. Barron.

Cap't Engr.

Comp & Supermt of M Academy

Appendix 2: Photographs

The West Point collection contains many books with attractive and interesting title pages, frontispieces, and other visual features which yield insights into the historical moments of the works. We hope to make our catalog come alive by displaying here a careful selection of a few of the more engaging pages. We cannot include all of the frontispieces and title pages with artwork in the collection, since that would expand the volume beyond reasonable size. Appendices 3 and 4 list the works in the collection which contain interesting illustrations.

There are 30 figures in this appendix, all photographs of features found in the collection volumes. There are portraits of authors or other notable mathematicians, title pages, symbolic and sometimes elaborate illustrations and designs, and interesting written notes. We believe that each of the following photographs is noteworthy because it is rare, symbolic, or attractive. A listing of the photographs is as follows:

Figure	Brief Description
1	Title Page from {Marolois 1628}
2	Frontispiece from {Saint Vincent 1647}
3	Signature of René François de Sluse from {Saint Vincent 1647}
4	Marginal notes from {Saint Vincent 1647}
5	Loose note from {Saint Vincent 1647}
6	Portrait of John Wallis from {Wallis 1699}
7	Portrait of John Wallis from {Wallis 1695}
8	Frontispiece from {Ozanam 1697}
9	Portrait of Galileo from {Viviani 1701}
10	Title Page from {Euclid 1703}
11	Frontispiece from {Euclid 1703}
12	Title Page from {L'Hospital 1723}
13	Frontispiece from {Gregory 1726}
14	Title Page from {Pemberton 1728}
15	Frontispiece from {Bardet de Villeneuve 1740}
16	Portrait of Nicholas Saunderson from {Saunderson 1740}
17	Title Page from {Euler 1744}
18	Portrait of Johannes Bernoulli from {Leibniz and Bernoulli 1745}
19	Title Page from {Maclaurin 1749}
20	Frontispiece from {Bürja 1788}
21	Frontispiece from {Edwards 1803}
22	Frontispiece from {Adams 1803} of the "Great Theodolite"
23	Title Page from {Bonnycastle 1818a}
24	Portrait of Gaspard Monge from {Vallée 1819}
25	Cover of {Euler 1796} showing "Thayer binding"
26	Advertisement from {Davies–*Descriptive Geometry* 1860}
27	Editorial comments (marginal notes) from {Davies–*Algebra* [1877]a}
28	Notes from {Davies–*Algebra and Trigonometry* [1890]c}
29	Advertisement from {Church–*Calculus* 1874}
30	Editorial comments (marginal notes) from {Church–*Descriptive Geometry* [1892]}

OEVVRES MATHEMATICQVES
DE
SAMVEL MAROLOIS,
Traictant de la
GEOMETRIE ET FORTIFICATION,
Reduictes en meilleur ordre, & corrigees d'un
nombre infiny de fautes escoulees aux
impressions precedentes:

La Geometrie par

THEODORE VERBEECK Mathematicien.

Et la Fortification par

FRANCOIS van SCHOTEN, Mathematicien &
Professeur és Fortifications & sciences qui en
despendent, en l'Vniversité de Leyde.

A AMSTERDAM,
Chez Guillaume Iansson Cæsius.
M. DC. XXVIII.

FIGURE 1. Title Page from {Marolois 1628} with interesting printer's device. Books on fortification were of great interest at West Point.

FIGURE 2. This magnificent frontispiece, from *Opus geometricum quadraturae circuli et sectionum coni* {Saint Vincent 1647} is full of allegories. In the foreground, Archimedes is using a diagram drawn in the sand to search for the area of a circle, while Euclid peers over his shoulder. Neptune's banner, "Plus ultra," characterizes this problem as geometry's greatest secret, but ancient scholars are obstructed by the Pillars of Hercules. The sunbeam's "Mutat quadrata rotundis" indicates that Saint Vincent believes that he has solved the mystery.

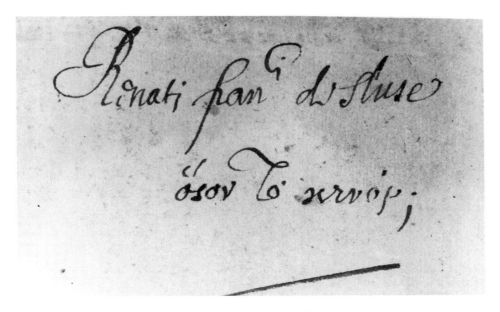

FIGURE 3. Signature of René François de Sluse (1629–1695), written on the endpaper of {Saint Vincent 1647}, indicates that he owned this volume. This signature raises the possibility that de Sluse is the author of some or all of the many comments written in the book (see Figure 4) or on the loose notes (see Figure 5).

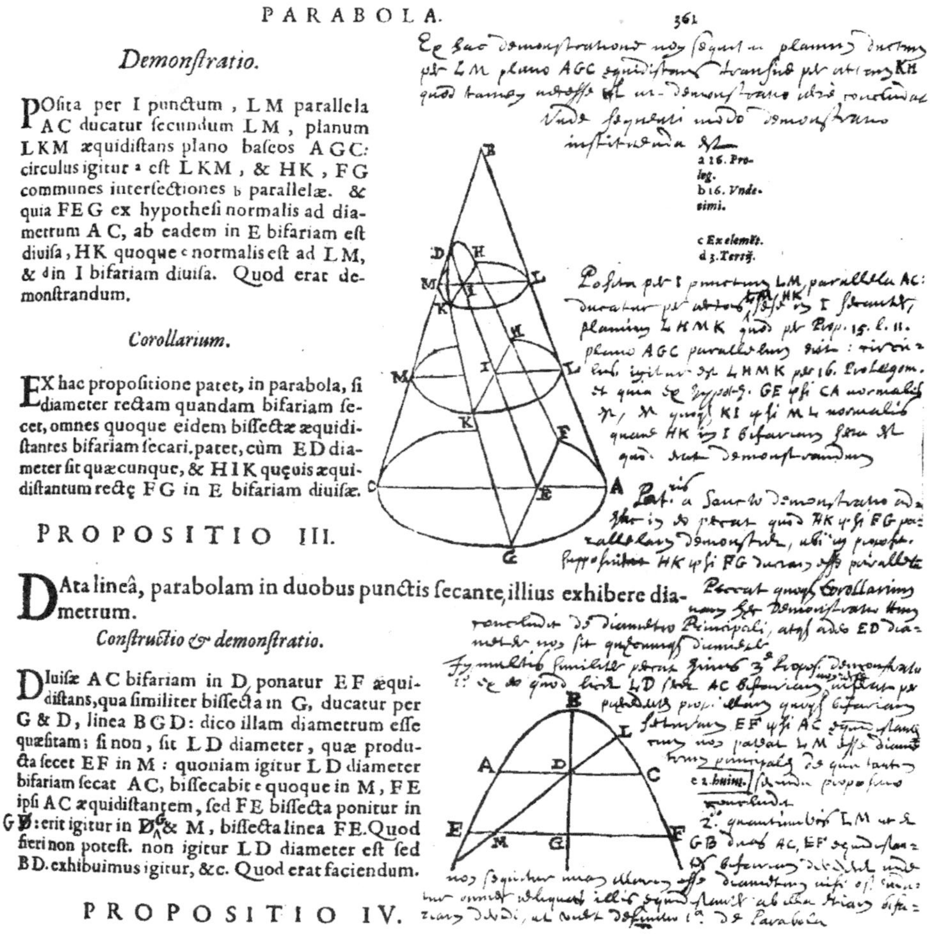

PARABOLA.   361

Demonstratio.

POsita per I punctum, LM parallela AC ducatur secundum LM, planum LKM æquidistans plano baseos AGC: circulus igitur ᵃ est LKM, & HK, FG communes intersectiones ᵇ parallelæ. & quia FEG ex hypothesi normalis ad diametrum AC, ab eadem in E bifariam est divisa, HK quoque ᶜ normalis est ad LM, & ᵈ in I bifariam divisa. Quod erat demonstrandum.

Corollarium.

EX hac propositione patet, in parabola, si diameter rectam quandam bifariam secet, omnes quoque eidem bissectæ æquidistantes bifariam secari. patet, cùm ED diameter sit quæcunque, & HIK quevis æquidistantum rectæ FG in E bifariam divisæ.

PROPOSITIO III.

DAta lineâ, parabolam in duobus punctis secante, illius exhibere diametrum.

Constructio & demonstratio.

DIuisæ AC bifariam in D, ponatur EF æquidistans, qua similiter bissecta in G, ducatur per G & D, linea BGD: dico illam diametrum esse quæsitam; si non, sit LD diameter, quæ producta secet EF in M : quoniam igitur LD diameter bifariam secat AC, bissecabit ᵉ quoque in M, FE ipsi AC æquidistantem, sed FE bissecta ponitur in G : erit igitur in D & M, bissecta linea FE. Quod fieri non potest. non igitur LD diameter est sed BD. exhibuimus igitur, &c. Quod erat faciendum.

PROPOSITIO IV.

(marginal printed references: a 16. Prob. leg. b 16. Undesimi. c Ex elem. d 3. Tertii.)

FIGURE 4. Page 361 of {Saint Vincent 1647} showing some of the marginal notes written in the book. There are numerous annotations of this type throughout the volume in the West Point collection.

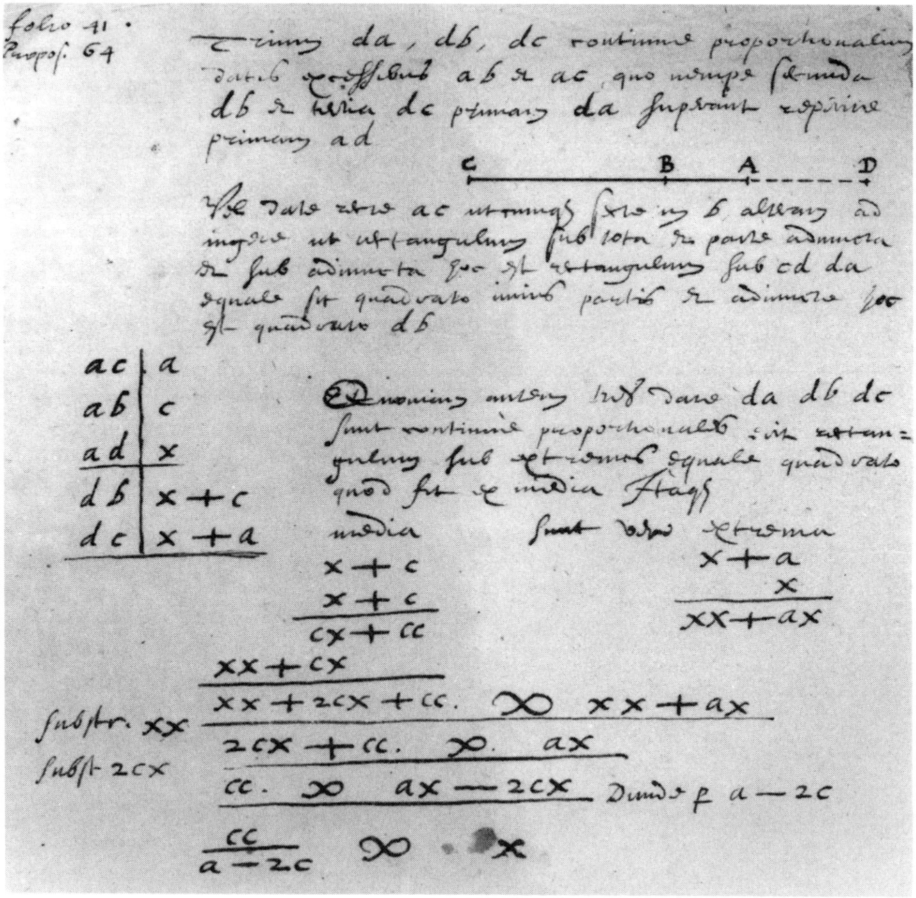

FIGURE 5. One of the loose notes found in {Saint Vincent 1647}. There are over 50 notes of this type inserted in the volume, and they appear to be written on 17th-century paper. {Saint Vincent 1647} is written in a classical geometrical style, making heavy use of geometrical proportions; whereas, the author of this note is translating these ideas into Cartesian notation.

FIGURE 6. An engraving by M. Burghers of John Wallis (1616-1703), Savilian Professor of Geometry at Oxford, from the third volume of {Wallis 1699}.

FIGURE 7. Detail of an engraving of John Wallis by David Loggan from the first volume of {Wallis 1695}.

FIGURE 8. Frontispiece from {Ozanam 1697}, volume 1 of *Cours de mathématique*, illustrating a variety of scientific instruments. Although Ozanam is best known for his work on recreational mathematics, this volume probably entered the West Point collection because it contained mathematics deemed useful in the military arts.

FIGURE 9. Engraving of a bust of Galileo at age 48 with reference to his membership in the Society of Lynxes, from {Viviani 1701}. Viviani was Galileo's last pupil.

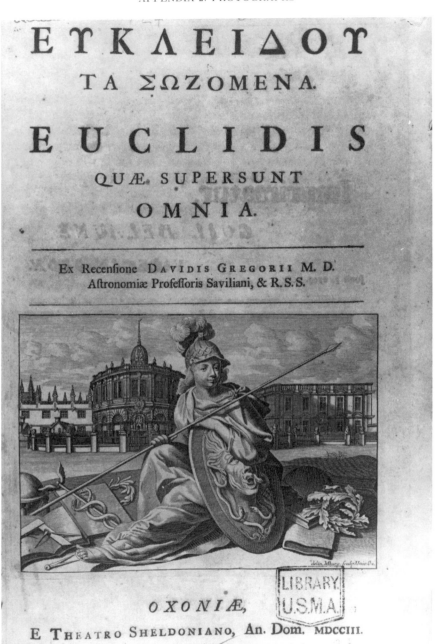

FIGURE 10. Title Page from {Euclid 1703}. This is the first edition of the collected extant works of Euclid, published at Oxford by Savilian Professor David Gregory. In the background of the etching is the Sheldonian Theatre. In the foreground are some tools of geometry.

FIGURE 11. Frontispiece from {Euclid 1703}. Three philosophers have made it to shore from the shipwreck and have found geometrical diagrams written in the sand.

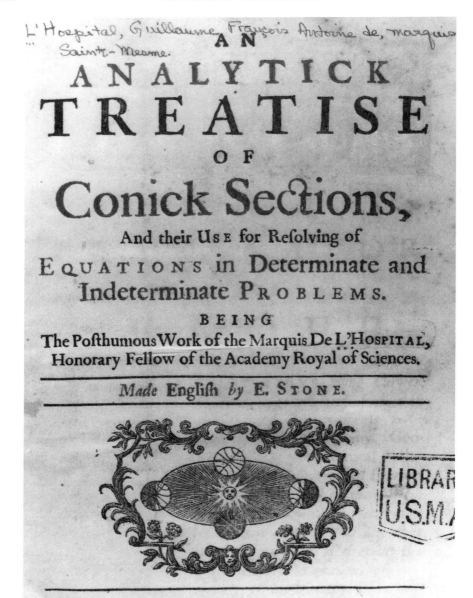

AN

ANALYTICK

TREATISE

OF

Conick Sections,

And their Use for Resolving of

EQUATIONS in Determinate and

Indeterminate PROBLEMS.

BEING

The Posthumous Work of the Marquis De L'HOSPITAL,
Honorary Fellow of the Academy Royal of Sciences.

Made English by E. STONE.

LONDON:

Printed for J. SENEX, in *Fleetstreet*; W. TAYLOR, in *Pater-
Noster-Row*; W. and J. INNYS, in St. *Paul's* Church-yard;
and J. OSBORN, in *Lombard-street*. MDCCXXIII.

FIGURE 12. Title Page from {L'Hospital 1723}. Note the presence
of the library stamp of the United States Military Academy, which
appears on most volumes in the collection.

FIGURE 13. Frontispiece from {Gregory 1726}, indicating the book's composition of both astronomy and geometry.

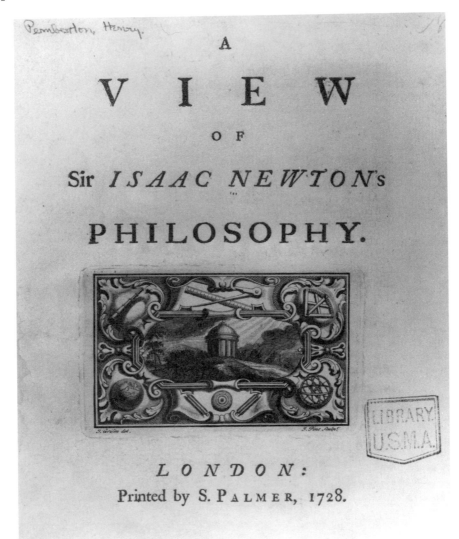

FIGURE 14. Title Page from {Pemberton 1728}. Henry Pemberton (1694–1771) was the British doctor and amateur mathematician who edited the third edition of Newton's *Principia*. {Pemberton 1728} is a popularization of Newton's ideas in natural philosophy (Newtonian mechanics).

FIGURE 15. Frontispiece from {Bardet de Villeneuve 1740}, which contains mathematics related to military operations and, therefore, of particular interest to West Point. Besides mathematics, navigation, architecture, fortification, and artillery are also considered part of "La Science Militaire."

NICHOLAS SAUNDERSON *119*
Lucaſian Profeſſor of Mathematicks in
the University of Cambridge
Died 19. Ap. 1739 Aged 56
J.Vanderbanck pinx.1718   From the Original painted for Martin Folkes esq.   G:Vander Gucht Sculp.

FIGURE 16. Portrait of Nicholas Saunderson (1682–1739), Lucasian Professor at Cambridge, from {Saunderson 1740}. Saunderson was blinded by smallpox at age one.

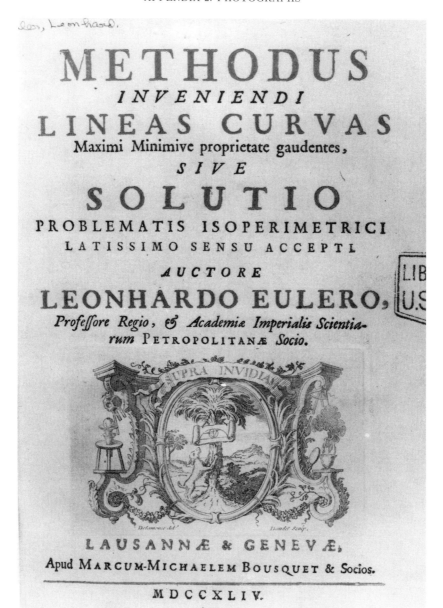

Euler, Leonhard.

METHODUS
INVENIENDI
LINEAS CURVAS
Maximi Minimive proprietate gaudentes,
SIVE
SOLUTIO
PROBLEMATIS ISOPERIMETRICI
LATISSIMO SENSU ACCEPTI
AUCTORE
LEONHARDO EULERO,
Professore Regio, & Academiæ Imperialis Scientiarum PETROPOLITANÆ *Socio.*

SUPRA INVIDIAM

LAUSANNÆ & GENEVÆ,
Apud MARCUM-MICHAELEM BOUSQUET & Socios.

MDCCXLIV.

FIGURE 17. Title Page from {Euler 1744}. This book is one of Euler's most famous works because it established the calculus of variations as a new and separate mathematical discipline. Note the printer's device illustrating the cycloid.

FIGURE 18. Engraving of Johannes Bernoulli from {Leibniz and Bernoulli 1745}. The diagram in his hand indicates that Bernoulli was the author of the brachystochrone problem (see the cycloid in Figure 17). The epigram by Voltaire reads: "His spirit saw truth, and his heart knew justice. He has brought honor to Switzerland and to humanity."

TRAITÉ

DES

FLUXIONS.

Par M. COLIN MACLAURIN, Profeſſeur
de Mathématique dans l'Univerſité d'Edimbourg,
de la Société Royale de Londres.

TRADUIT DE L'ANGLOIS,

Par le R. P. PEZENAS, Jéſuite, Profeſſeur Royal
d'Hydrographie à Marſeille, de l'Académie des
beaux Arts de Lyon.

TOME PREMIER.

A PARIS, QUAY DES AUGUSTINS,

Chez CHARLES-ANTOINE JOMBERT, Libraire du Roi pour l'Artillerie
& le Génie, au coin de la rue Gille-cœur, à l'Image Notre-Dame.

M. DCC. XLIX.

Avec Approbation & Privilege du Roi.

FIGURE 19. Title Page from {Maclaurin 1749}, a French translation of Maclaurin's *A Treatise on Fluxions* (1742), which is also in the West Point collection as {Maclaurin 1742}.

FIGURE 20. Frontispiece from {Bürja 1788}. The author is clearly indicating that Euler is to be added to the list of great mathematicians.

TAN TO CHE BASTA.

FIGURE 21. Frontispiece from {Edwards 1803}. This work on
linear perspective follows directly in the line of that of {Taylor
1811}. It also intends to be accessible to the practicing artist
with little mathematical background. The motto, freely translated from the Italian, "As Much as Required," and the figures
of the muse-teacher and the child-learner personify these pedagogical aims. The drawing board and palette are supplemented by
the attractive arrangement of dividers, brush and stylus, in the
medallion below, to promote Edwards' work to the practical artist.

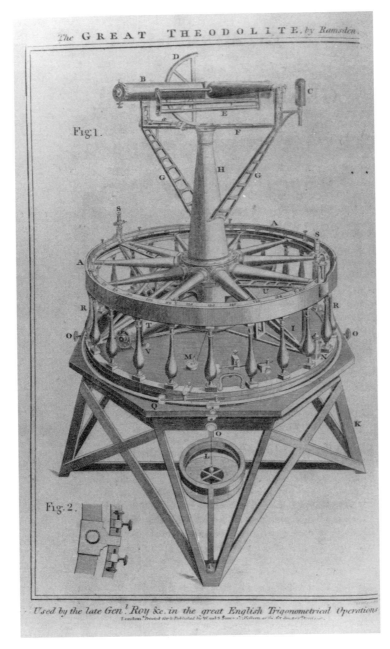

FIGURE 22. Frontispiece from {Adams 1803} of the "Great Theodolite." This volume has interest at West Point because it was used to teach surveying to military engineers.

A

TREATISE

ON

PLANE AND SPHERICAL

TRIGONOMETRY:

WITH THEIR MOST USEFUL PRACTICAL APPLICATIONS.

BY JOHN BONNYCASTLE,

PROFESSOR OF MATHEMATICS IN THE ROYAL MILITARY ACADEMY, WOOLWICH.

THE THIRD EDITION,

CORRECTED AND IMPROVED.

LONDON:

PRINTED FOR CADELL AND DAVIES; JOHN RICHARDSON; BALDWIN, CRADOCK, AND JOY; LAW AND WHITTAKER; AND JOHN ROBINSON.

1818.

FIGURE 23. Title Page from {Bonnycastle 1818a} with an engraving of the use of survey and navigation instruments.

FIGURE 24. Frontispiece of Gaspard Monge from {Vallée 1819}.
Monge was one of the principal founders of the École polytechnique (USMA was modeled after this institution) and the creator
of descriptive geometry. The USMA Library's collection is rich in
books on descriptive geometry.

FIGURE 25. Front cover of {Euler 1796} showing West Point's "Thayer binding." This is a full leather casing with the lettering and border decoration in gold leaf. Most of the books that Thayer and McRee bought in Paris and sent to West Point in 1816 were bound with a similar cover. A gift from the USMA Class of 1950 helped restore and display many of these "Thayer Collection" volumes.

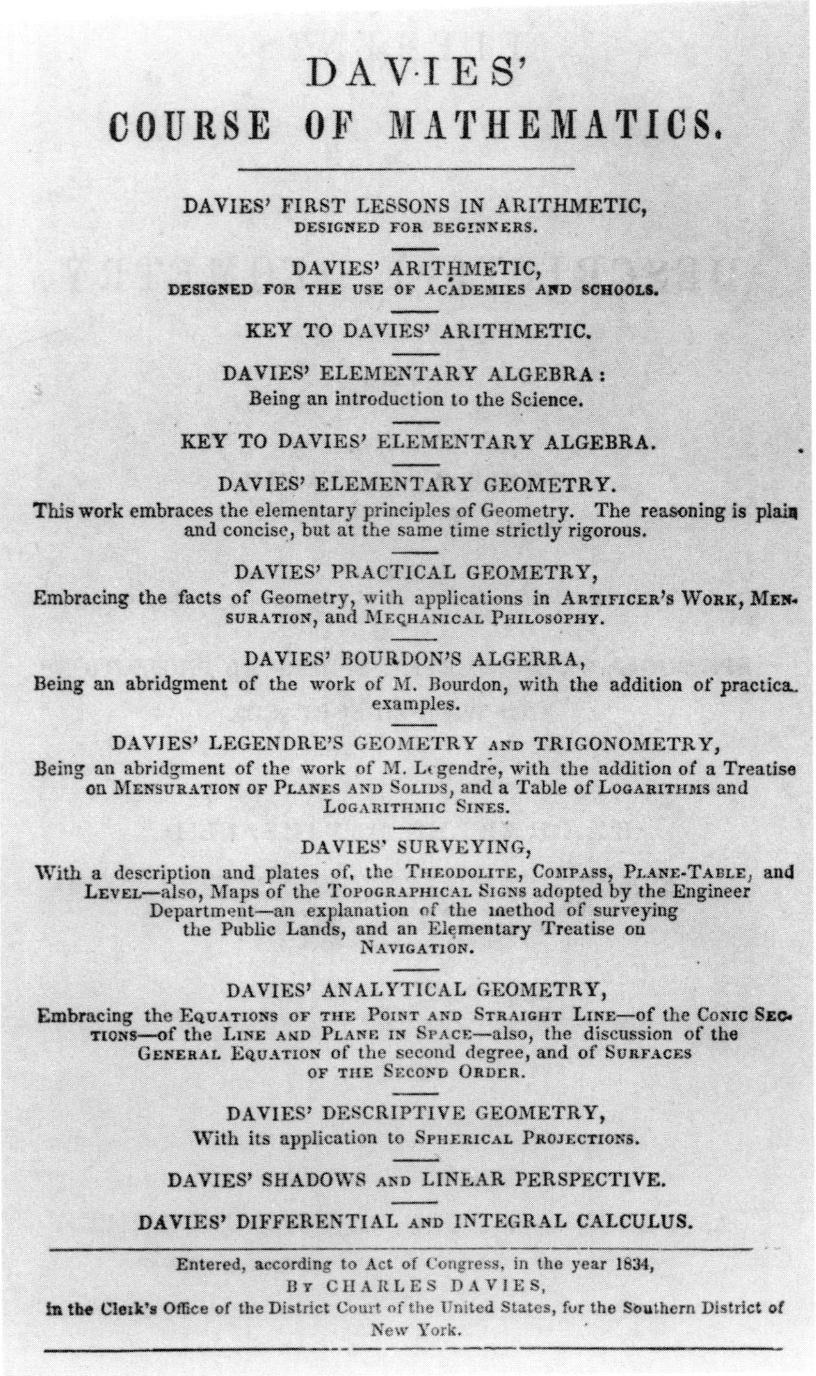

DAVIES'
COURSE OF MATHEMATICS.

DAVIES' FIRST LESSONS IN ARITHMETIC,
DESIGNED FOR BEGINNERS.

DAVIES' ARITHMETIC,
DESIGNED FOR THE USE OF ACADEMIES AND SCHOOLS.

KEY TO DAVIES' ARITHMETIC.

DAVIES' ELEMENTARY ALGEBRA:
Being an introduction to the Science.

KEY TO DAVIES' ELEMENTARY ALGEBRA.

DAVIES' ELEMENTARY GEOMETRY.
This work embraces the elementary principles of Geometry. The reasoning is plain and concise, but at the same time strictly rigorous.

DAVIES' PRACTICAL GEOMETRY,
Embracing the facts of Geometry, with applications in ARTIFICER'S WORK, MENSURATION, and MECHANICAL PHILOSOPHY.

DAVIES' BOURDON'S ALGERRA,
Being an abridgment of the work of M. Bourdon, with the addition of practical examples.

DAVIES' LEGENDRE'S GEOMETRY AND TRIGONOMETRY,
Being an abridgment of the work of M. Legendre, with the addition of a Treatise on MENSURATION OF PLANES AND SOLIDS, and a Table of LOGARITHMS and LOGARITHMIC SINES.

DAVIES' SURVEYING,
With a description and plates of, the THEODOLITE, COMPASS, PLANE-TABLE, and LEVEL—also, Maps of the TOPOGRAPHICAL SIGNS adopted by the Engineer Department—an explanation of the method of surveying the Public Lands, and an Elementary Treatise on NAVIGATION.

DAVIES' ANALYTICAL GEOMETRY,
Embracing the EQUATIONS OF THE POINT AND STRAIGHT LINE—of the CONIC SECTIONS—of the LINE AND PLANE IN SPACE—also, the discussion of the GENERAL EQUATION of the second degree, and of SURFACES OF THE SECOND ORDER.

DAVIES' DESCRIPTIVE GEOMETRY,
With its application to SPHERICAL PROJECTIONS.

DAVIES' SHADOWS AND LINEAR PERSPECTIVE.

DAVIES' DIFFERENTIAL AND INTEGRAL CALCULUS.

Entered, according to Act of Congress, in the year 1834,
BY CHARLES DAVIES,
in the Clerk's Office of the District Court of the United States, for the Southern District of New York.

FIGURE 26. Advertisement from {Davies–*Descriptive Geometry* 1860} listing many of Charles Davies' works.

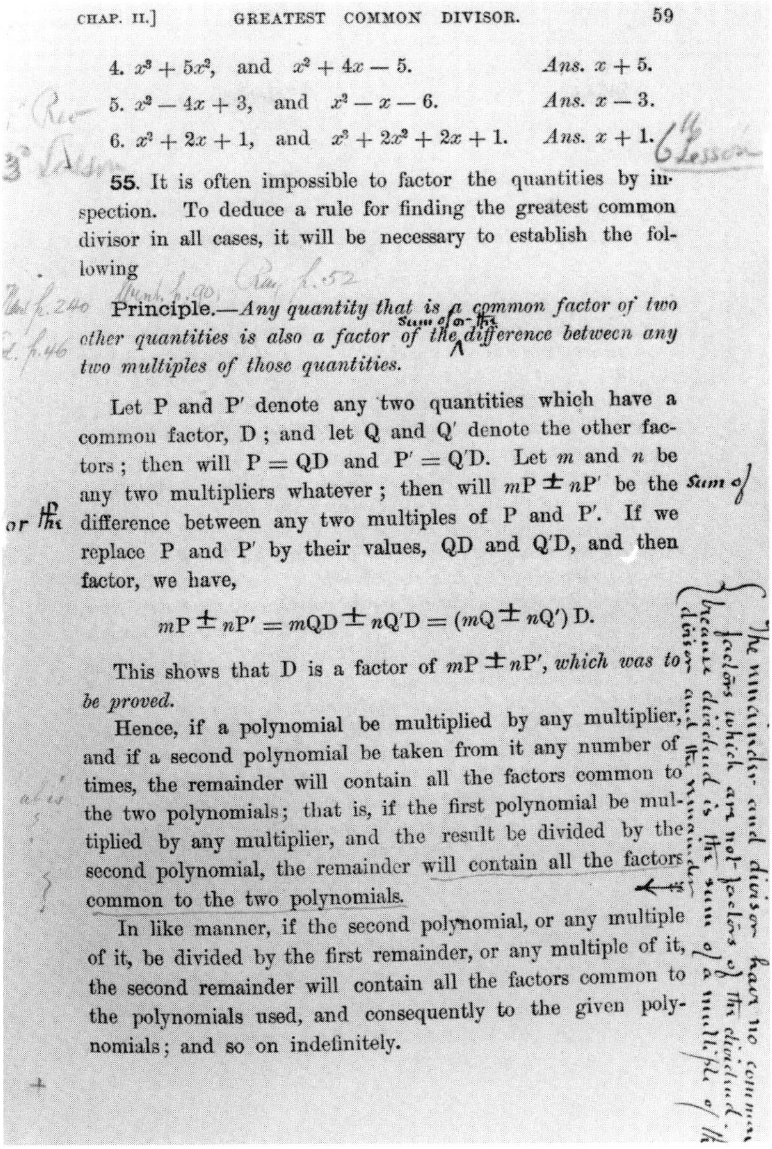

CHAP. II.]   GREATEST COMMON DIVISOR.   59

4. $x^3 + 5x^2$, and $x^2 + 4x - 5$. $Ans.$ $x + 5$.

5. $x^2 - 4x + 3$, and $x^2 - x - 6$. $Ans.$ $x - 3$.

6. $x^2 + 2x + 1$, and $x^3 + 2x^2 + 2x + 1$. $Ans.$ $x + 1$.

55. It is often impossible to factor the quantities by in-spection. To deduce a rule for finding the greatest common divisor in all cases, it will be necessary to establish the fol-lowing

Principle.—Any quantity that is a common factor of two other quantities is also a factor of the difference between any two multiples of those quantities.

Let P and P' denote any two quantities which have a common factor, D ; and let Q and Q' denote the other fac-tors ; then will P = QD and P' = Q'D. Let m and n be any two multipliers whatever ; then will mP $\pm$ nP' be the difference between any two multiples of P and P'. If we replace P and P' by their values, QD and Q'D, and then factor, we have,

$$m\mathrm{P} \pm n\mathrm{P}' = m\mathrm{QD} \pm n\mathrm{Q'D} = (m\mathrm{Q} \pm n\mathrm{Q'})\,\mathrm{D}.$$

This shows that D is a factor of mP $\pm$ nP', *which was to be proved.*

Hence, if a polynomial be multiplied by any multiplier, and if a second polynomial be taken from it any number of times, the remainder will contain all the factors common to the two polynomials; that is, if the first polynomial be mul-tiplied by any multiplier, and the result be divided by the second polynomial, the remainder will contain all the factors common to the two polynomials.

In like manner, if the second polynomial, or any multiple of it, be divided by the first remainder, or any multiple of it, the second remainder will contain all the factors common to the polynomials used, and consequently to the given poly-nomials; and so on indefinitely.

FIGURE 27. Page 59 of {Davies–*Algebra* [1877]a} showing editorial comments and lesson notations. The numerous annotations on this page were for cadet and faculty use. They are typical of notes contained in many West Point textbooks in the collection which had been used by cadets or faculty. Notice the reference to the "6th lesson" indicating in Academy jargon where the lesson assignment was to begin. Notice also the reference to similar texts (e.g., Newcomb, Ray, Todhunter).

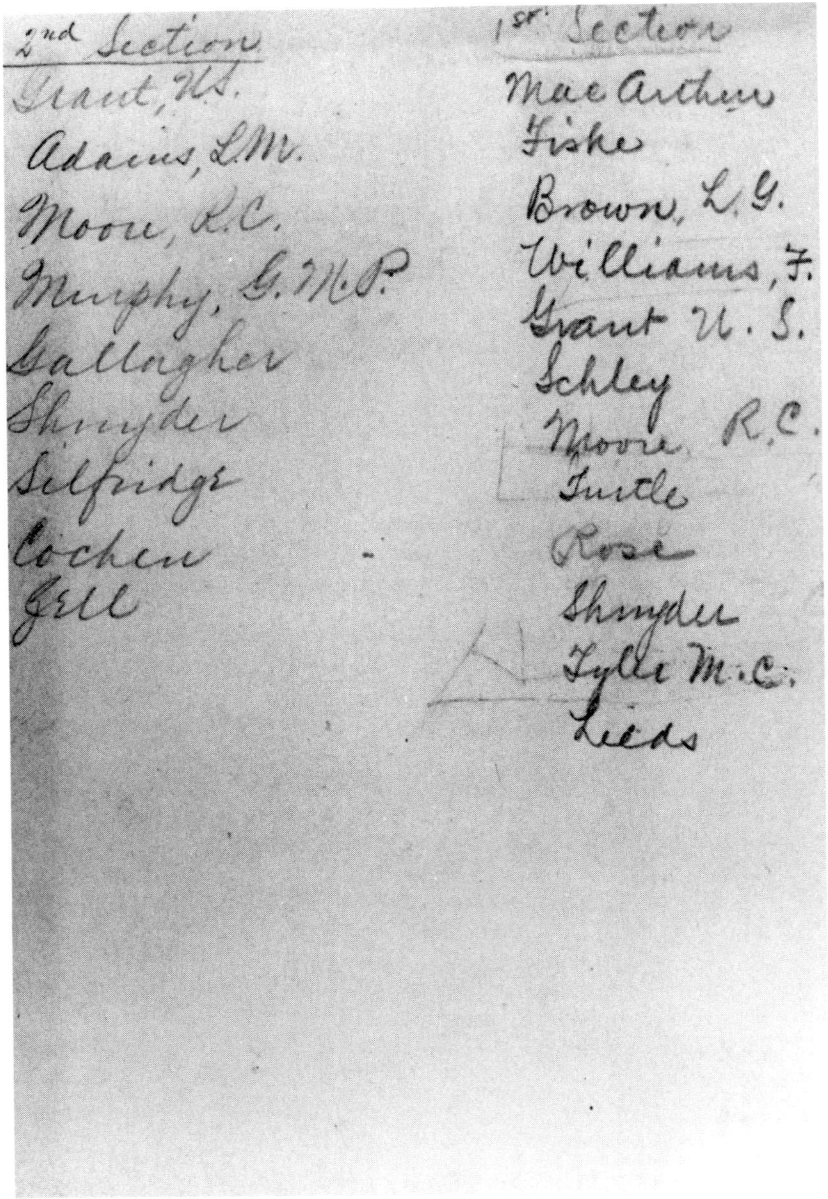

FIGURE 28. Leaf facing back cover from {Davies–*Algebra and Trigonometry* [1890]c}. This book belonged to Ulysses S. Grant, III (USMA 1903), grandson of President U. S. Grant (USMA 1843). From these lists, we know that Grant began the term as the top student of the second section and then moved up in the strictly linear order-of-merit of the Academy to the fifth student in the first (top) section. Grant's classmate, Douglas MacArthur (USMA 1903), is listed as top cadet in mathematics during the second sectioning (first seat of the first section).

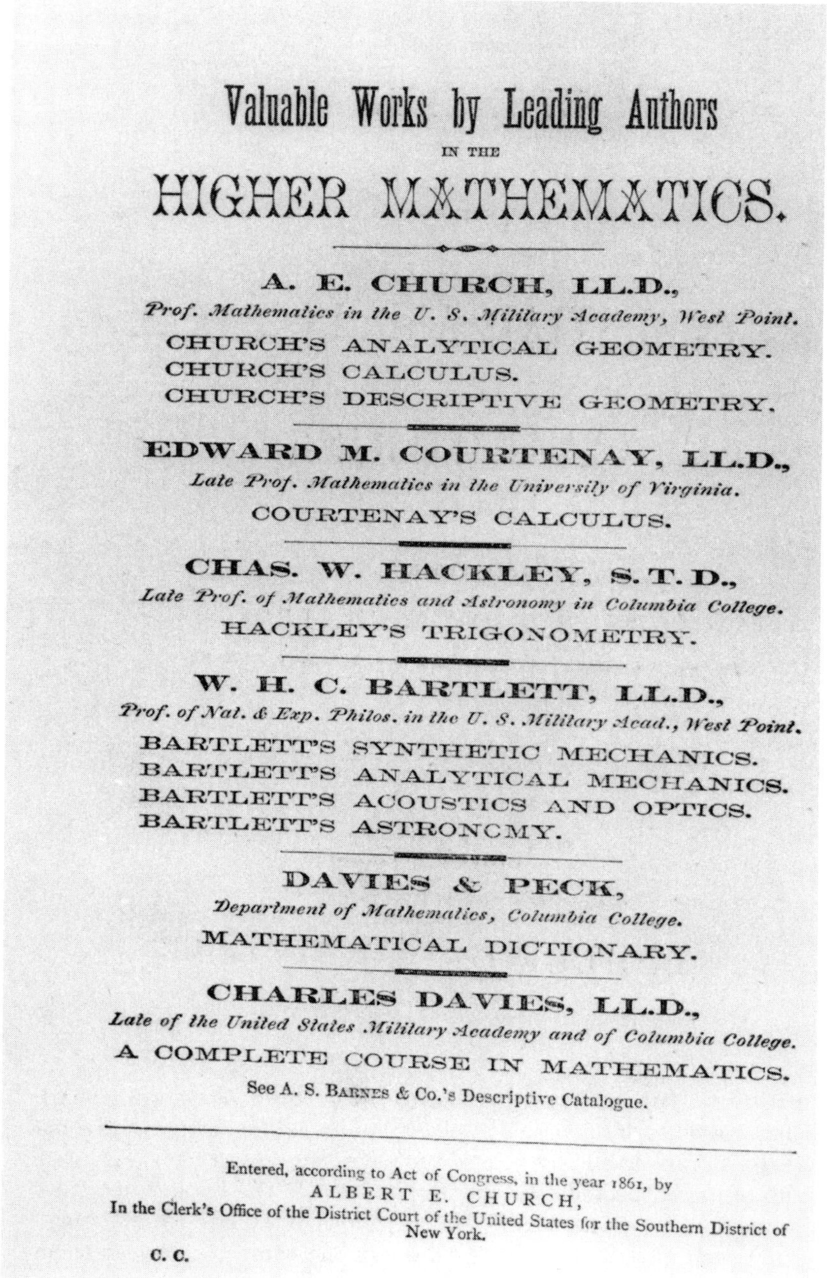

Valuable Works by Leading Authors

IN THE

HIGHER MATHEMATICS.

A. E. CHURCH, LL.D.,

Prof. Mathematics in the U. S. Military Academy, West Point.

CHURCH'S ANALYTICAL GEOMETRY.

CHURCH'S CALCULUS.

CHURCH'S DESCRIPTIVE GEOMETRY.

EDWARD M. COURTENAY, LL.D.,

Late Prof. Mathematics in the University of Virginia.

COURTENAY'S CALCULUS.

CHAS. W. HACKLEY, S. T. D.,

Late Prof. of Mathematics and Astronomy in Columbia College.

HACKLEY'S TRIGONOMETRY.

W. H. C. BARTLETT, LL.D.,

Prof. of Nat. & Exp. Philos. in the U. S. Military Acad., West Point.

BARTLETT'S SYNTHETIC MECHANICS.

BARTLETT'S ANALYTICAL MECHANICS.

BARTLETT'S ACOUSTICS AND OPTICS.

BARTLETT'S ASTRONCMY.

DAVIES & PECK,

Department of Mathematics, Columbia College.

MATHEMATICAL DICTIONARY.

CHARLES DAVIES, LL.D.,

Late of the United States Military Academy and of Columbia College.

A COMPLETE COURSE IN MATHEMATICS.

See A. S. BARNES & Co.'s Descriptive Catalogue.

Entered, according to Act of Congress, in the year 1861, by
ALBERT E. CHURCH,
In the Clerk's Office of the District Court of the United States for the Southern District of New York.

C. C.

FIGURE 29. Advertisement from {Church–*Calculus* 1874} listing works of West Point authors Albert Church (USMA 1828), Edward Courtenay (USMA 1821), Charles Hackley (USMA 1829), William Bartlett (USMA 1826), Charles Davies (USMA 1815), and William Peck (USMA 1844). All of these authors also taught at West Point for some period of time.

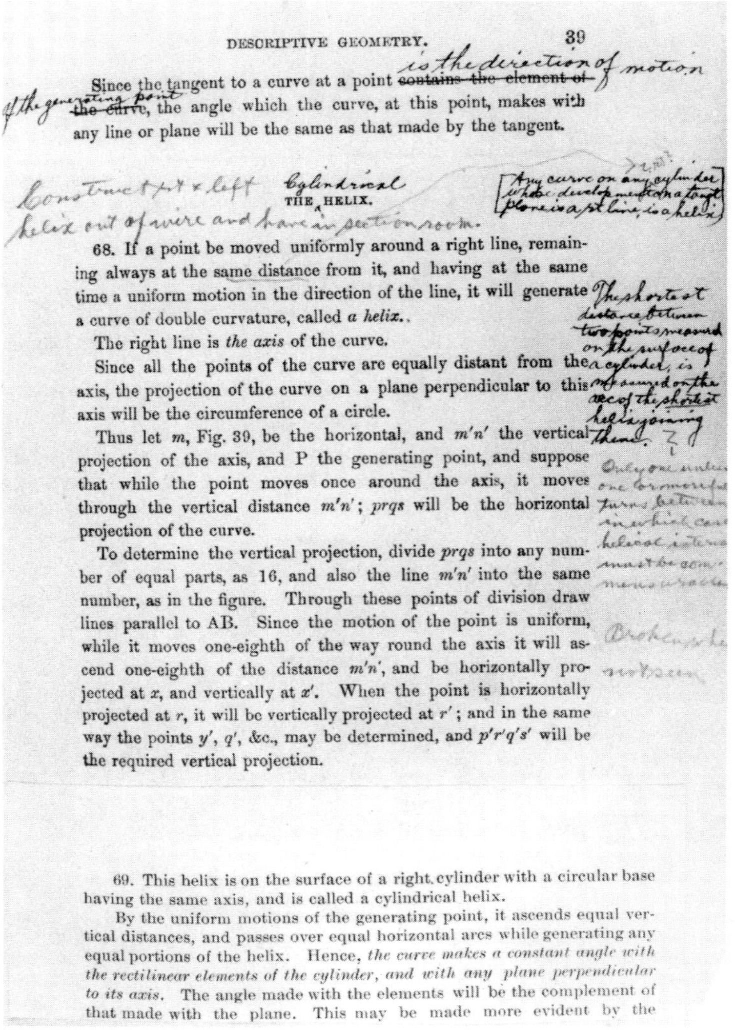

DESCRIPTIVE GEOMETRY. 39

[handwritten: is the direction of motion]

Since the tangent to a curve at a point ~~contains the element of~~ *[handwritten: of the generating point]* ~~the curve,~~ the angle which the curve, at this point, makes with any line or plane will be the same as that made by the tangent.

[handwritten: Construct a right & left cylindrical helix out of wire and have in section room.]

THE HELIX.

[handwritten: Any curve on any cylinder whose development on a tangent plane is a st. line, is a helix]

68. If a point be moved uniformly around a right line, remaining always at the same distance from it, and having at the same time a uniform motion in the direction of the line, it will generate a curve of double curvature, called *a helix.*.

[handwritten: The shortest distance between two points measured on the surface of a cylinder, is measured on the arc of the shortest helix joining them.]

The right line is *the axis* of the curve.

Since all the points of the curve are equally distant from the axis, the projection of the curve on a plane perpendicular to this axis will be the circumference of a circle.

Thus let *m*, Fig. 39, be the horizontal, and *m'n'* the vertical projection of the axis, and P the generating point, and suppose that while the point moves once around the axis, it moves through the vertical distance *m'n'*; *prqs* will be the horizontal projection of the curve.

[handwritten: Only one unless one or more full turns between in which case helix interns must be common or usu alcoholic.]

To determine the vertical projection, divide *prqs* into any number of equal parts, as 16, and also the line *m'n'* into the same number, as in the figure. Through these points of division draw lines parallel to AB. Since the motion of the point is uniform, while it moves one-eighth of the way round the axis it will ascend one-eighth of the distance *m'n'*, and be horizontally projected at *x*, and vertically at *x'*. When the point is horizontally projected at *r*, it will be vertically projected at *r'*; and in the same way the points *y'*, *q'*, &c., may be determined, and *p'r'q's'* will be the required vertical projection.

[handwritten: Broken when not seen]

69. This helix is on the surface of a right cylinder with a circular base having the same axis, and is called a cylindrical helix.

By the uniform motions of the generating point, it ascends equal vertical distances, and passes over equal horizontal arcs while generating any equal portions of the helix. Hence, *the curve makes a constant angle with the rectilinear elements of the cylinder, and with any plane perpendicular to its axis.* The angle made with the elements will be the complement of that made with the plane. This may be made more evident by the

FIGURE 30. Page 39 from {Church–*Descriptive Geometry* [1892]} showing editorial comments, clearly intended as changes for a future edition. At the bottom of the page, paragraph 69 is a printed slip of paper, pasted over the original text in the book. Such slips were common in the West Point textbooks of the late 19th century. They were distributed to all cadets in the course to correct their books with new, updated presentations.

Appendix 3: Portraits in the Collection

The following listing includes the portraits found in the books contained in this catalog. Portraits appearing in the photographs of the Figures contained in this volume are annotated. Some of the portraits appear in frontispieces and, therefore, are also listed in Appendix 4.

Portrait	Book
Abel, Niels Henrick	Peslouan 1906
Agnesi, Maria Gaetana	Rebière 1897
Arago, François	Lebon 1899
Archytas	Allman 1889
Barrow, Isaac	Barrow 1784
Bernoulli, Johann	Bernoulli 1742
Bernoulli, Johann (see Figure 18)	Leibniz and Bernoulli 1745
Bowdith, Mary	Bowditch 1840
Bowditch, Nathaniel	Bowditch 1840
Cardan, Jerome	Waters 1898
Carnot, Lazare	Carnot 1797
Châtelet, Gabrielle-Émilie	Rebière 1897
Copernicus, Nicolaus	Lebon 1899
DeMorgan, Augustus	DeMorgan 1902
Euler, Leonhard	Euler 1812
Euler, Leonhard	Tralles 1786
Fermat, Pierre de	Fermat 1891
Faye, H.	Lebon 1899
Fine, Oronce	Maupin 1898
Flamsteed, John	Flamsteed 1725
Fourier, Jean Baptiste	Fourier 1888
Frederick III	Bernoulli 1742
Galileo, Galilei	Lebon 1899
Galileo, Galilei (see Figure 9)	Viviani 1701
Germain, Sophie	Rebière 1897
Gulielmo, Leopold	Pappus 1660
Halley, Edmond	Halley 1752
Hamilton, William Rowan	Graves 1882
Herschel, John	Lebon 1899
Hill, George	Hill 1905
Hutton, Charles	Hutton 1812a
Huygens, Christian	Huygens 1734
Janssen, J.	Lebon 1899
Kepler, Johann	Kepler 1858-1871
Kepler, Johann	Lebon 1899
Kowalevsky, Sonja	Lebon 1899
Kowalevsky, Sonja	Rebière 1897

Portrait	Book
Ladd–Franklin, Christine	Rebière 1897
la Lande, Jerome de	Montucla 1798a
Laplace, Pierre Simon	Laplace 1813
Laplace, Pierre Simon	Lebon 1899
Leibniz, Wilhelm	Granville 1911
Leibniz, Wilhelm	Leibniz 1789
Le Verrier, Urbain.	Lebon 1899
Loewy, Maurice	Lebon 1899
Mackay, Andrew	Mackay 1812
de Moncada, D' Guleilel Raym.	Apollonius 1655
Monge, Gaspard (see Figure 24)	Vallée 1819
Montucla, Jean Étienne	Montucla 1798a
Newcomb, Simon	Lebon 1899
Newcomb, Simon	Newcomb 1903
Newton, Isaac	Granville 1911
Newton, Isaac	Lebon 1899
Newton, Isaac	Newton 1819
Pascal, Blaise	Pascal 1819
Perrier, F.	Lebon 1899
Picard, Émile	Lebon 1910
Poincaré, Henri	Lebon 1899
Rossi, Gaetanuo	Rossi 1804
Saint Vincent, Gregorius	Saint Vincent 1668
Saunderson, Nicholas (see Figure 16)	Saunderson 1740
Scott, Charlotte Angas	Rebière 1897
Somerville, Mary	Rebière 1897
Stokes, George	Stokes 1904
Tannery, Paul	Heiberg 1912
Tartaglia, Niccolò	Tartaglia 1592
Taylor, Brook	Taylor 1811
Thibault, Jean–Thomas	Thibault 1827
Tisserand, François Felix	Lebon 1899
Wallis, John (see Figure 6)	Wallis 1699
Wallis, John (see Figure 7)	Wallis 1695

Appendix 4: Frontispieces in the Collection

The following listing includes the frontispieces found in the books contained in this catalog. A brief description is given when possible. Frontispieces appearing in the photographs of the Figures contained in this volume are annotated. Frontispieces that contain portraits are also listed in Appendix 3.

Brief Description	Book
Great Theodolite (see Figure 22)	Adams 1803
Terrestrial Globe	Adams 1810
Burning mirrors, rainbow, mirrors, etc.	Alhazen 1572
Archytas	Allman 1889
Bust of Archimedes	Archimedes 1807
Battle scene (see Figure 15)	Bardet de Villeneuve 1740
Barrow	Barrow 1734
Map of Aigle	Biot 1802
"Queen" of Arithmetic	Bocklern 1690
Classical calculator	Bürja 1786
Great mathematicians (see Figure 20)	Bürja 1788
	Bürja 1791
Carnot	Carnot 1797
DeMorgan	DeMorgan 1902
Elaborate shield with lions	Digges 1590
	DuHamel 1701
Mother & child (see Figure 21)	Edwards 1803
Shipwreck (see Figure 11)	Euclid 1703
Euler	Euler 1812
Solar system	Euler 1835
Flamsteed	Flamsteed 1725
Fourier	Fourier 1888
Geometric figures	Frend 1796
Geometry and architecture	Frezier 1737
Newton	Granville 1911
Hamilton	Graves 1882
Statue of Newton	Gray 1907
Astronomy (see Figure 13)	Gregory 1726
Experiment	Guericke 1672
Halley	Halley 1752
Hill	Hill 1905
Astronomical observatory	Horrebow 1740
Hutton	Hutton 1812a
Huygens	Huygens 1724
House of astronomy	Kepler 1627
Device of planetary orbits	Kepler 1858-1871

Brief Description	Book
Laplace	Laplace 1813
Faye	Lebon 1899
Picard	Lebon 1910
American eagle and other coins	Lee 1797
Leibniz	Leibniz 1789
Johann Bernoulli	Leibniz and Bernoulli 1745
Mackay	Mackay 1812
Geometry and horology	Magdeleine 1665
Montucla	Montucla 1798a
la Lande	Montucla 1802
Cannons	Muller 1757
Newcomb	Newcomb 1903
Bust of Newton	Newton 1819
Woman & child (see Figure 8)	Ozanam 1697
Leopold Gulielmo	Pappus 1660
Pascal	Pascal 1819
Abel	Peslouan 1906
Proportional compasses	Robertson 1775
Rossi	Rossi 1804
Towers & angel (see Figure 3)	Saint Vincent 1647
Saint Vincent	Saint Vincent 1668
Saunderson (see Figure 16)	Sanderson 1740
Gunter's quadrant & sector	Sleeman 1805
Stokes	Stokes 1904
Taylor	Taylor 1811
Building in perspective	Thénot 1826
Thibault	Thibault 1827
Euler	Tralles 1786
Monge (see Figure 24)	Vallée 1819
Galileo (see Figure 9)	Viviani 1701
Wallis (see Figure 6)	Wallis 1695
Wallis (see Figure 7)	Wallis 1699
Cardan	Waters 1898